Shop Floor Control-
A Systems Perspective

From Deterministic Models
towards Agile Operations Management

Springer
Berlin
Heidelberg
New York
Barcelona
Budapest
Hong Kong
London
Milan
Paris
Santa Clara
Singapore
Tokyo

Eric Scherer (Ed.)

Shop Floor Control - A Systems Perspective

From Deterministic Models towards Agile Operations Management

With a Preface by
Kenneth Preiss

With 115 Figures
and 6 Tables

Springer

Dr. Eric Scherer
Institute for Industrial Management
and Engineering BWI
Swiss Federal Institute of Technolgy ETH
Zürichbergstraße 18
CH-8028 Zürich, Switzerland

ISBN-13:978-3-642-64349-1

Cataloging-in-Publication Data applied for
Die Deutsche Bibliothek - CIP-Einheitsaufnahme
Shop floor control - a systems perspective : from deterministic models towards agile operations management; with 6 tables / Eric Scherer (ed.). With a preface by Kenneth Preiss. - Berlin; Heidelberg; New York; Barcelona; Budapest; Hong Kong; London; Milan; Paris; Santa Clara; Singapore; Tokyo : Springer, 1998
ISBN-13:978-3-642-64349-1 e-ISBN-13:978-3-642-60313-6
DOI: 10.1007/978-3-642-60313-6

Softcover reprint of the hardcover 1st edition 1998

Hardcover-Design: Erich Kirchner, Heidelberg
SPIN 10656382 42/2202-5 4 3 2 1 0 - Printed on acid-free paper

„The more exact one plans, the more exact he fails."

Maxim of the week, Zürich Daily Press (Tagesanzeiger), July 6, 1996

Planning

We change things tomorrow, we change things today,
We change things, enraged, and we change things, gay.
We change things without a streak of meek,
We change things all seven days of the week.
We change things sometimes just for fun,
Or plan, or don't know what we've done!
We change things for good or just as a trial,
'Cause changing always means work for a while.
We change things, resigned, and without a fuss,
Because that is what is expected of us.
The old change things and so do the young,
We even change the things we've undone.
We change anything that we can alter,
Just to show that we will not falter.
And if our change was a piece of cake,
Then surely it can stand a slight remake.
And so we change things, all our days,
Anywhere we can start a new phase.
We change things now and without blinking,
This leaves no time at all for THINKING!
(The author reserves the right to revise).

As seen in a planner's office.

Foreword

The tools we use shape the thoughts we think. Our currently dominant concepts of organization derive from our mental pictures of machines. People talk about 'put on the brakes', 'put your foot on the gas', 'we will crash' and so on. The metaphor of machine penetrates thinking about society, organizations and relationships between people – as witness 'I am the big wheel but he is a cog in the machine'. This is more than just a turn of phrase. Our understanding of organizations, factories and manufacturing systems derives from that metaphor. This has many implications. For one, in a machine a movement of a controlling lever generates only one possible result, reliably, again and again. Whenever you turn the steering wheel of your car to the right, it turns right, and if it does not do so there can only be one reason, that the car is broken. Manufacturing or other business executives have the same instinct, that when they give an instruction the organization will go where they instruct it, always, and if it does not it needs repair or the replacement of some people. Furthermore, a machine should have a single operator or crew because two decision points can lead to conflicting actions which cause disaster. This is how we usually think of controlling the shop floor.

Industrial competition is now moving from being based upon muscle power and mechanical machinery to being based upon knowledge and networks. The most significant tool we wield today is the communication network. The machine has certainly not disappeared, but has become part of an interconnected system, not standing alone but in a network, integrated by communication. Networking is now becoming pervasive within systems of machines. For example, airplanes and cars are now connected to satellite navigation systems for lost pilots and drivers; machine tools and sewing machines are connected via computerized controls to networks which send design and scheduling information.

In a mechanical machine, a force applied at one point generates definite forces which we can calculate at any other point. There will only be one correct answer for that computed speed or force. This is not so in the Internet. There, when a message is sent from one point to another, we cannot know ahead of time the path it will take nor the time it will arrive. The best we can do is compute a probability of it going by one particular route and within a given time. In a network there is no certainty. Any single action has many possible results which can be predicted only on the basis of probabilities.

The shift we are living through, from industrial competition between companies at arms' length, to competition between coupled companies which exchange real time information in a dynamic world of constant change, is the change from Mass Production to Agility. It is a direct result of the development and spread of communication technology. This has changed the industrial agenda forever. And it is penetrating quickly. In a recent survey in the USA, three-fourths of over 500 manufacturers said that their major challenge was to become more agile. It took 35 years for 10 million people to be connected to telephones; it took 1 ¼ years for 10 million people to become connected to the world wide web. The networked agile world is hurtling toward us. This important book provides excellent guidance to those of us who need to manage and control within it.

As our defining tools change from the machine to the communication network, so do our concepts. Management of manufacturing in the age of the machine was by giving an order intended to push on elements in the organization to lead to one specific result in a given time. The development of communication networks has not only made available a new tool for management, but has created a new metaphor for the organization. The management problem has moved from being one of deterministic control, to being one of managing in an uncertain network. In the age of the network we manage by seeing the whole picture, integrating people, business processes and technology into one coordinated system. This cannot be achieved by managing the pieces separately, and is quite a new state for management of manufacturing and control of the shop floor.

This thoughtfully conceived book on shop floor control and scheduling illuminates these issues very well. It is a significant contribution to our understanding of how to manage manufacturing in the era in which agility and management of complexity have become the dominant competitive requirements. Dr. Scherer and his collaborators are to be congratulated for their contribution to this important subject.

Professor Kenneth Preiss
Ben Gurion University, Israel and
Agility Forum, Lehigh University, USA

From Deterministic Models towards Agile Operations Management

A look at research in the field of shop floor control shows that, on the one hand, shop floor control was and continues to be one of the most extensive domains of research in industrial management. Mainly under the banner of 'scheduling' hundreds, even thousands, of papers, books and theses have been published. A browse through renown scientific journals reveals the attention still given to shop floor control by researchers around the globe. On the other hand, however, current buzzwords and attractions in the literature on management, applied research and industrial innovation show that shop floor control seems to have become a topic of the past. The new focus is on strategic issues such as core competencies, leadership, process organization, enterprise networking, and general concepts like agility, virtuality, and fractalism.

So why publish a new book on shop floor control? A look ahead assures us that, in contrast to prevalent opinions, the manufacturing function maintains its importance with the increasing flexibilization, globalization and dynamization of the value-adding chain. Moreover, with the growing of networks and the ongoing diminishing of hierarchies, management functions at the operational level are gaining in importance. The 'wall' which hindered market dynamics from directly confronting the operational levels of enterprises has crumbled. Shop floor control is no more the front end of a hierarchical planning process, but has become a core function of any operational process chain. With requirements for shop floor control steadily rising, the solutions of the past have become more inadequate than ever. Researchers have to admit, that even though they expended tremendous energy on the problems of shop floor control, the past four decades since THOMPSON & MUTH[1] introduced the problem to the scientific community at large did not have much impact upon industrial reality. While some are still solving the '10 jobs x10 machines' problem and waiting for their thoroughly reviewed papers to be published, in industry such measures as group technology, KANBAN, teamwork and KAIZEN have proved to be feasible means of improving industrial performance at the shop floor level and, in many cases, have reduced the control problem somewhat.

To master the challenges of the future, there must be a change in our thinking paradigm. Manufacturing is not deterministic! What is more, manufacturing processes are beyond – traditional – control. As MELNIK, CARTER, DILTS & LYTH[2] have already noted, „the shop floor is not under control, it is out of control." The assumption held for the future is that the problem of systems design for shop floor control is no longer the problem of 'optimization'. The reductionistic paradigm of the optimization focus needs to be replaced by a holistic paradigm of agile activity, dynamic behavior and evolutionary development. This means that we can not continue to rely solely on mathematical modeling techniques but have to achieve a systems perspective that takes into account all aspects of shop floor control. The present volume provides a re-assessment of the problem of effective and efficient shop floor control. In doing so, the shop floor is considered to be that which in reality it is: a sociotechnical system where man, machines and computers interact according to formal rules and informal conventions. The objective of systems design for shop floor control is to achieve a system that is capable of managing the highly complex, dynamic processes of everyday manufacturing while simultaneously able to adapt evolutionary to changing requirements by itself.

The book does not aim to replace the many well-suited textbooks[3] on the problem. Rather, it provides the reader with additional views that allow a new and better understanding and should ultimately lead to organizational innovation and sustainable improvement of the situation in industry. For this purpose, the book is rooted in the study and experience of the 'real world' of everyday manufacturing and gathers together papers by authors from a vast background and many fields: industrial engineering, management science, work psychology, industrial sociology and business computing. All of the authors share their understanding of the shop floor as a socio-technical as well as cybernetic system. They also have industrial experience in research and consulting in common, and they share an appreciation of the human being as the most flexible and valuable resource in modern manufacturing.

Eric Scherer

Zürich, January 1998

---

1 MUTH, J.F., THOMPSON, G.L. (Eds.) (1963): Industrial scheduling. Prentice-Hall, Englewood Cliffs.

2 MELNIK, S.A., CARTER, P.L., DILTS, D.M., LYTH, D.M. (1985): Shop Floor Control. Dow Jones-Irwin, Homewood. p. vii.

3 for reading besides those mentioned above, the editor suggests:
BŁAZEWICZ, J. ET AL. (1996): Scheduling computer and manufacturing processes. Springer, Berlin.
BURBIDGE, J. L. (1978): The principles of production control. 4th Ed. MacDonald & Evans, Plymouth.
PINENDO, M. (1995): Scheduling: Theory, Algorithms and Systems. Prenctice Hall, Englewood Cliffs.
WIENDAHL, H.P. (1995): Load-oriented manufacturing control. Springer, Berlin.

About the Book

To achieve a systems perspective and holistic approach to shop floor control, the book's various contributors provide several different views. While all contributions were specifically invited and prepared for this publication, each still represents an individual view. To ensure commonality of the message, each paper was discussed in depth among the authors and the editor on several occasions, and each author had knowledge of the other contributors. Most of the papers were written originally in German and then carefully translated and reworked to conform to technical terminology. Most contributions have been additionally reviewed by researchers from English-speaking countries to ensure readability and reference to the state-of-the-art in research. As far as possible, the book draws upon literature available in the English language. Even so, several reference are based on the German-language literature, and several quotes have been translated for this book (in the text, no further mention will be made of this fact).

The book is divided into six parts. *Part One* provides a new *'Understanding of Shop Floor Control'*, and so it takes a *'Close look at Reality'* from three different perspectives: the qualitative functions and rules of shop floor control in everyday manufacturing (*Eric Scherer*), the non-linear dynamics of manufacturing processes and logistics (*Mirko Dobberstein*), and the rationality and politics of the management function in manufacturing (*Manfred Moldaschl*). *Part Two, 'Systems and Organizations'*, presents the foundations of systems theory and organizational design. In a first step it examines the problem of knowledge generation for systems design in the context of shop floor control and discusses the drawbacks of traditional research. The concepts of socio-technical and cybernetic systems theory is introduced, and the human characteristics in shop floor control are described as the keys to lasting improvement (*Eric Scherer*). In a second step, there is a review of the relation of system concepts and organizational design which serves as a basis for further reading (*Julia Kuark*).

For the practitioner, Part One provides an ideal starting point while the scientist probably may find the founding concepts in Part Two a good start for his reading. In *Part Three*, the question of *'Control Strategies for Complex Manufacturing Systems'* is raised. The possibilities of self-regulation as a means of reducing complexity is described (*Heinz Schüpbach*), and new models of control based on target systems, market principles and context control are introduced (*Volker Ahrens*).

Starting with *Part Four*, the book focuses on the two major resources for shop floor control: human beings and information technology. It analyzes *'Aspects of Decision Making'*. First, common approaches to the scheduling problem as the core decision making problem in shop floor control are reviewed and rated with regard to their suitability to a holistic approach (*Franz Stuber*). Then the concept of knowledge and experience as a link to the 'real world' of everyday manufacturing is introduced (*Jürgen Fleig and Robert Schneider*). *Part Five* describes the field of application of information technology (IT) on the shop floor. It focuses on *'Making Use of Information Technology: Man – Machine Interaction'* as a major means of achieving flexibility and adaptivity in an agile manufacturing system. The application of IT thereby is in an area of conflict between use and acceptance (*Eric Scherer*). Here the principles of supervisory control provide the foundations for future achievements (*Elizabeth Szelke*). Criteria for man-machine task allocation are discussed (*Eric Scherer and Martina Zölch*), and guidelines for the design of IT systems are outlined (*Franz Stuber*).

Finally, *Part Six* provides a look beyond standard textbooks and presents some signposts *'On the Road to Agility: Changing Organizational Structures'*. As a core concept for facilitating continuous improvement and sustainable change, aspects of shop floor control in teamwork-based organizations are presented (*Albert Hurtz*). Further insight is given into shop floor control as a task of organizational development (*Steffen Weik and Eric Scherer*). And finally, the question of education and the school-to-work transition are discussed as bases for improvement (*Martin Fischer and Franz Stuber*).

Acknowledgments

This books has its foundations in the activities of the Swiss Commission on Technology and Innovation (KTI/CTI). During the past years, the work of the editor and several authors was largely supported by the KTI research grant 'NuWIS' (Project No. 2923.1). A second basis has been the initiatives of the IFIP Working Group 5.7, 'Production Management Systems', in particular the special interest group on 'Knowledge-based Reactive Scheduling'. Many discussions and ideas originated there.

The editor likes to give thanks to all authors who contributed papers and were always willing to enter into discussion and make room for my requests. A special mention goes to Franz Stuber and Volker Ahrens, who acted as a helpful advisory board. Special thanks go to Kenneth Preiss for his spontaneous agreement to provide the foreword, to Paul Schönsleben, who supported my work during my years at the Swiss Federal Institute of Technology (ETH) in Zürich and made several useful suggestions, and to Eberhard Ulich, whose scientific concept of MTO (Human – Technology – Organization) forms one of the foundation for this book. A final thank you goes to Ellen Russon of East Sandwich, Massachusetts, for her great work as a translator. Her careful and superb translation work is in every sentence of this book.

Contents

Understanding Shop Floor Control – A Close Look at Reality

Part One

The Reality of Shop Floor Control – Approaches to Systems Innovation

Eric Scherer

One of the tasks of production planning and control (PPC) of an enterprise is to transform direct customer demands or marketing sales plans into production orders. It thus represents the interface between the external market of an enterprise and its internal, value-adding areas. The goal of PPC is to achieve the best possible correspondence between external demands and internal possibilities of the enterprise. As a rule, PPC is based on sequential-hierarchical procedures, which, step-wise, perform ever more detailed planning leading to implementation. It is the job of management at the operative level of the enterprise to make the final plan for the shop floor, in the form of a weekly or daily schedule for example, and to carry it out. Here management must take the actual, current situation in production and the availability of the individual resources required into account. But it must also examine whether or not the released orders and due dates, determined on the basis of sales predictions, are at all realistic. This task is called variously production activity control, operations management, or shop floor control. What these terms express in common is that the practice of control activities always has close structural links to other activities of operative management such as personnel management, interference management, maintenance and quality assurance; control places them in relation to one another. „The shop floor is understood as the locale of production activities" (NULLMEIER & BRINKOP 1991, p. 53), which shows the operative character of control activities and makes clear that shop floor control is a task within a physical environment.

1. Shop floor control as the operative task of PPC

The shop floor itself represents an elementary part of the enterprise (Figure 1). It is here that value-added activities takes place. In order to fulfill this task, many functions must be performed. The functions differ in their reference variables, such as frequency of repetition and time horizon, and can be summed up as 'logistical process chains' and 'technological process chains'. They form the basis of the organization of production. At the same time, production represents a sociotechnical system (see ULICH, 1994, p. 154ff). Within production, people perform various activities and actions. The organization forms the environment of the acting individual, with which the person interacts via the work task. In order to fulfill the task assigned, the person utilizes the machines and procedural techniques which form the technical subsystem (see SUSMAN 1976, p. 146ff)[1]. Here, depending upon the task, the person may be required to cooperate with or control the machine. At the same time, the person is embedded within a social network and stands in direct contact to other people, so that in addition to the technical subsystem, there is a social subsystem made up of „organization members having individual and group-specific needs both physical and psychological" (ALIOTH 1980, p. 26).

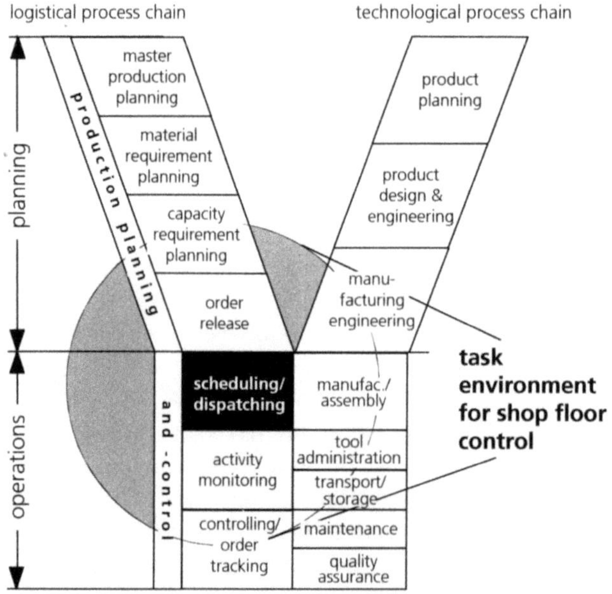

Fig. 1 SCHEER'S (1992) Y-model applied to shop floor control.

[1] The technical subsystem is not restricted to physical-material technologies alone, but rather is comprised of all mechanistic functions and elements of an organization, such as the method of order processing, whether or not they are supported by computers or other aids.

Within PPC, control represents the implementation of planning instructions – in the form of value-added processes. The shop floor takes on orders from the planning department where a rough capacity scheduling is performed. With this, the reference variable changes: while as a rule, the planning department processes one order after the other, the processing of numerous orders takes place on the shop floor in a parallel and simultaneous manner. Here it is the task of the shop floor to have an overview of total orders on hand and to achieve the most optimal processing possible[2]. Shop floor control thus stands within the field of tension between technology and logistics, that is, it must coordinate the economic goals of PPC with the technological conditions and capacities of production.

The example of a journey may serve, as an analogy, to describe the character of control more precisely. Prior to a trip, the traveler uses both a map and previous personal experience to plan a route from point A to destination B. Once initial planning and the necessary preparations have been completed, the traveler begins the journey ('carries it out'). During the course of the trip, the traveler must constantly adjust to actual situations and decide on new options in order to reach the destination. Every fork in the road demands that the traveler judge whether the real path before him corresponds to his planned route ('mapping'). Obstacles, such as construction sites, may force him to detour ('correction'). These activities serve the decision-making process and take place in a manner parallel to carrying out the journey. They are called control (Figure 2).

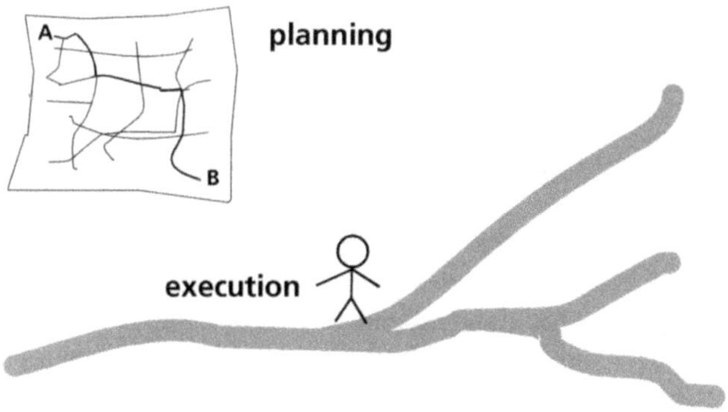

Fig. 2 The relation between planning and execution.

Control is thus constant surveillance and adaptation of original planning during the execution phase of a task. According to the situation at hand, this may be restricted to simple comparison of the plan and actual reality ('mapping'), or it may lead to correction of the original plans. The criterium for decisions is always the

2 This already shows the inadequacy of all too simple design approaches: shop floor control is not a business process, but rather a work task. It cannot be classified as a partial activity of an order or a business process and cannot be divided up sensibly among several orders. Shop floor control takes place continuously and cannot be fixed – temporally or locally – unambiguously.

goal that is to be reached, while the plan itself serves as an aid. In some cases, it can happen that the goal itself will change. Correspondingly, the original plan will be altered in its details to include newly acquired data. For our journey example, this might be adjustment of the original time plan at the actual start of the trip due to poor weather conditions. In the case of shop floor control, as a rule, finite scheduling refines and details the original master plan with regard to time (days rather than weeks) and capacity (specific work stations rather than whole shops) to cover a shorter time frame (for example, on Friday, as preparation for the coming week). This time scheduling aspect of shop floor control is shown in Figure 3 in comparison to production planning previous to the shop floor.

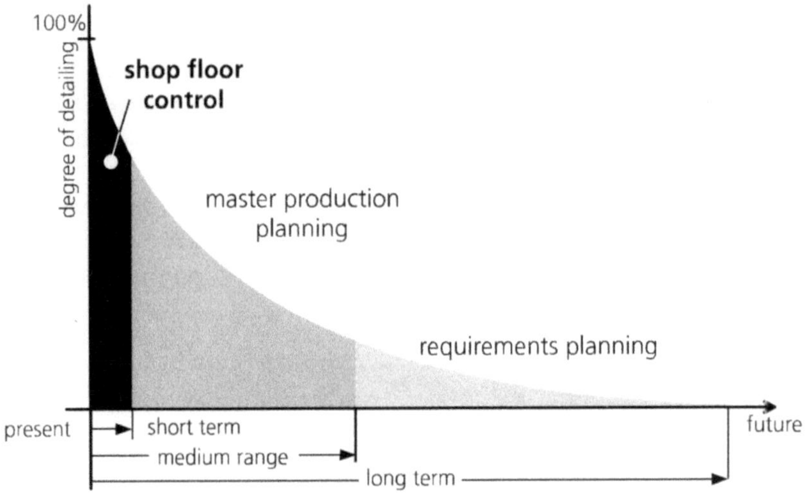

Fig. 3 The temporal range of planning activities in production (following KRÜGER 1991).

As a rule, research has viewed, described and developed shop floor control as a part of the logistical PPC system from a purely functional perspective and with a high degree of formalization. Traditional industrial engineering in English-language research, for example, is based almost exclusively upon purely mathematical optimization models (see MCKAY, SAFAYENI & BUZACOTT 1988, p. 87) and seeks optimization of a few, as a rule purely economic, objectives (see the contribution by STUBER in this volume). Within the reality of an enterprise, however, the issues in shop floor control are considerably more extensive and unfathomable. The field of shop floor control may only be understood by way of numerous aspects, levels of perspective and subsystems. In the world of practice, the subdivision shown in Figure 4 has proved to be serviceable.

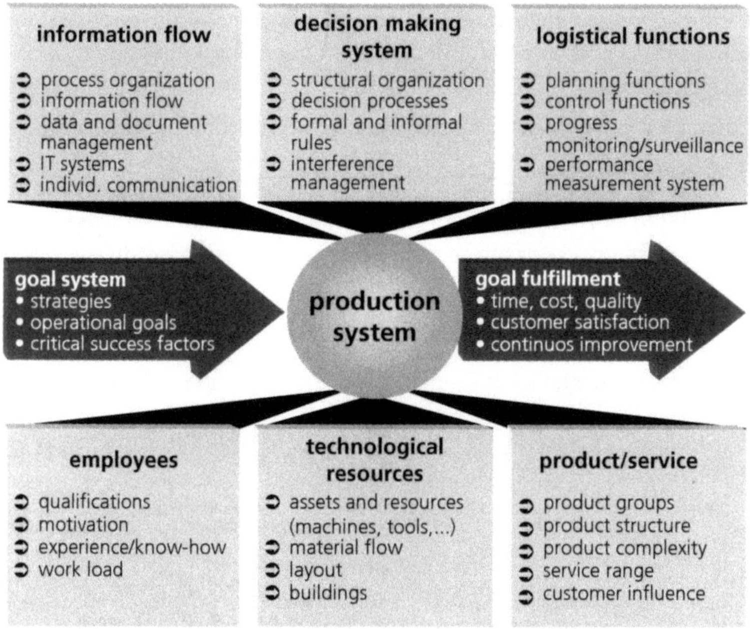

Fig. 4 Aspects for analysis, assessment, and design of logistic systems.

In considering these aspects of shop floor control, it is important to remember that an aspect-view serves as an aid, making it possible to explain the characteristics of a system from certain angles of vision. In reality, the individual aspects are tightly coupled and may not be considered as separate from one another for design.

2. Uncertainty and variation: Characteristics and behavior of production systems

The basis of effective control of a system is excellent knowledge of its characteristics and behavior. In the following, we will examine some characteristics, typically found in production practice, which are significant with regard to the design of control systems. The significance of and the relation among the characteristics described will vary according to the practical application at hand. Important is the recognition that the system to be controlled is very dependent upon the aspects to be controlled.

2.1. Heterogeneous and distributed structures

All enterprises are subject to the constant necessity to change strategies and structures in order to adapt to market conditions. Many production plants are characterized by heterogeneity, which results as a consequence of the adaptation of individual part-areas and aspects of production. Heterogeneity as a characteristic is manifested in several features:

- different technologies in machinery, such as manual, NC and CNC control,
- varying production resources, such as machines, persons, space,
- overlapping or inconsistently defined organizational units, shared utilization of resources, such as test and measurement equipment, or of central services, such as a tool room,
- varying requirements upon machine operators and their qualification,
- contradictory production goals, for example, minimization of stock and simultaneous increase in delivery service level.

This listing of various aspects of heterogeneity could go on and on. In the world of practice, one speaks of 'structures which grow'.

In consequence, heterogeneity soon affects the actual control system and applies to further factors. This is seen in, for example

- varying production strategies, e.g., KANBAN and make-to-order,
- non-uniform internal and external interfaces, such as the meaning of due dates (hard or within a certain range), order and batch sizes, delivery agreements, etc.

While heterogeneity guarantees flexibility on the market, it places high demands upon the control system. A heterogeneous production system can no longer be structured unambiguously through constant, universal criteria, and total reorganization is rarely – such as in the construction of a new factory – possible. This weakness is characteristic especially of the small and mid-size enterprise which, although it must offer a relatively wide palette of services, has only limited resources to utilize.

In addition to heterogeneity, production is characterized by a distributed, decentralized structure. Order processing extends as a rule over a range of various organizational units. Such subunits – a job shop, operational unit, or production island – will be termed 'cells' in the following. At the cell level, planning orders are implemented in the form of value-added processes within an operative and productive cell.

At the operational level of production, the division of work among individual cells is horizontal. The cells fulfill part-tasks along the processing of manufacturing orders. Individual cells in production handle simultaneously, parallel and sequentially one order or several, different orders. This causes, at the level of material flow, the growth of a distributed system, which is complemented by a similarly distributed and decentralized system at the level of information flow. There is coupling among individual cells in a distributed system which, through interfaces and buffers which develop, becomes dampened (Figure 5).

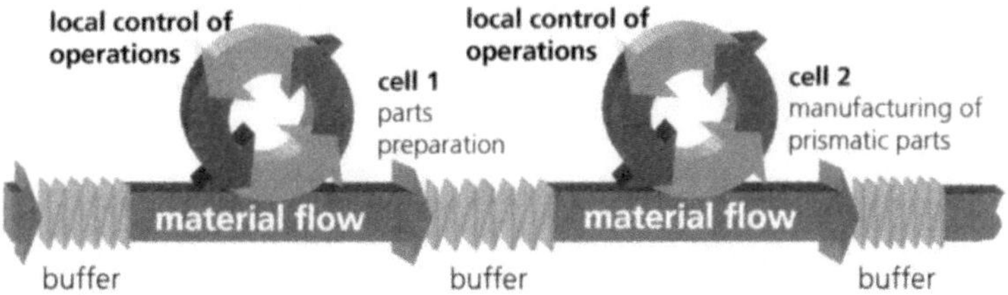

Fig. 5 Coupling of cells in manufacturing.

2.2. Resources and technical availability

In order to execute a single productive operation, a whole array of resources is required. Examples of such resources for an operation executed on a machine are material, set-up personnel and machine and tool operator. According to the type of machine, tool, or the order, additional resources will be required such as NC programs, gauges, fixtures, measurement instruments to set machines and tools, measurement devices for quality control, storage tanks, transport means, energy, and so on. To complete the operation, these resources are indispensable, whereby they will be needed simultaneously or in temporal sequence depending on the job at hand.

Because resources are usually available only in limited quantities (for economic reasons), there is competition for resources among the individual cells. An important task of shop floor control is thus to coordinate resource use and to prevent that at a certain point in time n, two cells will require a certain resource, while at point n+1, the resource would not be needed at all. In this way, the availability of resources becomes a central variable in predictive control.

Figure 6 presents an example case illustrating the importance of the availability of resources to reliable planning; for each operation, four resources are needed, five operations must be completed in succession, and for each defined resource, there is 90% availability at any point in time. This results in a planning reliability of 12.5%. This example shows clearly that any advance planning, even for short-term periods, is always beset with uncertainties. If the significantly higher degree of integration and the close coupling within production is taken into account, it becomes clear that planning is a very limited aid to control.

2.3. Variations and disturbances

Order processing is subject to many kinds of variations and disturbances. Variations and disturbances can occur in and be caused by both the information system and the physical system. Variations and disturbances may be classified according to two criteria:

- according to the occurrence and effect of the variation or disturbance
- according to the cause of the variation or disturbance.

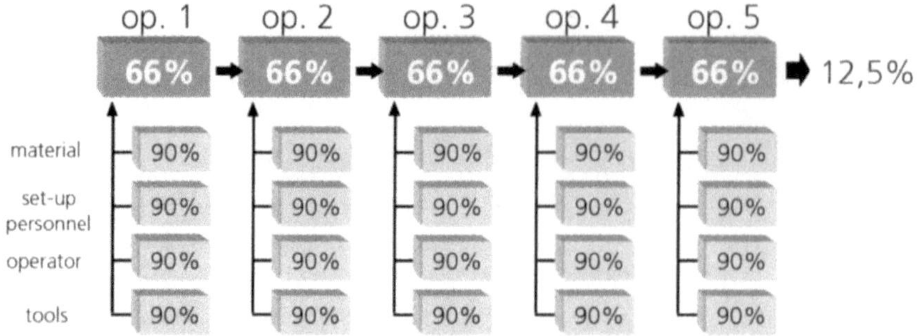

Fig. 6 Planning reliability when resources are not in complete availability.

The cause of variations and disturbances may also lie external to the actual boundaries of the system, in the form of, for example, false specifications in customer drawings. The classification above allows simplification of the unclear division between 'control system' and 'controlled system'. While the physical area of production must clearly be assigned to the controlled system, the information system is at once both control system and controlled system. Figure 7 shows possible types of variations and disturbances which can arise within shop floor control.

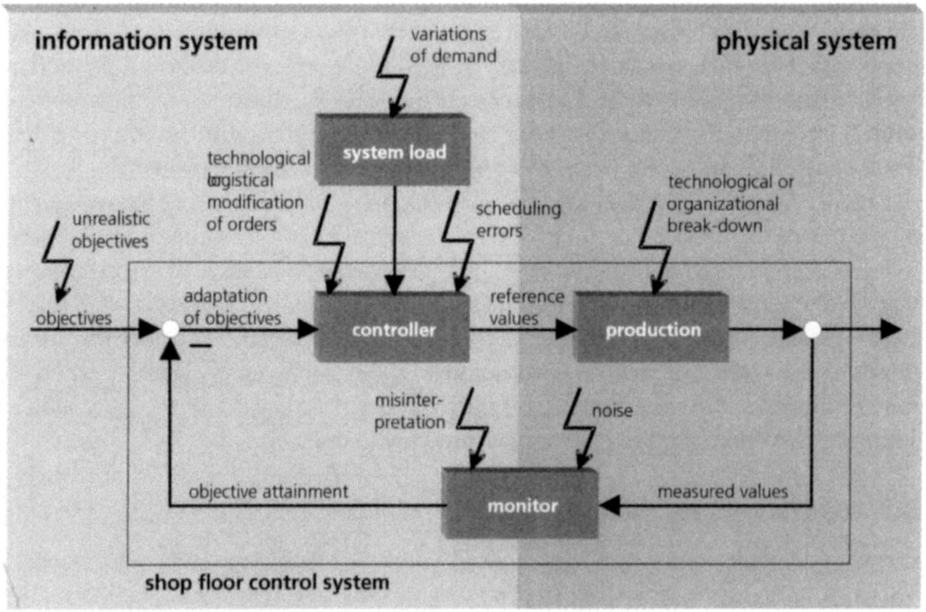

Fig.7 The structure of shop floor control as a feedback system and the classification of possible variations and disturbances (following SIMON 1995, p. 36).

Depending on the type of variation or disturbance, there are varying trouble-shooting strategies. Events which occur in the system's environment, such as a change in the customer order or a change in product specifications in engineering, lead first of all to a change in the system load on the shop floor. To this the shop floor must react flexibly. However, changes in the surround may cause considerable disturbance if they are not reported to shop floor control in a timely fashion. This points up the great importance of information coupling of the shop floor to the comprehensive information system of an enterprise. The information flow, which in the standard case will pass through many stations, must, in the case of changes, proceed rapidly and directly.

Control errors and misinterpretations of observed events are a direct measure of the quality of shop floor control as a decision-making instance, for both the function of control and the function of monitoring occur in plant reality through one and the same instance and with temporal overlapping. The quality of decision-making depends upon the qualifications of the employee responsible as well as the quality of the aids and information used. Lists, diagrams, indicators on the computer screen, as well as a clear structuring of production, may all contribute to minimization of such errors.

As several investigations have shown, at least a third to half of all disturbances in an enterprise occur within the physical subsystem: WESTKÄMPER & JESCHKE (1995) estimate that 49% of disturbances in the firm studied occurred in the production department. Possible disturbances included machine downtimes, errors in machine control and excessive scrap rates. Figure 8 shows the findings of their study of an enterprise having 90 employees manufacturing gears and transmissions both in make-to-order and make-to-stock production philosophy. The study classified errors according to basic functions within production, with a further detailed breakdown for the areas of manufacturing, planning and shop floor control.

A more detailed look reveals that the majority of errors and disturbances are not caused by the technical components of the production system, but rather can be traced back to organizational defects and individual mistakes. Disturbances minimize the usefulness of technical facilities, and they have a direct effect upon the availability of resources. Although the numbers will differ from firm to firm and among varying types of automated systems, it remains clear that disturbances have considerable economic significance. As WESTKÄMPER & JESCHKE (1995, p. 85) remark, „disturbances and errors seldom occur systematically. They more often occur by chance." This state of affairs is significant: the occurrence of a particular disturbance is statistically of minimal probability – errors are *unique*. They become statistically calculable only through very rough estimation. Therefore, errors and their effects on the system, and thus resulting control requirements, cannot be planned. Instead they require reactive, local strategies for solution. Disturbances are a dominating factor in shop floor control. In some cases, they become so

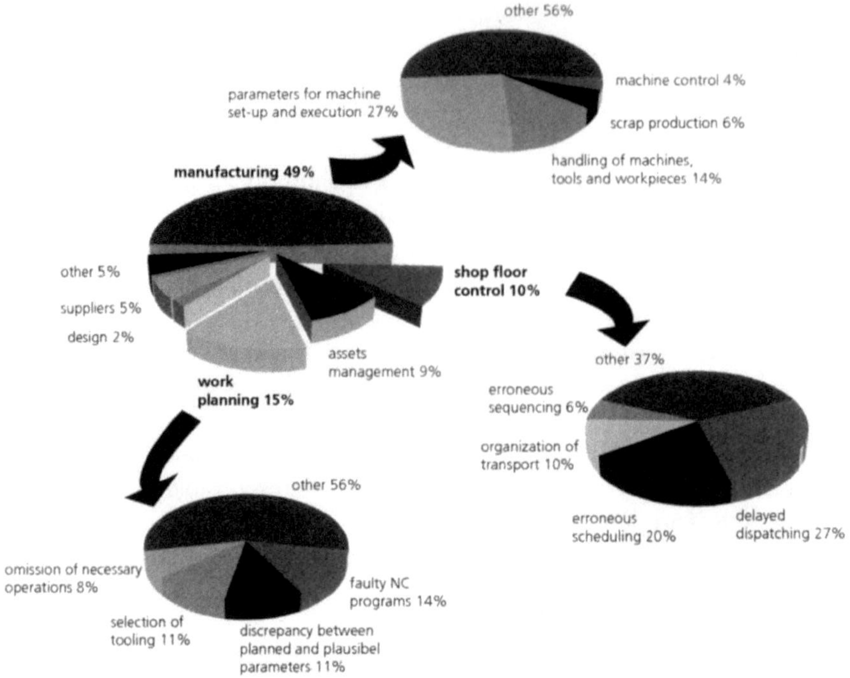

Fig. 8 Causes of errors, variations and disturbances in gear manufacturing (WESTKÄMPER & JESCHKE 1995).

weighty that the terms 'shop floor control' and 'interference management' may be used almost interchangeably.

3. Shop floor control between work task and formal organization

As a rule, traditional design approaches take as a reference the 'formal' organization of shop floor control. The real world of operations practice shows considerable deviation from this. More precise investigation requires that we study the reality of shop floor control, which is in part very informal. The main starting point is the role and importance of human beings. The person as the final decision-maker is an essential part of the control system. Within the framework of the social system, the person forms a network of relationships which is of great importance to shop floor control. The human view, and also human behavior, can differ considerably from the assumptions gained from a purely functional, formal perspective. Here lies one of the greatest shortcomings of research and development in the field of logistics in the 1980's: shop floor control was handled as a system in a technical sense. The role of employees in shop floor control and consequences for the work situation of employees received little consideration.

3.1. Formal and informal systems

As described above, shop floor control deals basically with a work task. Only to a small degree is this manifested in formalized form, such as in the form of weekly planning meetings of planners and floor supervisors. A large number of all control decisions are made within the situation and not in a formal context, as for example when supervisor and foreman meet over coffee and confer, or by a skilled worker when a machine goes down[3]. Here neither the location nor the point in time of the control decision can be determined in advance, and it can be seldom established what persons or roles within the shop floor take on control activities. Shop floor control yields to formal representation only in a very limited fashion, whether to order flow charts or to the assignment of tasks to a workplace in an 'organization chart'. This phenomenon is reinforced by 'the exception as a rule' and the uniqueness of possible trigger events, for these make formalization of even the basic rules of shop floor control difficult and always incomplete[4].

The functioning of shop floor control may be explained by the fact that individual persons view control of the system as their own constant or sporadic task, or by the fact that persons' productive actions represent explicit or implicit control-relevant decision-making. It is from their actions that an informal system develops alongside the formal system. Particularly in the case of shop floor control, which eludes final formalization due to its very nature, a strong informal system is built. This is shown clearly in the conditions surrounding the failure of deterministic planning approaches in practice. Despite the fact that this failure should lead to breakdown of the total system, it does not: things still function somehow. The formal system is complemented by an informal system[5] which supports, circumvents or, in some cases, counteracts it.

This reveals that the informal system of an enterprise is of particular significance with regard to shop floor control, because under disadvantageous circumstances, it can contribute to upholding and maintaining the entire system (see WIGHT 1974, p. 5ff., BECKER 1985, p. 87-87). It is important not to view the informal system as a symptom of a poorly functioning formal system, a view prevalent in practice (BECKER 1985, p. 88). Remember, for years there was a trend towards the 'paper-less factory', a practice which attempted to force people to use only 'correct', read 'formal', channels. The informal system is actually a symptom revealing that the powers of authority granted by the formal system are too little to allow effective task fulfillment, or that work activity is too tightly controlled by management (see EMERY & THORSRUD 1976, p. 17f). Close analysis indicates that the formal and informal systems complement each other. In the case of shop floor control, the weaknesses of the formal system actually serve to reinforce the

---

[3] Detailed descriptions and case studies of the behavior of shop floor controllers in practice are found in the works of MCKAY (MCKAY ET AL. 1988, 1989 & 1992).

[4] This may also be a reason for the failure of KI-based software in computer-aided control.

[5] A concrete example of the formation of such an informal system through self-organization is presented by RIIS & FRICK (1991).

informal system (MALSCH 1987, p. 82), and thus serve indirectly to improve the total system.

Therefore, any reorganization efforts must recognize the important existence of the informal system – and its nature as a system – and include it in analysis. But this is particularly difficult, as the representatives of plant interests involved in reorganization take on a most disadvantageous constellation: on the one hand, management sees in the informal system an undermining of the formal system, and on the other hand, representatives of the operations level do not have conscious knowledge of the existence of informal system of which they themselves are also representatives.

3.2. Computer application and work organization

The work situation of the shop is characterized by a double man-machine division of function. On the one hand, the person works on machines and facilities which aid him in the fulfillment of his primary task, the production of parts and products. On the other hand, the person utilizes computer systems for planning, monitoring and control of his work. Thus, from the viewpoint of the use of technology on the shop floor, there is a triangle of man, production facilities and computer-cased PPC system. The person is a part of both the information and physical systems (Figure 9).

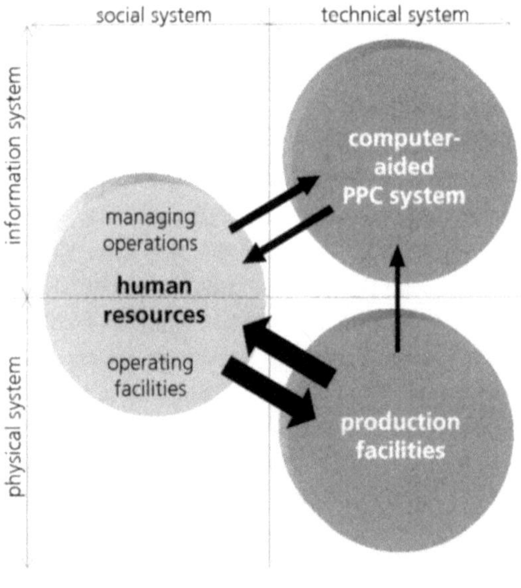

Fig. 9 The shop floor employee in the field of tension of hybrid systems.

Close examination shows an evident, further cause of the failure of deterministic approaches: truly deterministic approaches must rely upon computer-aided processing. Even when data are increasingly available, they are not, or not quickly enough, in a form that can be utilized by the computer system. In consequence, the calculations performed by the system are never up-to-date. The information coupling of production facilities and computer-aided PPC must, at least in part, take place through human beings in order to be complete. Through this, the expenditure required to feed the computer system with the necessary data in a crucial situations – where decisions are urgent – leads to considerable delays and costs.

3.3. Models of planning and the reality of execution

Shop floor control is characterized by the duality of the terms 'planning' and 'control'. Planning takes place on the basis of assumptions. A model serves to provide this basis, the model being a conception or schema which describes operations flow and relationships in production and logistics in an abstract, generalized form. In plant practice, such models take the form of tables, lists, master data compilations, etc. The planning model for shop floor control is easily represented by computer systems. In consequence, the model as represented in the computer system has a considerable influence on the thinking of those working with it. But the planning model is general in function and in its logical essence and is thus simple. However, due to the extensive character of shop floor control throughout the entire enterprise, the overall model on the computer system is just as complex as the production processes it represents. This leads to planning taking on its own 'reality', which corresponds to the world of the model (see KROHN & KÜPPERS 1990, p. 112).

In practice, the great importance of models in planning tasks results in a 'gap' : while planning is based upon a succinctly defined model, the actual reality of the shop floor is completely different. Abstract models are not adequate, because the shop floor is a physical, material work system (see WEICK 1990, p. 8). Consequently, thinking and work in central planning departments, whose employees work almost exclusively at computer workplaces, are dominated by the model and its logic. This leads to the formation of its own 'reality of planning' which deviates significantly from, or even opposes, the 'reality of the shop floor' (MANSKE & WOBBE-OHLENBURG 1985, p. 402). In practice, the dual realities are linked only by means of instructions and feedback. The effects caused by this 'gap' in the PPC are evident in many areas (Figure 10) and are well known (see WIENDAHL & AHRENS 1995, p. 4).

The effects of this gap are particularly strong where planning department and shop floor are also separated spatially and in terms of structural organization. These effects are also very apparent – as the author was to observe frequently – in the context of reorganization projects: adequate discussion of problems between planners and executors often fails due to their differing views of reality.

Although this gap is found in many enterprises, it is hardly taken into consideration in system design and, in particular, in software development. Today's

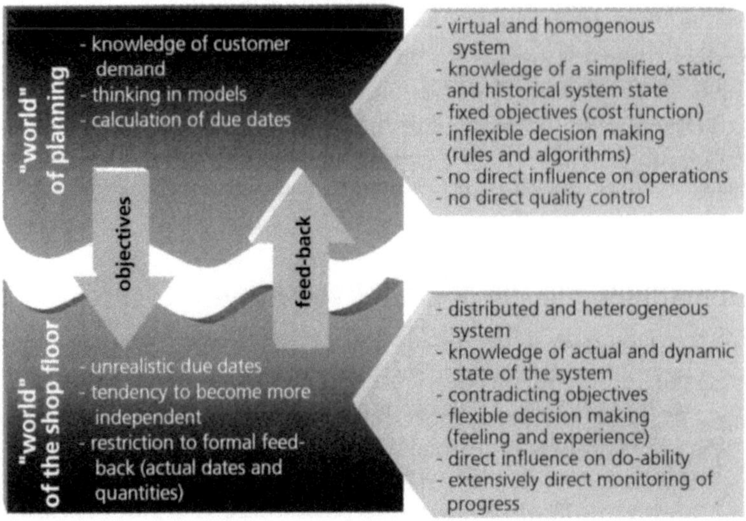

Fig. 10 View of planning on the shop floor – view of the shop floor in planning.

tendencies towards the refinement of deterministic models has served to even increase the gap. Mathematically precise models, which still today represent the basis of planning, contradict in many cases the cognitive conceptions of shop floor workers who execute the planned work activities and who have a primary grasp of the results of their work (see MALSCH 1987, p. 86). This 'information interference' between highly exact, model-based planning on a computer system and human perceptions in the physical production environment leads to the situation where counteractive control to maintain the system becomes increasingly impossible (see MELNIK & CARTER 1987, p. 43). The total control system is finally robbed of any local flexibility.

The reason that such a gap could arise within shop floor control is, among other things, the vertical division of labor between planning and execution – and with this also between proactive and reactive control – which has been reinforced by computer systems (see DÖRR 1991, p. 137f). At the same time, the traditional distance between work in the planning department and work on the shop floor has been greatly reduced by the computer, and through this, the autonomy of the operative areas – gained previously through the 'proper channels' – has declined. All this makes the task of bridging the gap between planning and operations imperative. However, it must not be forgotten that independent planning is still necessary, as it must serve as the interface to customer and supplier markets, and thus, leads a priori to contradictory interests. The temporal dimension must also be considered: while planning – both temporally and spatially – may be settled closer to production operative activities, this is not possible for planning which, for example, must schedule raw materials having nine months' delivery lead times. Planning depends upon a model-based reality.

3.4. The significance of goal systems

Goals are of particular significance in shop floor control, as they represent the objective for control decisions. A manufacturing order, for example, describes a task by giving instructions on what is to be done and how often this is to be done. The order is complemented by an order-specific goal in the form of a restriction „by May 22nd“ or – as in rush orders – by an optimization goal such as „rush - as soon as possible“. Overlapping the order-specific goal system are the typical goals of shop floor control, such as 'capacity utilization', 'minimization of lead time', 'minimization of inventory' and 'increase delivery reliability', but there are also enterprise goals such as 'best quality'. As a central concern of efforts to decentralize control authority, objectives are viewed as an important means of giving a common orientation to individual, local control activities towards a superordinate goal of the enterprise (see WARNECKE 1992, p. 142 & p. 165-172).

In mathematical planning theory, economically-oriented goal systems predominate, which as a rule consist of the variables mentioned above ('capacity utilization', 'lead times', 'inventory' and 'delivery reliability'). Accordingly, these are integrated into the goal system of the computer-based PPC system and represent optimization objectives. It is the task of shop floor control to compare these to technological determinants and, correspondingly, to follow technology-oriented goals, such as reduction of set-up times by means of clever sequencing of orders, or to take technological limiting factors into account, such as when processing the same alloys within the same blocks of time due to the necessary separate disposal. Thus, as a rule, technological goal variables represent restrictions and are integrated into decision-making without the aid of the computer[6].

Economic objectives of the enterprise and technological goals have a very formal character and may usually be interpreted with clear references and in a relatively uniform manner. That is, the significance of a certain objective is generally clear. At the level of the sociotechnical system, the formal goal system is expanded to include individual interpretation. The responsible scheduler attempts to take the social situation within the shop floor into consideration. A typical example of this is the widely accepted 'Friday rule': 'Never begin demanding jobs on Friday afternoon'. Individual employees themselves take on orders and interpret them as personal tasks. They add on personal goals to their work. The employee is not always completely conscious of all individual goals.

Whereas in the literature and in reorganization projects economic considerations take the forefront, the sociotechnical character of production almost always remains unconsidered in the structuring of goal systems. As a rule, the traditional „dilemma of production control“ (WIENDAHL 1995, 5) – referring to the conflict among lead times, on-time performance, capacity utilization and inventories – is the focus. The second, actually more far-reaching dilemma – shop

6 Curiously, the restricting nature of technological constraints was not introduced into scheduling theory until 1980 (see FOX 1987).

floor control in the field of tension among economic, technological and individual objectives – receives little attention (Figure 11).

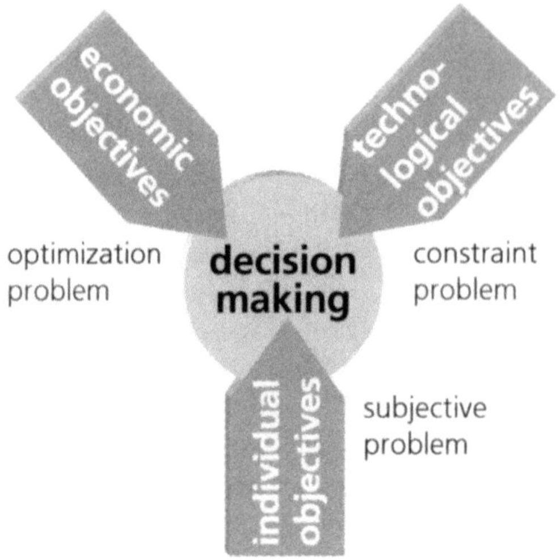

Fig. 11 The dilemma of shop floor control – A reinterpretation.

Here again, the 'gap' between the reality of the model and the reality of the shop floor is revealed: what in the model appears to be a serious problem may be of little relevance in reality, and vice versa.

Objectives are subject to interpretation and are thus individual and dependent upon *what is to be decided* and upon *whomever is responsible* for decision-making, that is, goals do not necessarily have to accord with the formal goals of the organization. Within the field of shop floor control particularly, where contradictory and conflicting goals predominate, this circumstance becomes even more significant: because direct control of goal achievement is not possible, individual goals relevant to a certain decision can be easily hidden from someone on the outside, such as a superior or a representative of the planning department. The decision-maker, as 'master of goal interpretation', thus automatically gains a very high degree of autonomy, even if this is not granted by the formal system. Even though this autonomy will partly depend upon the individual's 'desire to cover up' and 'ability to cover up', the generalization holds that decision makers within shop floor control are highly – though informally – autonomous. This autonomy is significantly less restricted by the formal organization than by structural and technological constraints, which limit decision options considerably.

Individually-formed objectives are thus of especial importance for the enterprise, as they greatly influence the form and quality of operative activities and in this way have an effect on the total system. At the same time, for shop floor control – and particularly in distributed, heterogeneous systems – formal goals cannot simply be forced from the top down. The call for stronger inclusion of the reality of the shop floor as a sociotechnical system thus becomes a clear demand upon the design of goal systems. In order to obtain greater acceptance by employees, the formal system must be explained, and employees should be able to adapt it for use in their own work areas. Acceptance is required also of the fact that the goal system can not be free of contradictions – as WARNECKE (1992, p. 138) demands for the „fractal company" – and that it cannot serve only enterprise objectives Rather, the ability of decision makers within distributed production to compromise must be fostered and increased in targeted fashion.

3.5. Control decisions in practice

The majority of control decisions are made on the spot upon the basis of an unexpected triggering event and are therefore very difficult to formalize. In such cases, an order affected by the event is usually chosen as the starting point for decisions and planned anew, with other orders being more or less clearly taken into account. But even regularly occurring control decisions – such as weekly planning by the supervisor – diverge considerably from planning theory and derived sequencing rules– for example FIFO (first in – first out), EDT (earliest due date), SPT (shortest processing time) and so on. Here again, the way humans proceed is individual and hardly formalizable. After all, control in the case of reactive decisions takes place under great time pressure, and even proactive decisions must be made in a minimal amount of time.

Figure 12 shows as an example the representation of a possible decision-making process in the world of practice. In contrast to reactive control, this example shows proactive components. In other words, here there is a certain amount of time to plan in advance, and there are considerable degrees of freedom for decision-making. It is important to note that in the example there are several simultaneous orders, and that they must be planned for almost simultaneous processing or in small iterations. Information at the outset comes from the master schedules from central planning and by means of a priority list produced by a computer-based PPC system. Priority lists are – as observed by the author in many factories – not given very serious consideration by shop floor employees. However, while they may not explicitly determine the results of sequence planning, they do influence the sequence of planning decisions themselves. As a first step generally, the feasibility in the current situation is assessed: 'Are machines, materials, NC programs, etc. available?'. There follows a rather more social evaluation: 'Do we really want to start work on this now, or do we wait for the shift change?' As a next step, the aspect of cost minimization gains significance, and finally, there is attention paid to minimizing individual work loads – in set-up work, for example. Work load and required effort clearly play a more important role with regard to employees and one's own department than it does with regard to machines and other departments – the issue is 'just so long as it does not require more from our department!'

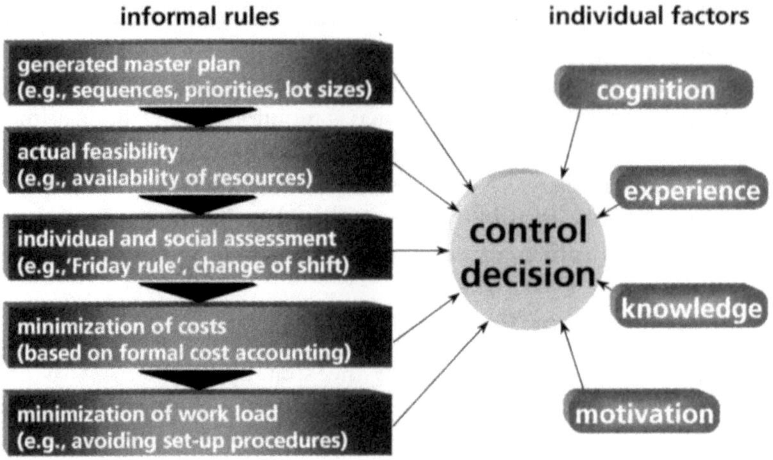

Fig. 12 Determining factors and rules for control decisions at the level of the shop floor (a typical example).

Further, the perception, experience, knowledge and motivation of decision makers plays an important role in control decisions (see BÖHLE & ROSE 1992 on the concept of experience knowledge). Knowledge is acquired formally and can be transmitted by means of training. Experience grows on the basis of reflection upon one's own work and is not formalizable to the same degree.

Experience and decision authority

„Adaptations and interventions guided by experience are necessary [to control] in production, particularly when external influences [...] demand immediate and unplanned-for changes in the production situation" (FLEIG & SCHNEIDER 1995, p. 6). The knowledge of experience referred to is in part intuitive in nature and accessible in concrete situations. At the same time, it is highly personal – bound to individuals – and it cannot easily be transmitted. Experience grows through work activity. Actors learn continuously on the basis of the results of their actions and thus gain increasing experience.

The prerequisite is that actors be confronted with the results of their work. This can take place by means of feedback, for example. Here the problematic nature, in numerous aspects, of the task field in shop floor control is shown:

1 The controlling individual is excluded from an effective or explicit feedback loop due to the formal system:

- The individual – as an employee in central planning, for example – receives only 'muffled' feedback due to spatial, temporal and structural-organizational separation.
- The individual does not even perceive that he has made a control-relevant decision and, at the same time, because he is not embedded within the formal system, he is excluded from formal feedback.

Another structural characteristic of control proves to be even more problematic:

2 The effect of an individual control decision cannot be discerned, because

- Control is performed by several controllers continuously and simultaneously within the same system boundaries
- The effects of a control decision cannot be delimited in a temporal sense or in their influence.

It is mainly the structural determinants under which control in production takes place that are revealed to be the most problematic. These determinants prevent the actor from forming anything more than only a very limited picture of the effects of his actions. As a consequence, experience can be false.

However, true experience may be furthered by a work and task design which encourages and promotes the growth of experience. In addition to numerous other criteria, such as degrees of freedom, demands and the like, this means that in shop floor control, a decision-maker wherever possible should be settled in close proximity – both spatial and organizational – to the area influenced by his decision. Expressed another way, any person who must deal with the consequences of a decision should be given authority to decide and responsibility for that decision. In the case of shop floor control, this means that this aspect of experience-promoting work design alone shows that control should take place in the closest possible proximity to shop floor operations.

Information and perception

In a particular decision-making situation, perceptions are an essential determinant of decisions. Here, perception is dependent upon, on the one hand, the information principally available, and on the other hand, dependent upon certain framework conditions which will determine whether perception takes place at all and to what degree. Practice shows clearly that the information utilized by the human decision maker goes far beyond the information that is supplied by the formal system. In on-the-spot decisions in particular, individual bits of information play an important role. This sort of information has to be gathered by or is already known to the decision-maker. For example, often information is required as to the current processing status of a work order on a certain machine. This information is considerable more familiar to and more simply accessible to a decision-maker integrated within the shop floor, who works in direct proximity to operations or is himself in part an operator, than to a decision maker far removed from the shop floor, who attempts to gain an overview from within his office. It must be noted that the shop floor-integrated decision maker will have the information at his disposal much more quickly – even when it must first be gathered. This is due both to his routinized proficiency and to his familiarity with the social and organizational conditions on the shop floor. At the same time, a shop floor-integrated decision-maker is as a rule more capable of testing and judging pieces of information with regard to their significance and truth.

4. Systems design for shop floor control

The numerous strategies for reorganization and new software technologies notwithstanding, the problem of shop floor control, as the central task of operative production management, remains unsolved. Rush orders, missing parts lists, backorder lists and 'due date chasers' still belong to the everyday in many enterprises[7]. Until today, „the shop floor is not under control, it is out of control“ (MELNIK ET AL. 1985, vii).

4.1. The logic of failure

The reasons for the failure of past approaches are manifold, but close examination allows us to reduce them to a few problem areas. As the basis of the 'logic of failure' presented here, we can view the lasting disregard of the reality of shop floor control in many popular design approaches:

- Disregard of the systems character of shop floor control:
  Solutions are based upon simple and quantitative models based on linear causal chains. The diverse characteristics of the shop floor as a sociotechnical system are not taken into account.
- Dominance of electronic data processing as the means of design:
  The design and implementation of data processing-based systems for shop floor control is treated as equivalent to the design of shop floor control itself. The significant issues of organizational design are ignored.
- Inadequate project management and inadequate organizational development:
  The design of the formal system is treated as equivalent to the design of the total system. It should not be forgotten here that the development logics of technical systems design differ from the logics of social systems design. Individual interests and powers of authority as well as enterprise-political and social aspects are overlooked.

4.2. The systems reality of shop floor control

Production is a complex, sociotechnical system. Heterogeneity and distributed, decentralized structure, limited and uncertain resource availability, as well as variations and disturbances are the significant characteristics here which contribute to the complexity of the task of the planning and control of production. Shop floor control in the sense of a technical control system or – as it could be described in view of the underlying management principle – through means of a technocratic approach, becomes an impossibility. In designing shop floor control, the following must be taken into account:

7 The "Hurry-up department" of a New England company in the 1890s, that PLOSSL (1985) describes, seems to be persisting in many enterprises until today.

- The structure of production is by means of its heterogeneity and decentralization so complex that it can no longer be controlled through a universal and clearly defined shop floor control.
- Shop floor control in the broader sense itself forms a heterogeneous system, which is not oriented goal-wise only towards the control of production, but rather is a system in which various interests and interpretations act in combination.

Added to this is the fact that production is a distributed system involving distributed decision-making and the implementation of decisions. Many decisions made within the framework of local control have an effect upon other cells distributed within the system. In such cases, coordination of local control is required. Problems which arise typically in this regard are:

- the short temporal periods in which joint decision-making is to be achieved
- the limited degrees of freedom for altering original planning
- fuzzy knowledge of the potential consequences of decisions and of the cells fundamentally affected by decisions
- contradictory goals of the affected cells.

In the reality of the shop floor, happenings deviate rapidly from original planning. These deviations become greater and more extensive in scope through time, because a disturbance, downtime of a machine, for example, produces further disturbances, such as the loading of an alternative machine which is then now longer available for its originally planned utilization. In principle, buffers, in the form of material inventory for example, represent one possibility of restricting the chain reaction of disturbances. However, the use of buffers as a control aid rapidly becomes uneconomic. An attempt to constantly adapt planning is also not realistically feasible, as this would cause uneconomical high expense in terms of data entry.

It is thus not generally possible to arrive at a completely optimal, satisfactory solution. Instead, originally fixed master data, such as production dates, must be questioned and loosened up – for example, in conjunction with the client through new scheduling. The options of decision-making in a dynamic field like production are very limited and, due to the high complexity of the total system and the limited degrees of freedom, often not even recognizable.

4.3. Approaches to systems innovation

The circumstances described put control in its traditional interpretation under question. Shop floor control is not only responsible for optimal functioning of shop floor operations; it has the much more fundamental task of making sure that things function at all. With this, the 'maintenance' of the total dynamic system becomes the task of control.

The task of control is thus reactive and situational, that is, the decision to select a certain procedure must made rapidly and with only limited knowledge of the situation. In control engineering terms, it is here that the limits of the translation of closed loops is demonstrated: at the level of operations, action must be taken in many cases without feedback. Control is then, of necessity, guided by intuition. This is why experience knowledge and informal processes become central variables in the design of control systems.

With this, we now understand the particular importance of shop floor control: it is confronted directly with system behavior, and further, it determines system behavior directly. Control becomes a propelling variable within PPC in its entirety. Planning – as a unit of both function and organization – at best serves to support the system; it determines system behavior only in an indirect way.

The preceding discussion relativizes the view of control which predominates in the literature. Control is itself as heterogeneous and as difficult to grasp as the production processes to be controlled. In conclusion, the relevant characteristics of shop floor control are summed up as follows:

1. Shop floor control is a task within a sociotechnical system, that is, there exist multiple views/realities.
2. Shop floor control takes place within the field of tension between the formal and informal systems
3. Control as an activity can neither be defined nor localized unambiguously, but rather is effected by the actions of employees.
4. Shop floor control is a task within a networked system. This means that action in one area always has an influence upon the events in other areas.
5. Control is parallel to and closely linked to the physical system. Planning, in contrast, is sequential and its link to the physical system is weak.
6. Control takes place within a very heterogeneous information field.
7. For control, only a small proportion of total work time on the shop floor is required, but it is of great structural relevance.
8. Control is not a business process, but rather a work activity.

The nature of shop floor control as a sociotechnical system must be taken into account consistently. This will provide the basis for lasting systems innovation.

5. References

ALIOTH, A. (1980): Entwicklung und Einführung alternativer Arbeitsformen. Verlag Hans Huber, Bern.

BECKER, R.A. (1985): All factories are not the same. Interfaces, 15:3, p. 85-93.

BÖHLE, F., ROSE, H. (1992): Technik und Erfahrung. Arbeit in hochautomatisierten Systemen. Campus, Frankfurt.

DÖRR, G. (1991): Die Lücken der Arbeitsorganisation – Neue Kontroll- und Kooperationsformen durch computergestützte Reorganisation im Maschinenbau. Edition Sigma, Berlin.

EMERY, F., THORSRUD, E. (1982): Democracy at work. Martinus Nijhoff Social Science Division, Leiden.

FLEIG, J., SCHNEIDER, R. (1995): Erfahrung und Technik in der Produktion. Springer, Berlin.

FOX, M.S. (1987): Constraint-directed search: a case study of job-scheduling. Morgan Kaufmann, San Franscisco.

KROHN, W., KÜPPERS, G. (1990): Selbstreferenz und Planung. In: NIEDERSEN, U., POHLMANN, L. (Eds.): Selbstorganisation. Duncker & Humblot. Berlin. p. 109-127.

MALSCH, T. (1987): Die Informatisierung des betrieblichen Erfahrungswissens und der „Imperialismus der instrumentellen Vernunft". Zeitschrift für Soziologie, 6:2, p. 77-91.

MANSKE, F., WOBBE-OHLENBURG, W. (1985): Fertigungssteuerung im Maschinenbau aus der Sicht von Unternehmensleitung und Werkstattpersonal, Teil 1. VDI-Z, 127:11, p. 395-402.

MCKAY, K.N., SAFAYENI, F.R., BUZACOTT, J.A. (1988): Job-shop scheduling theory: What is relevant? Interfaces, 18:4, p. 84-90.

MCKAY, K.N., BUZACOTT, J.A., SAFAYENI, F.R. (1989): The scheduler's knowledge of uncertainty: The missing link. In: BROWNE, J. (Ed.): Knowledge-Based Production Management Systems. Elsevier, Amsterdam. p. 171-189.

MCKAY, K.N., BUZACOTT, J.A., CHARNESS, N., SAFAYENI, F.R. (1992): The scheduler's predictive expertise: An interdisciplinary perspective. In: DOUKIDIS, G.I., PAUL, R.J. (Eds.): Artificial Intelligence in Operational Research. Macmillan, Houndmills and London. p. 139-150.

MCKAY, K., MURGIANO, C.J. (1991): Decision support for planning and scheduling within the automated factory. Proceedings of AUTOFACT '91 conference. SME, Dearborn.

MELNIK, S.A., CARTER, P.L., DILTS, D.M., LYTH, D.M. (1985): Shop floor control. Dow Jones-Irwin, Homewood.

MELNIK, S.A., CARTER, P.L. (1987): Shop floor control – Principles and practices and case studies. APICS, Falls Church.

NULLMEIER, E., BRINKOP, T. (1991): Die Reduktion der Fertigung auf Informationsverarbeitung, oder: der Einzug des Büros in die Werkstatt. In: KUTSCHA, J. (Ed.): Wirksamkeit von Massnahmen zur Gestaltung von Büroarbeit. Wissenschaftliche Diskussionspapiere, Heft 7. Bundesinstitut für Berufsbildung, Berlin. p. 51-72.

PLOSSL, G.W. (1985): Production and inventory control. Principles and techniques. 2nd Ed. Prentice-Hall, Englewood Cliffs.

RIIS, J.O., FRICK, J. (1991): Organizational learning: A neglected dimension of Production Management Systems Design. In: ELORANTA, E. (Ed.): Advances in Production Management Systems. Elsevier, Amsterdam. p. 141-149.

SCHEER, A.-W. (1992): Computer Integrated manufacturing – Towards the factory of the future. 2nd Ed. Springer, Berlin.

SIMON, D. (1995): Fertigungsregelung durch zielgrössenorientierte Planung und logistisches Störungsmanagement. Springer, Berlin.

SUSMAN, G (1976): Autonomy at work: A socio-technical analysis of participative management. Praeger, New York.

ULICH, E. (1994): Arbeitspsychologie. 3rd Ed. vdf/Poeschel, Zürich/Stuttgart.

WARNECKE, H.-J. (1992): The Fractal Company – A revolution in corporate culture. Springer, Berlin.

WEICK, K.E. (1990): Technology as equivoque: Sensemaking in new technologies. In: GOODMAN, P.S., SPROULL, L.S. (Eds.): Technology and organizations. Joosey-Bass, San Francisco. p. 1-44.

WESTKÄMPER, E., JESCHKE, K. (1995): Massnahmen zur Null-Fehler-Produktion im Maschinenbau. VDI-Z, 137:3/4, p. 84-87.

WIENDAHL, H.P. (1995): Load-oriented manufacturing control. Springer, Berlin.

WIENDAHL, H.P., AHRENS, V. (1995): Planning and control in self-organized production systems. WGP Annals Production Engineering, II:2.

WIGHT, O.W. (1974): Production and inventory management in the computer age. CBI Publishing, Boston.

Control on the Verge of Chaos – Analysis of Dynamic Production Structures

Mirko Dobberstein

Production is a dynamic system. From the perspective of production research, the instrumental-cybernetic approach of J. W. FORRESTER (1961) is significant. It is based essentially on work with feedback networks of state variables of an observed system. FORRESTER'S approach to computer-aided, cybernetic systems theory became known as „industrial dynamics“. Industrial dynamics represents the attempt to use systems analysis to describe and analyze the dynamics of networked, feedback-looped structures of an enterprise. The basic assumption is that the dynamic behavior of an enterprise can be determined principally by an underlying structure (which can be modeled), which is itself mainly determined by internal temporal shifts in the downstream flow of materials and in the upstream flow of information. The structure comprises not so much physical aspects of machine and production processes, but rather quantitatively measurable and qualitatively describable factors that affect important decisions in the planning and control of production systems (ROBERTS 1985). Industrial dynamics thus links the three most important dimensions of socio-technical systems: *technology* (machining, manufacturing technology, plant buildings, etc.), *organization* (explicit and implicit constraints, information systems, decision making hierarchy, etc.) and *human beings* (employees, knowledge, customers) and studies their dynamic ‘interaction’.

1. Approaching production as a dynamic system

The basic structure always contains various causes of systems-immanent, non-linear reinforcing effects, temporal delays and feedback. Interlocking feedback loops can lead to fluctuating (randomly varying) system behavior which results in high production costs, warehousing costs, employment variations (undertime, overtime) and so on. The fluctuations are not exclusively dependent upon external factors, but rather are frequently caused by interrelations inherent to the system (FORRESTER 1961).

At the forefront of systems analysis stand the higher connections in the enterprise which determine the links and the dynamic behavior of employees and the flow of orders, materials, capital and information. Feedback loops in information flows which serve as the basis for decision-making within the enterprise must be pinpointed. In doing this, system boundaries have to be drawn which go beyond the company itself. System boundaries enclose not only the company, but also developments in economic and market activity (or customer behavior), dependencies upon suppliers and upon market conditions. Meaningful system delineation may be performed by tracing back all relevant cause and effect chains of individual systems flow. Figure 1 is a simple schema of cause-and-effect connections in the dynamic system of manufacturing, warehouse, distribution and market. Industrial dynamics attempts to discover, formulate and analyze in detail these connections, in order, to allow decision-making for planning and control of the system.

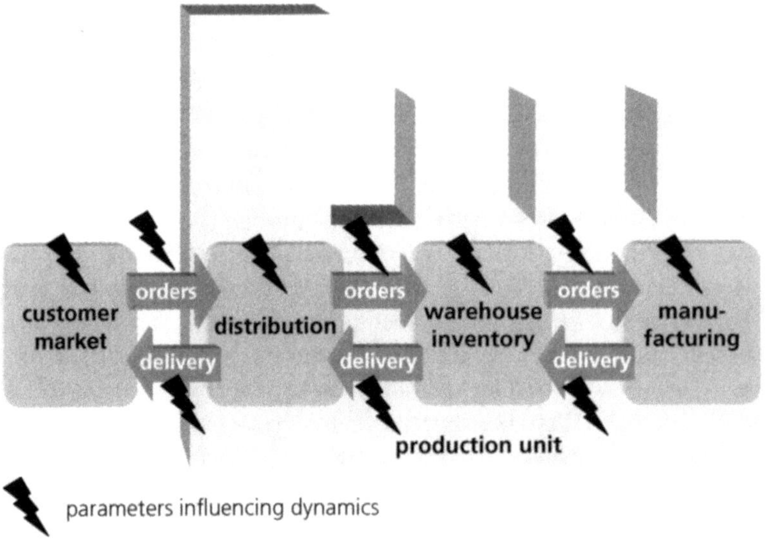

Fig. 1 Simplified example of superordinate cause-and-effect connections (following FORRESTER 1961).

Questions central to analysis include the following examples:

- How do small fluctuations in sales lead to large effects in production?

- How does the system behave in reference to varying sales rates (stochastically distributed, periodic, constant)?
- What are the strategies that lead to internal oscillations, to chaotic system behavior and to stable system behavior?

Systems theory profits from the findings of the theory of non-linear, dynamic systems, as here the assumption was confirmed that the behavior of complex systems is for the most part determined by a few control parameters (HAKEN 1991, 1993 & 1995). Through the aid of the theory of non-linear systems, the assumption can be confirmed that fluctuations of system behavior are often dependent upon endogenous factors (FORRESTER 1961). More specifically, systems theory approaches to economics has gained new impetus through the findings on non-linear dynamics (HOMMES 1991). RASMUSSEN & MOSEKILDE (1988) already demonstrated scientifically that deterministic chaos exists in (certain) systems theory models of manufacturing-distribution-customer systems. According to TOWILL & DEL VECCHIO (1994), systems theory is particularly suited to the investigation of the dynamic behavior of supply chains, in which several subsystems are linked sequentially by the demand for products. In such supply chains, small fluctuations in customer demand become further reinforced by control strategies, independent of each other, within individual subsystems (Figure 2 & 3).

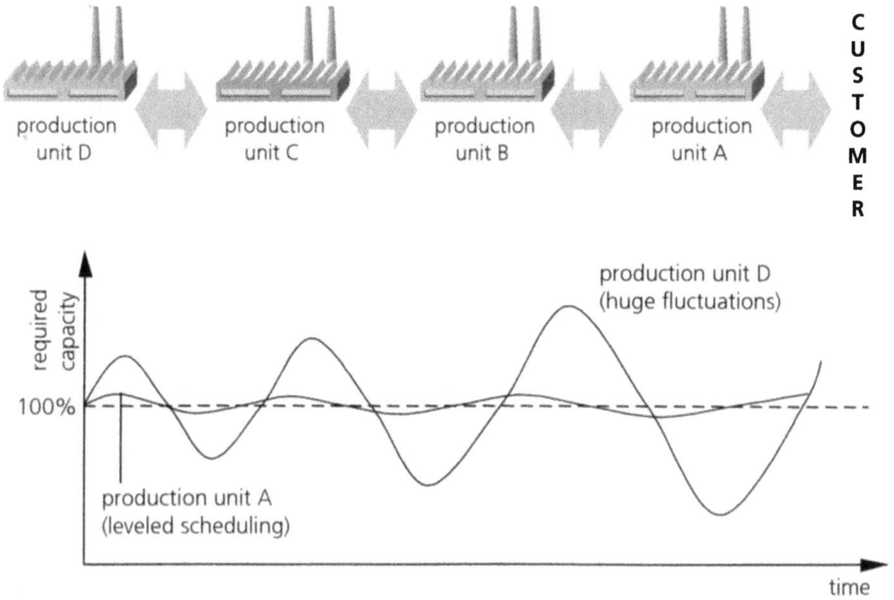

Fig. 2 Amplification of demand as transmitted along a logistic chain (HOULIHAN 1987).

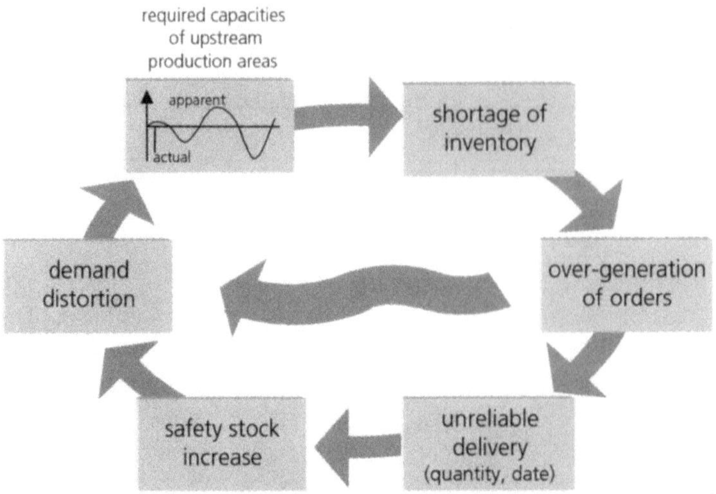

Fig. 3 Amplification effect as described by HOULIHAN (1987).

When the demand for products is transmitted over a whole series of company domains, each having its own regulating strategies, distortion of the actual demand becomes greater from sub-system to sub-system. The reason is that uncertainties regarding the actually required quantity of product increase, the quality and accessibility of information decrease, reserve inventories are produced, and so on (HOULIHAN 1987). TOWILL & DEL VECCHIO (1994) point out that this increasing effect – the so-called „amplification effect" – is normally determined by much shorter time periods than would be expected (weeks/months as opposed to months/years) and that the effect is very frequently manifested in industry.

Cybernetics approaches towards structuring production and towards solving the control problem are based upon a feedback systems model. Crucial to the efficiency of such feedback-oriented approaches is the speed of reaction. In complex dynamic systems, reaction time is often not sufficient. For this reason, the scientific field of non-linear systems has had success utilizing other methods of planning and control of dynamic systems, which base upon active influencing of the system (DRESSLER 1994, HÜBLER 1992). Transposing this principle to the control of production systems, production control must act preventively, in a proactive rather than reactive manner, to past events (BAGANZ 1994, SCHERER & DOBBERSTEIN 1996).

2. Complexity in production and its effects

In general, production systems and the processes and decision-making situations within it are rated as complex (EVERSHEIM 1994, WARNECKE 1992, CHRYSSOLOURIS 1994). Complexity hinders understanding of relations within production systems and presents an obstacle to goal-oriented design of control structures. WIENDAHL & SCHOLTISSEK (1994) describe the difficulties involved in mastering the complexity in production as follows:

> „Yet the complexity of systems, in particular in this case the complexity of production systems, is the crucial cause of many management problems in industrial companies.“ (WIENDAHL & SCHOLTISSEK 1994).

To design control systems for complex production systems, a clear understanding of the complexity is therefore a requirement.

This raises the issue of the forms of complexity which arise in production systems and production control systems. The finding that complexity is a determining factor in view of production systems is not novel (EVERSHEIM 1994, SCHÜPBACH 1994). What has changed is the way is which complexity is viewed. Complexity was long regarded merely as an additional, aggravating condition to be dealt with. For this reason, mainly linear or linearized models and 'Ceteris-Paribus'[1] perspectives stood in the foreground (HAYEK 1972). However, recently the influence of systems theory had caused a change in our understanding of complexity. Complexity is now seen as one of the causes of the difficulties of mastering production processes (SCHERER & DOBBERSTEIN 1996, SCHÜPBACH 1994). Processes often termed chaotic are caused by the complex structure of production systems. Complexity is responsible for the dynamics which make production control and the planning and prediction of processes so difficult. Due to this, complexity can no longer be seen as an accompanying difficulty or condition, but rather must be viewed as an essential feature to be included into design.

In order to explain and grasp the complexity of production systems, we will fall back upon analogies to other complex and dynamic systems. Through this, we aim to achieve the following:

- better analysis and explanation of the structure and behavior of production systems,
- derivation of a method of structuring control systems and action regulations for operations.

2.1. Various descriptions of the concept of complexity

As a rule, an observer describes a system as complex when it can no longer be described easily. The perception of complexity is here based upon a subjective feeling. In the scientific view, complexity does not, however, depend simply upon a system being complicated and difficult to understand due to numerous elements

1 'Ceteris-paribus' describes the assumption that during analysis only one variable is changing while the others remain constant. Thereby an independence of influences is assumed.

and relationships, but rather particularly depends upon the dynamics, frequency and strength with which system and environmental factors change through the course of time (RÖPKE 1977).

The example of a simple double pendulum illustrates, however, that many approaches do not accurately describe complexity in the sense of dynamics. A double pendulum consists only of two simple part-systems and a simple connector (Figure 4). Although the total system may be described exactly by means of physical laws of motion, the orbit curve of the motion may be predicted only for a very short time period. A strong enough push[2] through an outside force the system performs chaotic, i.e., without regularity. The causes of this are found in the strong non-linearity of the relationships.

If we examine the controllability or predictability of the double pendulum, this system is to be seen as a complex system. In contrast to systems with many elements, its complexity is nor based upon the complexity of the system, but rather upon the type of dynamic coupling of the part-systems. Both part-systems are of themselves non-linear and thus sensitive to external influences. In addition, the two part-systems are coupled by a mechanism which transmits the dynamics of the one part-system to the other, directly and without damping (considering idealized conditions).

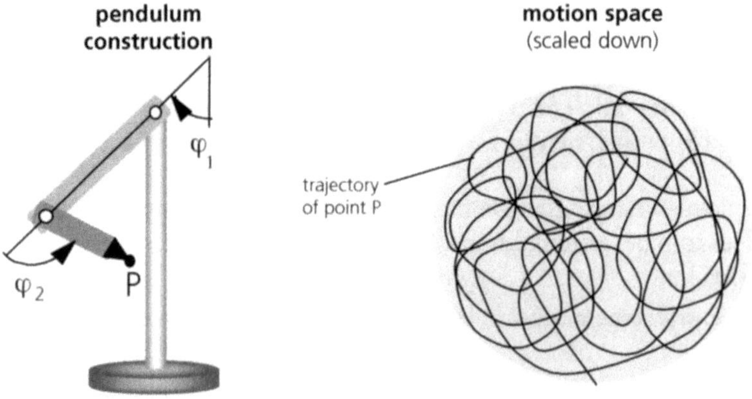

Fig. 4 Pendulum as an example of a system simple in structure but complex in its dynamics (see RICHTER 1993).

The inner pendulum swings from a fixed hanging point and rotates; its position is characterized by the angle φ_1. The second pendulum is hung within the first, and its position is relative to the first with an angle of φ_2. In addition to the angles, angle velocity is required to characterize the motion state of the system. When swings exceed the linear range, the double pendulum can move into 'chaotic' orbits.

---

2 When motions are very small, the system remains nearly linear.

Similar systems constellations are also found in logistics systems. Economic production demands that the capacity load of all resources is as high as possible. This incontrovertible constraint results in only small buffers, or tolerance, of dependency among part-areas. Through this, the dynamics of individual areas are easily transmitted to other areas and cascade through the system. This occurs wherever many processes stand in strict dependency upon each other – as, for example, technology-determined manufacturing sequences – and many different and scarce resources are needed at the same time and different orders stand in competition. In one-of-a-kind and small-series production, these are the conditions which almost always exist. The introduction of the just-in-time (JIT) philosophy often led to these conditions of dynamic effects for serial production as well. The JIT requirements of automobile producers in particular cause great logistics difficulties for automotive suppliers (SCHUMANN ET AL. 1994). Set-up times and production times are often longer than the reaction time available, if delivery dates, quantities or indeed the articles themselves are changed at short notice.

The considerations show that for the analysis and description of logistics systems, the concept of complexity must include the aspect of dynamics as an essential complexity feature. Approaches in this vein are found in so-called „new cybernetics" (PROBST 1981). One important finding is that structure and behavior are very closely linked, affect each other mutually and cannot be clearly separated:

> „Within the nonclassical paradigm, the behavior makes the structure and vice-versa; hence they are not distinct." (PASK 1978).

The behavior of complex systems is dependent upon the interaction of influencing variables, positive and negative feedback, intermeshed feedback systems, etc. In addition, behavior is non-linear, fluctuating, exponential in growth and can reverse itself (FORRESTER 1968). Thus, complex systems can be predicted only with difficulty – or not at all. FORRESTER shows repeatedly that it is this very unexpectedness of behavior, possible instability with small perturbations, but stability with large perturbations, shift from negative to positive feedback, growth acceleration, and so on, which characterize complex dynamic systems. Complexity, therefore, is determined by two factors:

1. Complicated structure:
   Number of subsystems, number and variety of elements, type and frequency of connections, etc.

2. Dynamic behavior:
   Number of possible state transitions, speed and strength of change, character of the relations between cause and effect, etc.

Whereas it is easy to find features of complexity from a structural perspective, which describe complexity not only qualitatively but in part also quantitatively, it is difficult to find descriptive and measurement variables for dynamics.

One generally valid measure of complexity is ASHBY's concept of *variety* (ASHBY 1958 & 1972). Variety indicates the number of possible system states a system can enter. According to ASHBY (1958), BEER (1966) and PASK (1978), variety is dependent upon

- the number of elements,
- the relationships and
- the variety of the elements themselves , e.g., number of possible system states of the elements.

Variety is an important aspect in connection with systems control. If we consider that the control of logistics processes is largely determined by decisions, variety seems to be a measure well suited to determining complexity. The measure yields information on the temporal intervals within which decisions must be made, on the situations that call for decisions, on the possible consequences of decisions, etc. Through answering these questions, we can establish the degrees of capability and capacity that a control system will require.

And yet, variety is an abstract measure which allows us to view principal aspects of complex dynamic systems. For the concrete design of control structures, variety is not very practicable and is thus only indirectly suited. In production systems, the set of all possible states is, as a rule, neither known nor ascertainable.

Beyond the necessity of concretizing the measurement variable of variety, there is a principal difficulty in determining the complexity of systems. BEER (1966) points out that the measurement of complexity is dependent upon the observer and corresponding system formation. Depending on one's perspective on the system and the objectives followed, systems can be variously delimited, whereby delimitations should not be seen as equivalent to objects (GOMEZ, MALIK & OELLER 1975). The measurement of complexity is thus subject to subjective criteria of systems model building.

2.2. Classifying complexity

Objective and quantitative evaluation of complexity according to characteristic variables is not possible. For this reason, complexity can be assessed through the aid of classification.

An essential criterium of production systems control is the predictability of the future course of the system. Only where predictability is high, reliable is planning possible. Operational production planning must, with its degree of detailing and its planning range, be oriented towards the principal possibilities of predictability. Predictability is thus an important criterium in classification.

In general, systems can show varying, dynamic degrees of complexity, which result from the number of parameters, motion equations, state variables and so on, which are needed to describe the system. Following CRAMER (1993) and BEER (1959), three fundamental degrees of complexity may be distinguished for deterministic[3] and stochastic[4] systems, whereby the quantitative increase in complexity brings forth qualitatively new system characteristics (Table 1).

	deterministic systems	stochastic systems
subcritical complexity	may be described by equations; exact predictions possible; only few elements and relationships.	simple system but showing no predictable behavior; although there are few elements and relationships, behavior can be predicted only statistically.
critical complexity	systems form sub- and part-systems (usually many); predictability subject to practical but not absolute limits; behavior principally predictable; superposition principle holds.	system describable to a large extent, but allowing only general, statistical predictions.
fundamental complexity	systems which – despite deterministic initial conditions – result in non-deterministic or deterministic-chaotic states; prognosis not only practically impossible, but also principally impossible; superposition principle no longer holds.	system can never be described completely; no unambiguous determination of behavior possible.

Tab. 1 Classification of complexity following CRAMER (1993) and BEER (1959).

In real systems, usually the three degrees of complexity coexist. The principle differentiation between deterministic and stochastic systems is thus, in most cases, no longer possible. Still, the degree of complexity places an important requirement upon the design of production control and the choice of appropriate control methods. These must be oriented towards the prevailing degree of complexity.

The frequent fundamental complexity of real systems requires description which transcends the formulation of deterministic relationships. A purely deterministic perspective, such as that assumed by production planning and control (PPC) in most enterprises, does not do justice to the fundamental complexity of real systems. Most production planning and control approaches treats real systems as if they evidenced subcritical complexity.

Due to the non-deterministic character of fundamentally complex systems, the temporal and structural phase shift from critical to fundamental complexity may not be determined. What can be established is that fundamental complexity can be caused by neither the effect of external, stochastic influences alone (stochastic system) nor solely through a very high number of parameters describing the system (only practical limits, no principal ones) (RASBAND 1990). Fundamental-complex systems are, in contrast, usually systems which, like modern production systems,

[3] Deterministic systems can be described by exact laws (equation or differential equation systems).
[4] Stochastic systems can not be described by laws, but only though the aid of statistical descriptions.

evidence an intermeshed structure of subsystems having numerous feedback loops and non-linear structures.

2.3. Taking stock of complexity approaches

A general, unambiguous and certain definition of complexity appears to be impossible, because the facets of complexity description are too numerous. Description and definition must therefore always be subject to the particular objective. But some essential aspects may serve as the basis for designing production control structures:

1 Complexity is an independent characteristic of production systems and must accordingly be viewed as a design feature.

2 Along with structural features such as number and type of elements and relationships, variety and dynamics are fundamental components of complexity which have a decisive influence upon the controllability of production systems.

3 Production control must be oriented towards the degree of system complexity. Control structures which neglect this aspect will not succeed.

Successful control of production shows itself in the quality and temporal scope of predictability. Only if a minimal degree of predictability exists may delivery dates and delivery quantities be fulfilled with certain reliability. For this reason, the aspect of dynamic behavior is particularly important with regard to design of production control. This means, on the one hand, that false assumptions as to the degree of complexity present in a production system will result in non-functioning control structures. On the other hand, it is here, within this aspect, that the real potential for the design of control lies. As long as there is fundamental complexity, prediction of the course of production is impossible, and thus the controllability of the system must be assessed as low. Therefore, structuring must aim towards reducing the degree of complexity faced by an controller.

Up to now, there have been no serviceable concepts or models in production technology which explain the analysis and handling of real life phenomena comprehensively. A promising approach is to consider the findings of the theory of non-linear dynamic systems.

3. Fundaments of the theory of non-linear systems (chaos theory)

The theory of non-linear dynamic systems is generally known as 'chaos research' or 'chaos theory'. Chaos theory is based upon a holistic 'perception' of complex systems and seeks to do justice to the fundamental-complex nature of real systems. The term 'chaos theory' serves as an umbrella covering a multitude of heterogeneous fields of research (Figure 5), which in part have only few aspects in common. However, all of the areas do deal explicitly with the description and treatment of complex systems.

Popular associations to the term chaos suggest the meaning of lawless, statistical confusion, or the dissolution of all order. In contrast, the modern scientific conception of chaos refers to the irregular, temporal behavior of the variables characterizing a system, whose irregularity stems from internal relationships (EBELING 1986). The state of a system is not 'chaotic', but rather, is established exactly by the determining relationships. This is called the 'inner stochasticity' of a deterministic system. For this reason, in a scientific sense, one speaks of deterministic chaos (PEITGEN, JÜRGENS & SAUPE 1992). In deterministic chaos, the proverbial 'chaos' and 'order' stand in a complementary relationship (RICHTER 1993).

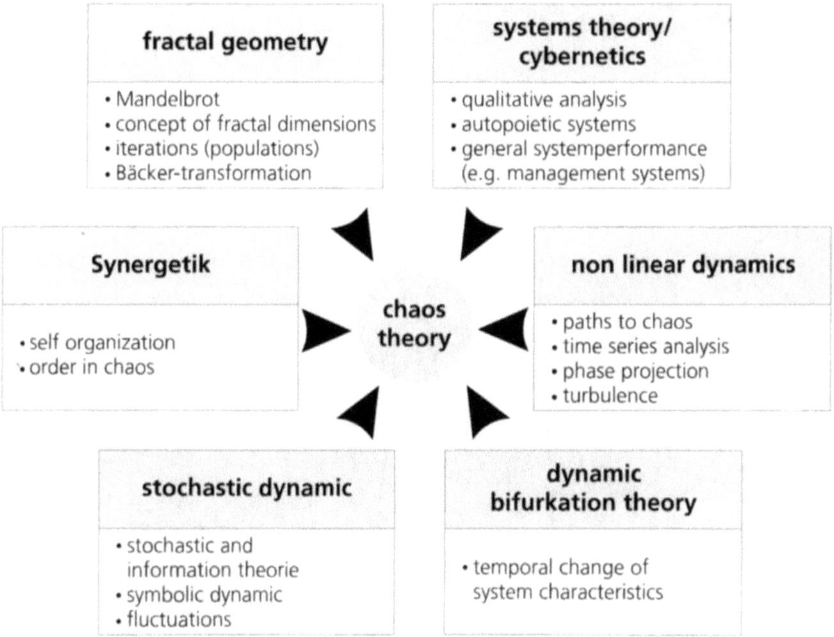

Fig. 5 Research fields in chaos theory.

The behavior of chaotic systems is subject to deterministic regularity, but the development of the system is not predictable due to its sensitive dependency upon initial conditions (MAYER-KRESS 1992). Sensitive dependency on initial parameters is understood as the vulnerability of the system to deviations from initial values. The smallest variations in these values lead irrevocably to totally different, unforeseeable developments (BRIGGS & PEAT 1990). In order to predict the temporal behavior of such systems on the basis of the given deterministic equations, the initial parameters would have to be set with theoretically infinite – mathematical – precision.

> The inexactness in chaotic systems is not due to any practical insufficiencies of methods or aiding tools, e.g., insufficient exactitude of computers. Rather, inexactness is the principal nature of such systems and is not even theoretically resolvable (BREUER 1993). Planning failure in such cases is not the result of inadequate description of the system, but rather the result of the system's 'uncertain regularity' (KUNICK & STEEB 1986).

Figure 6 illustrates sensitive dependency upon initial parameters in the Hénon system, which consists of two, coupled iteration functions. Two curves —dotted and solid lines – show system behavior for initial values in close proximity to one another (+ 0.00001). It is easy to see that the curves, which seem identical at the outset, clearly deviate from one another after a few iterations.

iteration	initial conditions	
$x_n + 1 = 1 - 1.4 * x^2_n + 0.3 * y_n$	—— $x_0 = 0$;	$y_0 = 0$
$y_{n+1} = x_n$	- - $x_0 = 0.00001$;	$y_0 = 0$

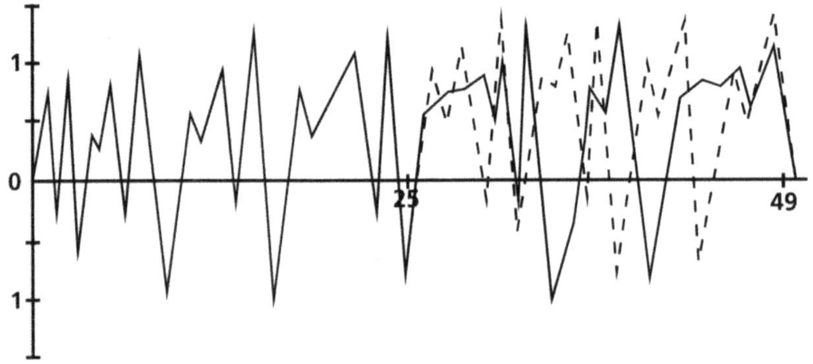

Fig. 6 The sensitive dependence of initial conditions in the Hénon system.

As we examine complex processes in production systems, let us establish that investigations based upon the findings of chaotic systems have nothing to do with any 'chaotic' conditions in a factory. Rather, the chaos theory perspective allows us to evaluate the seemingly irregular temporal behavior of system state variables, such as orders in hand, inventory, capacity or processing times. The relationships among these variables are often known, but due to the complexity of the processes within production systems, no unambiguous predictions as to their temporal development can be made.

3.1. Change – A scientific perspective

The findings of chaos theory have led to a new and altered scientific perspective. The differences in the chaos theory view can be best shown by means of comparing it to the methods previously used by empirical research to study real systems. In addition, we will differentiate the relations among *necessity*, *random events*, *possibility* and *impossibility* with regard to general investigation and analysis of production processes.

Traditionally, empirical research attempted to describe the behavior of real systems by deterministic laws (BRIGGS & PEAT 1990). The starting assumption was the existence of strongly causal cause-and-effect relationships. If the laws and regularities and the initial state of the object were known, then system behavior could be calculated for any point in time in the future. Expressed another way: the universal possibility of prognosis is based upon the necessity of the deterministic law (KÜPPERS 1993). The fact that the precision of the actual calculation is always limited was ignored, because it was assumed that the prognosis would remain within a verifiable range of certainty.

Sharply delimited from these were such processes for which 'it was not yet possible' to formulate a deterministic law. Experiments studying such processes cannot be replicated. The processes can be described only by means of statistical methods or probability theory (for example, the atomic model). Although conjectures can be made as to the 'probable' behavior of the system, deterministic prognosis of the system's behavior is not possible. In other words, random events determines system behavior, and random events are the reason that exact prediction is impossible.

In the special case of logistics processes, the basic assumption has been that once a certain degree of complexity and dynamics is reached, the processes are stochastic. A typical statement made upon this basis is one such as this: if measure X can be implemented, then lead time Y can be expected (AHRENS & JANSSEN 1994).

PPC systems were developed with the aim to improve planning and calculation of the proverbial 'chaos' in production systems. Expectations were, and remain, high that this path of integration will lead to resolution of conflict among objectives of the 'magical rectangle': short lead time, high delivery reliability, low inventory and high capacity utilization. Rather than to seek the underlying laws of the behavior of the system, the real system was subordinated to deterministic laws. In the world of practice, this has caused immeasurable difficulties, and the results have remained far below expectations.

Industrial experience has shown that real systems may not be forced into the mold of given scheduling plans, or at least, that the expected degree of predictability, prognosis, and plan-ability of system behavior has not been not achieved: 'Planning limps along far behind the actual developments'. What is unknown is whether this is due to unrealistic planning expectations or due to inadequate planning instruments (MANSKE 1986).

3.2. Causality in chaotic systems

In deterministic chaos, random events and necessity represent two sides of one and the same system development. Deterministic-chaotic systems are subject to laws, but their development – due to sensitive dependency upon initial conditions – is predictable only for short time periods and, in the long-term, must be seen as completely unpredictable (RUELLE 1980 & 1989). Thus on the one hand, there is in reality the possibility of prognosis. On the other hand, the question must always be raised as to the point at which systems behavior can no longer be predicted, that is, at what point the system is determined by random events[5]. Put another way, for deterministic-chaotic processes, it holds that:

- laws determine the principal impossibility of prediction and
- random events determine the principal possibility of prediction.

Upon this background, it is no longer possible to separate sharply necessity and random events on the one hand, and possibility and impossibility on the other. In contrast to separating, chaos theory introduces *chiasmus*[6], a relation which links together lawfulness, random events, predictability and unpredictability.

In describing the stability of dynamic systems, we can distinguish among various forms of causality (Figure 7). In strongly deterministic systems, e.g., simple mechanical systems, the principle of strong causality holds (KÜPPERS 1993). With strong causality, similar causes result in similar effects. In the case of deterministic chaos, however, the law of strong causality is broken due to the sensitivity of such systems. Even small variations in cause may, on the basis of exponential growth, lead to completely different effects. But the principle of weak causality does continue to hold for chaotic systems. The same causes, if they can be determined with mathematical precision, still lead to the same effects.

The fact that causality has only limited validity in complex, dynamic production systems has extensive consequences for all planning procedures and decisions. Control activities may no longer be linked to single events, but instead, must be oriented towards the course of a process. One possible solution is to move from reactive to proactive control (SCHERER & DOBBERSTEIN 1996).

---

5 In production systems: when measure *X* is implemented, the relevant lead times have the values *a, b, c,* ... up to point in time *Y*. Beyond point in time *Y*, any prediction will be inexact (AHRENS & JANSSEN 1994).

6 *Chiasmus,* from the Greek *chiasmos,* or crosspiece.

strong causal linkage

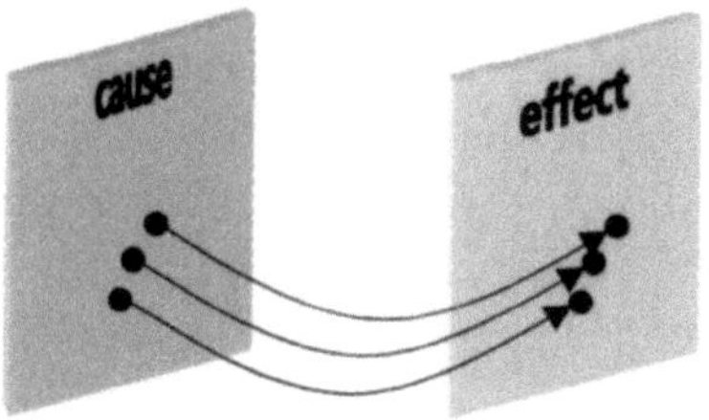

- same causes lead to same results
- similar causes lead to similar results
- minor causes lead to minor results
- major causes lead to major results

weak causal linkage

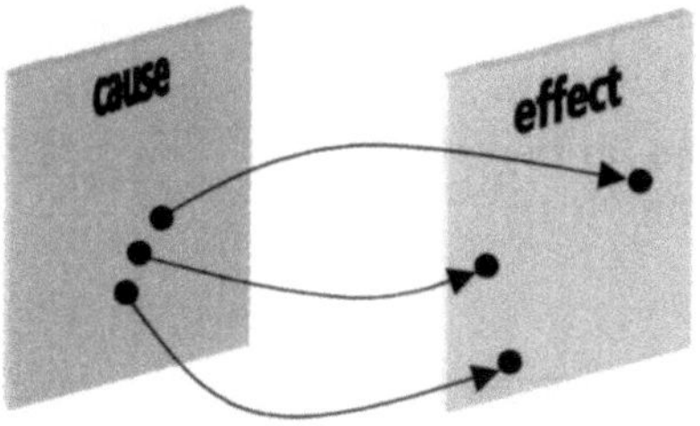

- same causes lead to same results
- similar causes may lead to similar results
- minor causes may lead to major results

Fig. 7 Forms of causality.

4. Analysis of the dynamics in production systems

According to the particular perspective taken, systems may be classified as 'deterministic', 'stochastic' or 'deterministic-chaotic'. In the general case, however, assignment to one of these classifications is not at all unambiguous. Real systems usually contain a varying mix of all these components (HAKEN 1995). In order to influence production systems in a targeted manner, it is necessary to have knowledge of their primary characteristics.

It has been established that in production systems the behavior of material and information flow is frequently irregular, that is, unpredictable. The question must be settled as to whether this irregularity is caused by deterministic-chaotic phenomena or, as assumed up to now, by stochastic influences, or most probably, by a combination of the two. This raises the issue of how, on the basis of data, conclusions may be drawn about the underlying nature of the dynamic system. There are two, principally different ways to gain a deeper understanding of such systems:

- The 'reduction approach' attempts to improve understanding of parts of the system.
- The 'complex approach' aims towards understanding and improvement of the total system (MORFILL & SCHEINGRABER 1994).

A complex approach reaches beyond the determination of statistical values characterizing the mean behavior of series of data. Here, previously inaccessible information on the structure and the dynamics of the processes underlying the time-series is extracted. This additional information may be used not only for the diagnosis of data sequences, but also to allow considerable improvement in our understanding of real production processes. In the following, methods of analysis used in the 'complex approach' will be introduced. These methods aid the non-linear analysis of logistical state variables.

The basis of all model design and analysis is the determination of the relevant state variables. State variables are those variables which – at any point in time – determine the state of the system completely and unambiguously. From these, all other system variables must be derived, through algebraic or logical expressions. State variables span a so-called phase projection, in that the states of the system may be recorded over time, in a time series (Figure 8).

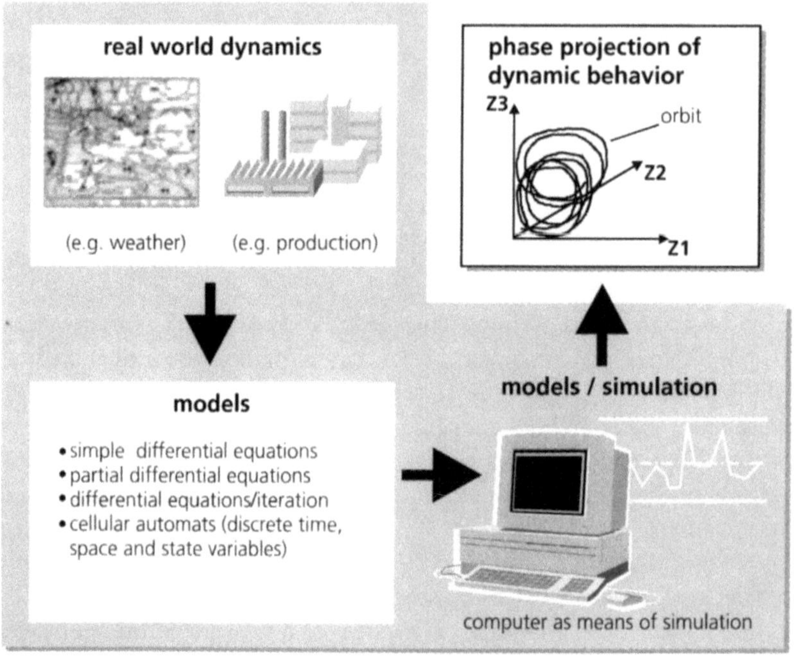

Fig. 8 Mapping of real systems in the phase projection.

The evolution of the system in the phase projection of state variables over time traces a continuous trajectory for a continuous process, or an orbit for a discrete process. The phase projection is one of the most useful concepts within non-linear dynamics. It allows systems characteristics to be described through non-linear phenomena (HAKEN 1983). Thus, here again, the degree to which systems are composed of deterministic, deterministic-chaotic or stochastic characteristics can be analyzed.

In principle, the trajectory of a non-linear system in the phase projection can have one of the following four modes (THOMPSON & STEWART 1986) (Figure 9):

1. Asymptotic stable or unstable:
   The state of a system develops towards a fixed point or away from a fixed point.

2. Periodic:
   The system state oscillates periodically on a so-called bounding cycle.

3. Quasi periodic:
   The system state oscillates with higher periodicity on a *k*-dimensional torus.

4. Deterministic-chaotic:
   The system state develops into a „strange attractor" – also called „chaotic attractor" (RUELLE 1980, RICHTER 1993). In this case the attractor appears in a set of a fractional Hausdorff-Besicovitch dimension (BRIGGS & PEAT 1990). Its shape can no longer be described by conventional geometry. But the structure does show spatial and temporal symmetries (fractal structure).

If the geometry of the phase projection is known, we can predict the sectors in which system behavior will be deterministic and the sectors in which behavior will be deterministic-chaotic. Within the phase projection, two sectors may be distinguished (WORG 1993): (1) the *'attractor'* describes the sector to which a set of trajectories with different initial conditions aims ($t \rightarrow \infty$). (2) Those sectors from which the trajectories drawn to the attractor originate are described as the *'attracted set'*.

The dynamic mapping of state variables which allows calculation of state $x(t+\Delta t)$ at point in time $t+\Delta t$ from state $x(t)$ at point in time t reflects the causality of processes within production systems. There are two fundamental possibilities for the nature of this causality in reality: (1) If the mapping of $x(t)$ and $x(t+\Delta t)$ is *unambiguous*, i.e., if the mapping rules are known, any future state can be derived unambiguously from the current state, and a deterministic relation exists over time. The relation may be constant, periodic or quasi-periodic. (2) If the character of the mapping is *not unambiguous*, meaning that there are several possible future system states, then the dynamics are either stochastic or deterministic-chaotic. In determining this, it must be taken into account that a production system may include a whole spectrum of qualitatively different types of behavior.

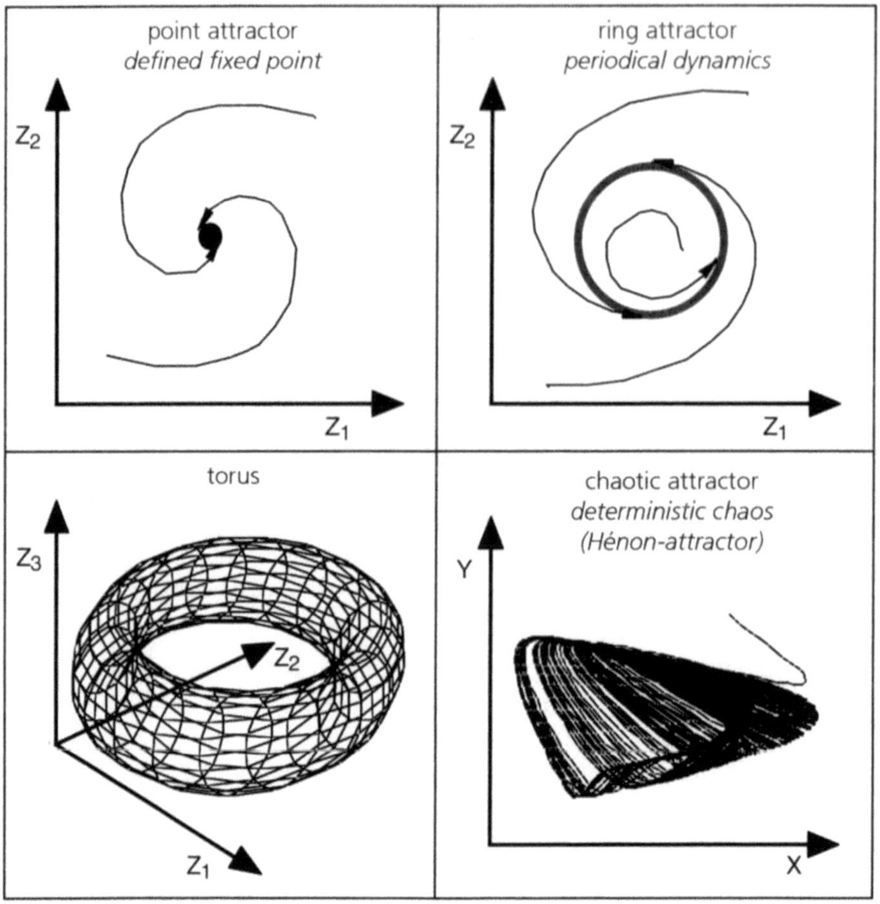

Fig. 9 Examples of phase projection structures (see also Figure 5: Hénon system).

If deterministic chaos is found, the question is raised as to how many dimensions in the phase projection will be required in order to describe the system unambiguously. Within such systems, regularities are very difficult to discern, which makes these systems unpredictable. In the case of deterministic chaos, no clear threshold exists between stochastic behavior and causal relations. On the one hand, stochastic developments are basically deterministic but so highly dimensional that they cannot be described. On the other hand, even deterministic systems show stochastic behavior which, in reality, is additionally overlapped by fluctuations.

Practical application of phase projection analysis must take into account the following conditions:

- In order to analyze temporal behavior, a certain minimal amount of data will be required.
- One basic problem in analyzing complex dynamical systems lies in the necessary reduction in description of the underlying dynamics through few state variables. In many real systems, the true state variables describing dynamics are not known or are very difficult – or impossible – to determine. In these cases, information on dynamic system behavior can also be derived from the temporal behavior of only one system-defining state variable. This, of course, has decisive practical advantages. To do this, it is necessary to determine the so-called 'substitution space' – a state space reconstructed from the time series. However, two significant parameters are required which are not easily determined:
  - ➔ the dimension dim_E of the substitution space,
  - ➔ the temporal lag of the reconstructed phase projection.
- With noisy data, appropriate methods of reducing noise must be implemented prior to performing non-linear analysis, in order to prevent distortion of the results. But here we run into the problem of the evaluation of noise, because in certain cases, fluctuations have a significant influence upon system dynamics.
- The origins of non-linear methods of analysis lay in the mathematical analysis of continuous data which result from differential or system-of-difference equations. In order to perform exact mathematical analysis of deterministic-chaotic data, it would be necessary to describe systems by means of differentiable systems of equations. But also special methods have been developed for the analysis of temporally discrete time series and applied successfully to experiments.

4.1. Methods of analyzing non-linear dynamic systems

Methods for the analysis of non-linear dynamical systems (Figure 10). They base upon a representation of time series and trajectories in the phase projection. The structure of the phase projection itself already yields significant information on system behavior. With the help of special graphing methods, the structure of the phase projection can be examined in greater detail. These methods serve to produce a visual classification of dynamics. However, only qualitative analysis allows for a stringent evaluation of whether the system concerned shows deterministic chaos or mainly stochastic processes.

The long-time behavior of deterministic-chaotic processes is represented in the phase projection by 'strange attractors', whereby deterministic-chaotic trajectories behave in such a way that perturbations show exponential growth. This limits the predictability of the temporal course of dynamics, or makes them impossible from

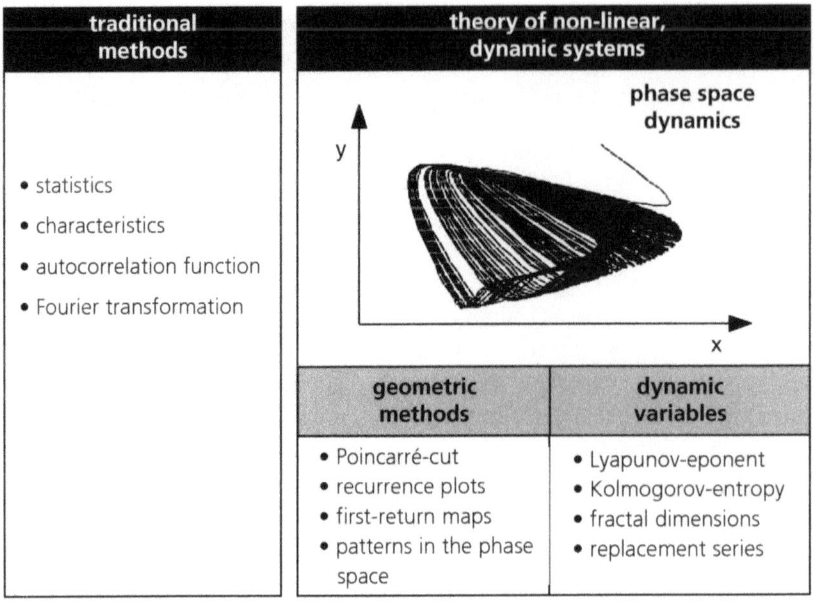

Fig. 10 Methods of analyzing non-linear dynamic systems.

a certain point in time on. The irregular behavior can be quantified by the Lyapunov exponent and the Kolmogorov entropy. In addition to dynamics, the geometric structure of the attractor can also be evaluated. For this, the fractal dimension is used.

Lyapunov exponent, Kolmogorov entropy and fractal dimension are the most important variables in classifying dynamic processes. They are called dynamic variables because, from an experimental perspective, they take on meaning only when they vary with change in system control parameters (BUZUG 1994). Through these variables, the route to chaos can be traced and analyzed quantitatively. This also means that a system is not characterized by a single 'system value', but rather by a whole spectrum of values. This spectrum reflects a complete picture of all possible dynamic system developments.

4.2. Graphic methods

Making portraits in phase projection – of time series and trajectories – provides visualization which helps us to gain a picture of the system's underlying dynamics. If the temporal development of logistics state variables could be represented graphically in a phase projection, as a picture in three dimensions of the state space, e.g., spanned by inventory, processing time and capacity vectors, dynamic shop floor activity could be visually classified and interpreted.

The trajectories of non-linear systems can, after an initialization phase, return to an attractor, whose structure can be examined (RICHTER 1993). If the system approaches a steady state, the attractor is called a 'fixed point', while periodic motions correspond to 'bounding cycles'. The attractor of quasi-periodic movement with n-frequencies is called an *n*-torus. Deterministic-chaotic dynamics in dissipative systems return to „strange attractors" (BRIGGS & PEAT 1990). Strange attractors, as compared to regular attractors, can show very complicated geometric forms, but are indeed structures when they are compared to the diffuse 'clouds' of points of the data of stochastic processes. In this way, the existence of an attractor corresponds to correlations in the phase projection (EBELING 1986).

The search for attractor structures represents a considerable achievement in expanding the methods of time series analysis. The phase projection portrait exposes a complete picture of all possible attractor structures of a system which can result from change in its control parameters. Using phase projection maps, changes in system behavior – for example, the route into chaos – can be followed and analyzed. Phase projection portraits of real systems usually show all described forms of attractors. In other words, a system is not defined by a single attractor, but by the dynamic change of attractor structures in dependency upon control parameters. The objective is to gain such a phase projection portrait of production systems, with which control of the system could be achieved by varying control parameters in desired, that is not deterministic-chaotic, phase projection regions. Phase projection portraits, which at first yield static pictures of dynamics, are animated through video sequences or special computer software, whereby dynamic behavior becomes visible (WOLF ET AL. 1985).

Reconstruction of the substitution space

A deterministic view of production systems starts from the assumption that all state variables can be identified and determined from the start. This is not possible, at present, for real production systems, so that there can be no investigation of the original phase projection. Often, however, at least one logistics state variable can be observed. This is enough to allow reconstruction of the attractor structure in the original phase projection through the so-called substitution space. This requires that delay time τ and the dimension of imbedding dim_E be given and that for each point in time a dim_E dimensional state vector be formed (DRESSLER 1994):

$$x(t) = (x(t), x(t-\tau), \ldots, x(t-(\dim_E - 1) \cdot \tau))$$

This creates a picture of the set of measurements on one orbit which lies in a multi-dimensional substitution space. TAKENS has proved that the 'experimental' attractor in substitution space yields a true reflection – a differentiable equivalent – of the original attractor (TAKENS 1982). Choice of the dimension of imbedding and the delay time will allow each point of the attractor in substitution space to clearly characterize the system state.

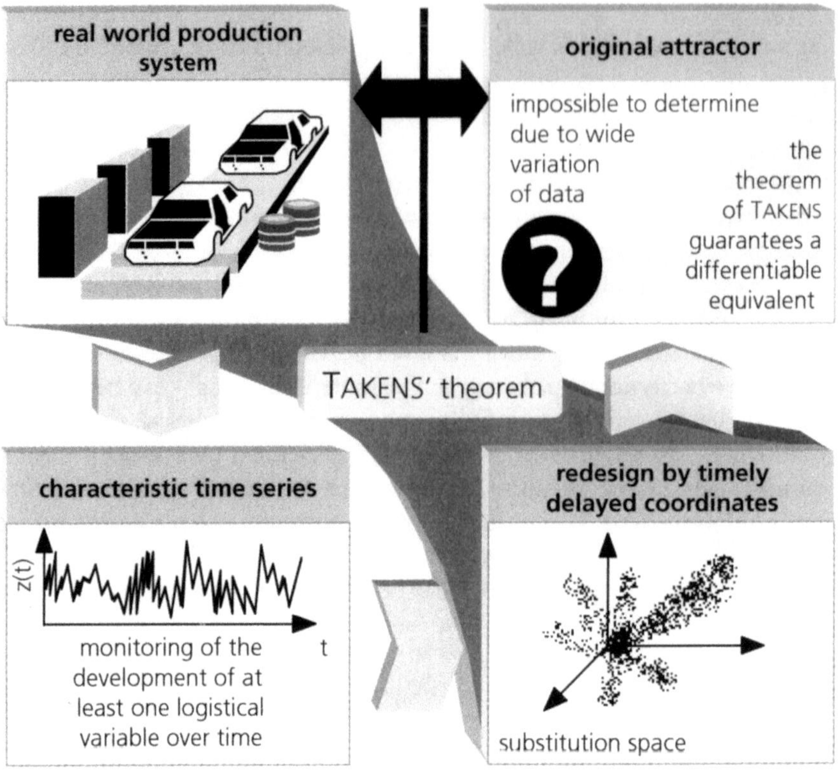

Fig. 11 Schematic representation of the reconstruction of dynamics from a time series of a logistics time variable.

Figure 11 shows how the attractor in substitution space relates to the attractor in the original phase projection. Collecting data on logistics state variables is, initially, the only access to the dynamics of real production systems. The original attractor remains hidden. The path of reconstruction proceeds over the recording of the real data of at least one characteristic state variable and then usually leads to a complicated time series. The structure of the attractor is revealed through successive plotting of the delayed coordinates of this state variable. To assure through TAKENS' technique that the data is imbedded in a phase projection of enough dimensions, the imbedding dimension dim_E of the Euclidean space must be at least $2n + 1 \leq dim_E$, whereby n corresponds to the dimension of the original attractor. That is in the worst of cases, and often $dim_{EE} = n$ will suffice (WORG 1993). Then there is an (invertable) mapping between the attractor in substitution space and the attractor in the original phase projection, which can be interpreted as a coordinate transformation. This is not explicitly known, but its 'theoretical existence' alone justifies investigation of the system through the attractor in substitution space (DRESSLER 1994). The resulting attractor is usually somewhat distorted through the imbedding, but important characteristics, such as the

Lyapunov exponent or Kolmogorov entropy, are, to certain limits, independent of the number of coordinates (DRESSLER 1994). Within the framework of this paper, this technique of imbedding data to reconstruct phase projection is used to retrieve information on system dynamics from a graphic representation.

4.3. Quantitative methods

Lyapunov exponent

The stability of the dynamics of logistic state variables with small perturbations and changes of control parameters, or environmental conditions, is a matter of great interest. A production system should be organized in such a way that a continuous flow of materials, low lead times and greatest possible capacity utilization are assured. High capacity utilization requires that inventory show no unpredictable fluctuations. System behavior must be sufficiently stable to prevent breakdowns in planning and control of the production system. If internal or external influences change the desired state of the system, that state must be quickly regained. The variable which describes this behavior generally for non-linear, dynamical systems is the Lyapunov exponent (λ) (SCHUSTER 1989). The numbers provide a picture of the properties which lead to different types of stability and instability (BUZUG 1994):

- Lyapunov stability:
  With small interferences, the state of the system remains within the finite region of a reference state ($t \rightarrow \alpha$).
- Asymptotic stability:
  The state of the system at first deviates with small interferences, but then pulls back towards the reference state.
- Instability:
  Small interferences cause the system to pull away from the reference state; that is, there is an interference in every surround of the state whose amplitude does not remain forever smaller than a certain interval. The fluctuation grows exponentially.

According to these considerations, both Lyapunov stability and asymptotic stability are desirable with regard to the dynamic behavior of production systems. Instability, on the other hand, is as a rule to be avoided, as in such a state, no real planning or control is possible, and the system itself reacts sensitively and unpredictably to small changes. Lyapunov exponents describe the mean expansion or contraction rates of conflicting effects (SCHUSTER 1989). Thus they are a quantitative measure of the exponential divergence or convergence of neighboring system states in the phase projection.

Lyapunov exponents are defined as growth rates of infinitesimal volume elements. The values of Lyapunov exponents (λ) have the following meaning (KUNICK & STEEB 1986):

- $\lambda<0$:
  Disturbances decrease exponentially; this means, that neighboring points in the phase projection converge (asymptotic stability).
- $\lambda=0$:
  On the average, disturbances of the system remain at the same level (Lyapunov stability).
- $\lambda>0$:
  Small disturbances increase exponentially: that is, nearby points in the phase projection diverge (instability).

The behavior of the system is always characterized by several Lyapunov exponents. The number of Lypanov exponents is determined by the number of state variables in the phase projection. The attractor structure in three-dimensional phase projection is determined by three Lyapunov exponents. In the case of strange attractors, one of the exponents is zero ($\lambda_1=0$), one is negative ($\lambda_2<0$) and one is positive ($\lambda_3>0$). This means that in one direction, changes in the system stay at the same level, in another the disturbances decrease and in a third the disturbances grow exponentially. This is the way the 'strange' dynamic behavior of the system comes about, which causes it to pull towards an attractor, i.e., asymptotic stability, but also -- within this attractor region -- infinitely diverge, i.e., instability.

For the analysis of logistics state variables, the problem is that there are only discrete time series at hand. Systems of differential equations, which would describe system development exactly, are generally not explicitly known. Strictly defined, Lyapunov exponents refer to differential equation systems, but techniques have been developed that allow calculation of Lyapunov exponents on the basis of discrete time series data. Two methods are the most common here (VASTANO & KOSTELICH 1986):

- Approximation of the discrete data series by means of linearized flow mapping,
- Calculation of the largest Lyapunov exponent according to a method by WOLF ET AL. (1985).

The first methods allows determination of all Lyapunov exponents, yielding a more exact picture of dynamics. The principle is to approximate system development using linearized mapping of its flow in the phase projection and to calculate all Lyapunov exponents on this basis (SANO & SAWADA 1985, ECKMANN ET AL. 1986). But although this method gives us a more comprehensive picture of the dynamics on the attractor, it has some practical disadvantages. The linear approximation will not suffice if the degree of non-linearity, or the curve of the attractor, is very large, or if the number of data points is very small. There have been attempts in such cases to use polynomials raised to higher powers. This approach was used, for example, by AHRENS & JANSSEN (1994) to investigate production systems. Here the reliability of the calculated Lyapunov exponents is strongly dependent upon several parameters, making this method inappropriate for the determination of the exponents where qualitative system dynamics are not known from the start. And in investigating real production systems today, these are surely unknown.

While WOLF's method is also dependent upon the selection of some parameters, it is considerably more practical in application due to its simplicity. It allows for calculation of only the largest Lyapunov exponent. This is enough, however, to allow us to decide whether the system shows deterministic chaos ($\lambda_{max}>0$) or not ($\lambda_{max}<0$).

In summary, it is possible, with the appropriate algorithms, to determine at least the greatest Lyapunov exponent on the basis of experimental time series data. The exponent is a measure of the 'inner stochasticity' of systems which aids differentiation between stochastic and deterministic-chaotic relations in the dynamics of logistics state variables. In the case of unknown attractor dynamics, substitution space must be reconstructed first.

Kolmogorov entropy

Kolmogorov entropy is a measure which bridges information theory, invented in the late 1940s by SHANNON & WEAVER (1949), and the theory of non-linear dynamical systems. The concept of information I is very closely related to the concept of the Lyapunov exponent λ.

If we observe a square, for example (Figure 12), whose subdivisions are supposed to represent possible system states, with an increasingly number of subdivisions there is an increase in the amount of system information gained which can be an aid in determining the actual state of the system (RASBAND 1990). This is equivalent to saying that uncertainty about the actual system state grows with the number of possible system states, or that 'knowledge' of the state system decreases. Information entropy reflects this relationship with mathematical precision (SHANNON & WEAVER 1949):

$$I = \lg \Omega = -\sum p_i \lg p_i$$

Kolmogorov entropy (K) is a measure of the production of information (dI/dt) and, thus, for the increase in uncertainty in a system (WORG 1993). In a deterministic-chaotic system with $K>0$, the number of possible system states, which develop from an initial state, increases with time (exponential divergence of trajectories $\lambda > 0$; Figure 12). Information theory would say that the information which is yielded by determining the system increases. This change can be seen as a gain in information, as two different initial states, which at first are indistinguishable, develop through infinite time into distinguishable system states (JACKSON 1990).

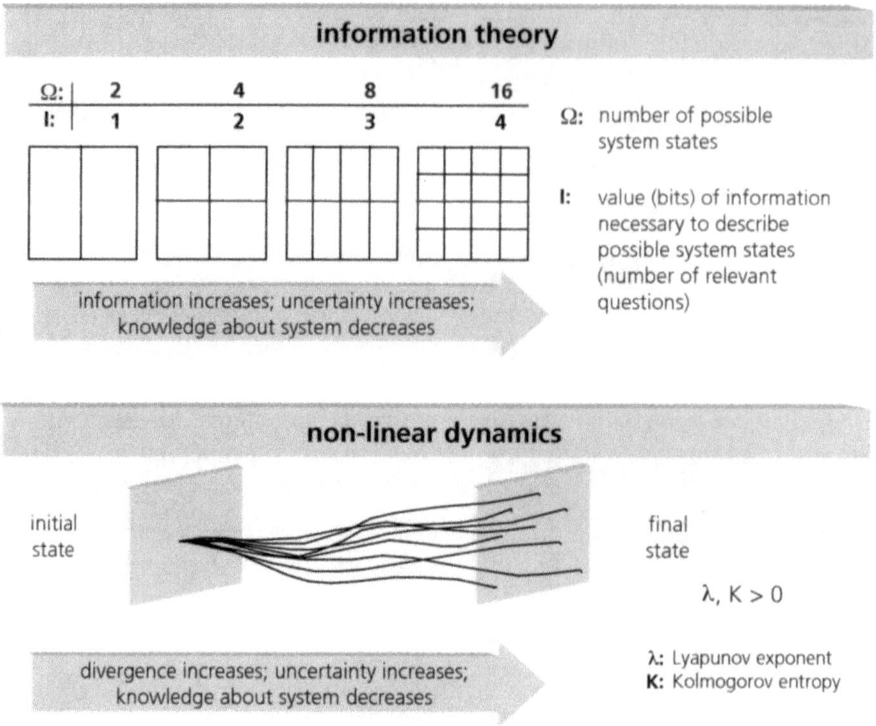

Fig. 12 Relation between information theory and non-linear dynamics.

The system thus produces information. If no subsequent measurements are made, more and more 'knowledge' about the system is lost, despite the underlying determinism, until ultimately, the system appears to the observer to be completely random. What the Kolmogorov exponent produces is a natural measure of the time scale over which 'knowledge' on a deterministic-chaotic system decreases. It stands in a close relation to the positive Lyapunov exponent, which is a measure of the exponential growth of disturbances in a system. For nearly all systems the Kolmogorov exponent will result from the sum of all positive Lyapunov exponents of a system (ECKMANN ET AL. 1986). The task of identifying the factors within a production organization which determine deterministic chaos corresponds to the problem of finding those terms and boundary conditions which lead to positive Kolmogorov entropy of the system (TÖNSHOFF & GLÖCKNER 1994).

Through the aid of Kolmogorov entropy, HORNS (1989) evaluates the planning horizons of finite scheduling generally and specifically. The basis is a simple simulation model (Petri nets) of a portal robot, which is shown in Figure 13. The portal robot transports pallets to and from a total of six buffers. Two of the buffers are connected to two different work stations by additional transfer robots. A section of the finite schedule shows how many time units are planned per operation and the sequence in which operations should be performed. When the portal robot is called to supply stations 1 and 2 with materials and to transport finished workpieces A and B away, the temporal overlapping of these events creates competition.

The starting point of the investigations is the question of the principle limits to predictability of temporal and causal chains of work operations – using deterministic simulation studies – if deterministic-chaotic phenomena appear. Efficient finite scheduling depends upon accurate prediction of the temporal-causal sequences of the work steps as they really occur in dependency on loading strategies. Simulation is a useful technique in evaluating various strategies with reference to objectives, such as low inventory, high capacity utilization, flexibility and so on. Machine loading in a job shop is usually planned for a period of one day or one shift, i.e., short-term control. Here, order sequences as well as sequences of set-up, machine utilization and work pieces are calculated according to the given objective.

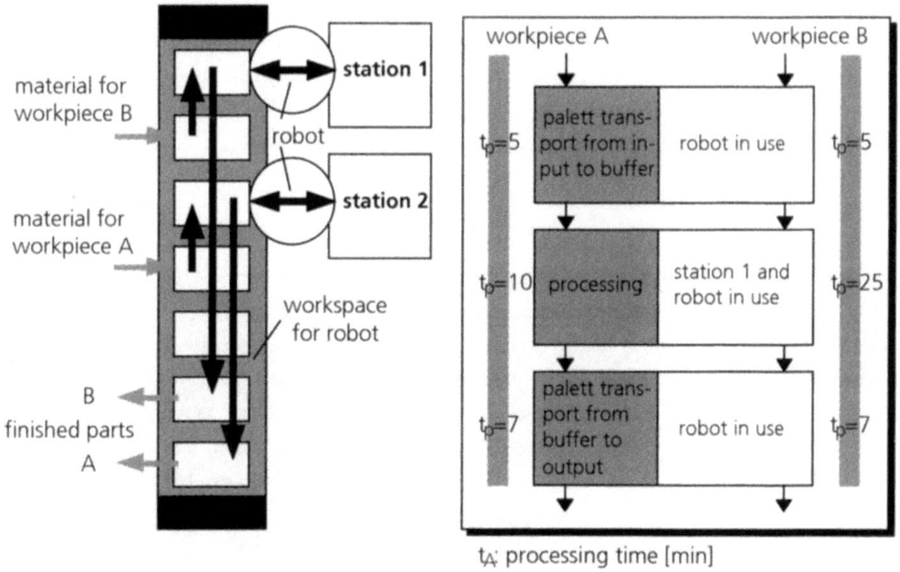

Fig. 13 Schematic structure of the production process with sections of the corresponding sequence plan as investigated by HORNS (1989).

In carrying out simulation studies, determining the times for processing, set up, queuing and transportation are very important. The precision of this data must be very high. If this prerequisite is not fulfilled, – for example, in the case of job shop production of newly designed products which have no comparative values – finite scheduling does not make sense. But even collecting the data under real conditions in 'normal cases' on the shop floor is a difficult task. Uncertainty about the actual length of the time units is unavoidably contained in every deterministic simulation of production processes. The length of the various time units has a direct influence upon the complex, temporal-causal relationships, so that uncertainties become reinforced through the processes' own dynamics. With this, 'knowledge' of the system is automatically lost through the course of simulation time.

In evaluating the strength of findings from such simulation studies, therefore, we must ask how this loss of information may be estimated. HORNS (1989) was able to show that this is indeed possible on a quantitative basis through the aid of the concept of Kolmogorov entropy. Kolmogorov entropy can be interpreted as a measure of the mean information loss when system state is calculated from the previous state, which is the case with so-called event-driven and discrete simulations. The goal is to find out the number of simulation steps it takes for Kolmogorov entropy to reach the information value determining the initial state. This is the same as saying that the simulation model has 'forgotten' its initial state. Each further observation on the basis of the model will now have no deterministic foundation.

With an appropriate choice of constants in the formula for Kolmogorov entropy, the loss of information can be expressed in units (bits). According to FARMER (1982), there is an explicit connection between the time ($\Delta\tau$) elapsed until the simulation model has 'forgotten' its initial state (Ψ_0)and the dimension of the attractor (D_I), Kolmogorov entropy (K) and uncertainty in determining the initial state (ζ), which is laid down the planning data (HORNS 1989):

$$\Delta\tau = \left(d(\zeta)\frac{D_I}{K} \right)$$

The information dimension and Kolmogorov entropy are calculated by HORNS (1989) on a temporally discrete orbit which is established by means of a simulation model produced with Petri nets. The state of the system is established by means of a vector, which at each step of the simulation reflects the temporal-causal relations of the total system, or total Petri net. In this way, the simulation model – represented by the vector – is analyzed directly, and not via the detour of evaluating ascertained values of inventory, capacity, and so on (HORNS 1989). In order to analyze the temporal development of this vector, orbits of 1500 points for various standard deviations of 'ideal' process parameters in a range of 0 to 0.45 time units – representing the uncertainties in ascertaining the planning data – were determined and analyzed. Positive standard deviations yielded positive values for Kolmogorov entropy $K \approx 0.36$ *(bits)*, which means that with these simulated uncertainties on real parameters, the observer loses 0.36 bits of information with each step of the simulation. If we assume that the initial state could be determined with a precision of 3 bits, the time elapsed before the system state forgets its initial state would be approximately $\Delta\tau \approx 40$ time units. Thus, it was demonstrated quantitatively for this simulation study that there are principle limits to the predictability of temporal-causal work sequences determined by deterministic-chaotic phenomena ($K>0$). However, in this case the limits are so wide that deterministic analysis of various loading strategies is not jeopardized. Upon this background, optimal strategies for loading the robots may now be determined without the danger that imprecision in establishing initial data will lead to distorted simulation results.

Although HORNS' (1989) quantitative findings remain at present approximate values, as Kolmogorov entropy and information dimension had to be calculated on the basis of non-validated estimates, this in no way affects the qualitative findings of his work.

Fractal dimension

The fractal dimension allows quantitative examination of the geometric structure of attractors in the phase projection. Strange attractors of non-linear dynamical systems show self-similarity in structure, that is, certain forms are similar – theoretically! – on different scales. Ascertaining these scaling properties is of interest in two respects: on the one hand, the fractal dimension is a measure of the degrees of freedom of the dynamic system and, on the other hand, insight into the geometric structure of complex attractors is gained (CRILLY, EARNSCHAW & JONES 1991).

In analyzing the dynamic behavior of production systems, the number of the degrees of freedom which determine the behavior of logistics state variables is of great significance. If it can be shown that the irregular behavior of logistics state variables is dependent upon only a few state variables, and so may be imbedded in a phase projection of few dimensions, this would prove that there is an underlying determinism of the dynamics of the system. And as the minimal number of values is known, we can further, and in a more targeted manner, seek the state variables of the system that are required in order to describe the system.

Figure 14 gives an overview of the most common methods of determining the fractal dimension. It reveals the close relation among individual concepts for analyzing deterministic-chaotic systems. Lyapunov exponents take on a central role, for there are methods available for practical application. And on the basis of Lyapunov exponents, nearly all other dynamic variables can be derived, at the least as estimates (KAPPLAN & YORKE 1979). The most common method of ascertaining the fractal dimension is determining correlation. Correlation is understood as a relation between events which is not random. The usual correlation function provides a measure of the mutual dependency of two or more states after a certain period of time, or after a certain number of successive data points (MORFILL & SCHEINGRABER 1993).

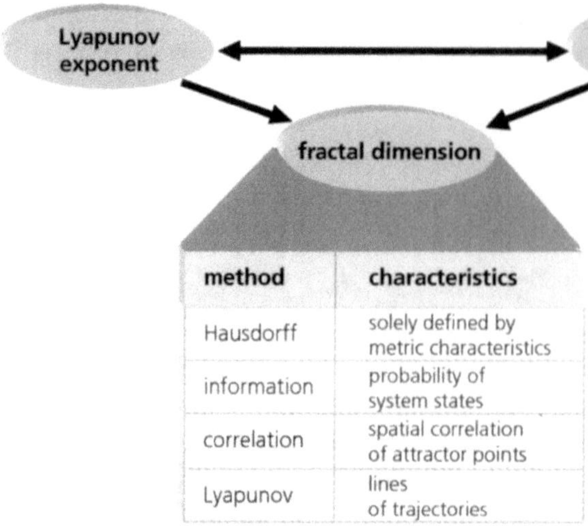

Fig. 14 Relation between dynamic variables and methods of determining fractal dimension.

In investigating logistics state variables, the correlation of system states is of great significance. If this can be shown for production systems, even though the temporal development of the values initially seems irregular, previous – static – concepts of the evaluation of production systems will have to be reconsidered. The correlation dimension reveals the number of state variables on which the dynamic behavior of the production system depends.

5. Simulation of dynamic systems

The consequences of dynamic effects and influences in production systems can not, however, be evaluated using static methods. Meaningful evaluations take the temporal-dynamic development in logistics networks into account. There are basically two distinguishable types of procedures:

- Data collection in real systems and
- data generation through modeling.

As compared to measuring real production systems, the model-aided procedure has a number of advantages:

- Experiments, e.g., planning scenarios, can be conducted with the model which could not be applied to real systems without serious changes or without high risk (GÜNZEL 1993).
- Models allow us to collect many, various data at different measuring points in the model, which in real systems can be gathered only with considerable difficulty or not at all, because methods are lacking.

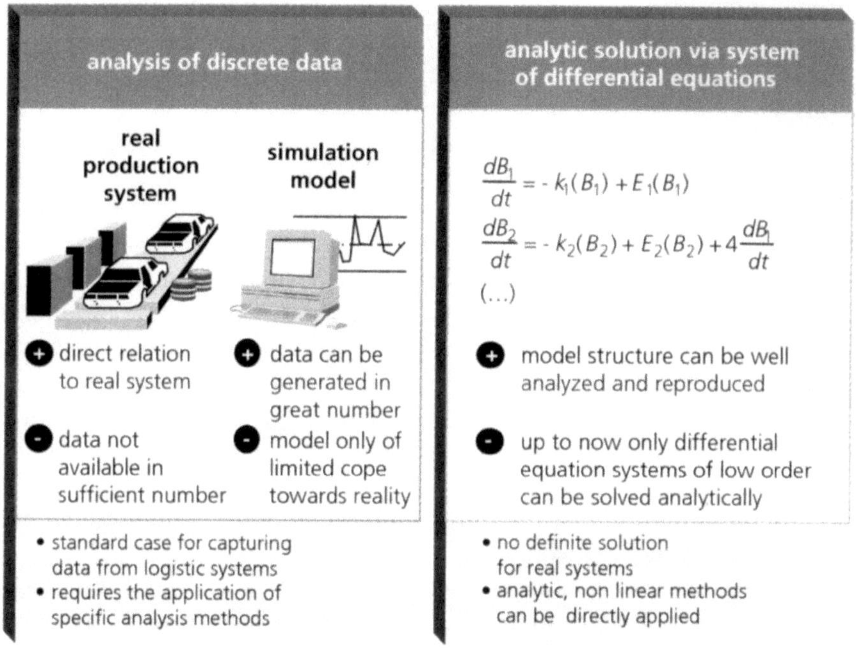

Fig. 15 Procedures in analyzing dynamic production systems.

Dynamic systems can be modeled by means of differential equations or discrete models (Figure 15). While in logistics systems the behavior of single subsystems can often be described, their complex effects upon each other maintain unknown. For this reason, modeling through differential equations is frequently extremely difficult or not possible. It is here that discrete simulation shows its advantages. Simulation allows production processes and information and decision-making processes to be reproduced. The results of simulation are data which represent the states of the system and the course of the processes over time. This data may be evaluated through the aid of statistical methods or chaos theory methods. Over and above that, simulations represent the processes in real systems in time lapse. As a rule, the processes occur faster on the computer than in reality. In this way, simulation produces bodies of data of any size within short periods of time.

The production of large bodies of data is a particularly important aspect with regard to evaluating dynamics. The quality and reliability of results of time series analysis of dynamic systems is crucially dependent upon the extensiveness and large volume of the available data (MORFILL & SCHEINGRABER 1994). And precisely because processes in real production systems are influenced by permanently overlapping interferences and single random events, large bodies of data must be examined, so that system-determining structures may be differentiated from interferences and random influences.

The necessity to study large bodies of data leads to the demand that systems being investigated should show a quasi stationary state over the analysis time period. Basic structural changes or changes in conditions have a clear effect on the dynamic processes within production systems. They may lead to new system structures and thus to new behavioral patterns in the system. For example, change in the product program, production technology or segmentation of the real system during the period of analysis can completely distort the findings of a long-running analysis of dynamics, making it impossible to identify the original dynamic characteristics. But because changes must be reckoned with when monitoring real production systems over longer time periods (for example, periods longer than one year), time series analyses may not be performed on real systems without difficulty. At the least, it is difficult to know whether structuring is attuned to the actual demands if the demands change during the period of analysis. Simulation allows this problem to be avoided. Following the initialization phase of the simulation, quasi stationary state may be, as a rule, assumed. By extending the length of the simulation, the necessary amount of data for analysis and evaluation can be collected without changes in the conditions.

The question remains as to what knowledge can be derived from these procedures. Certainly, a detailed simulation of logistics systems over a long time period does not tell us anything about long-term development. This is because, for one thing, critical system changes cannot be taken into account. Complex dynamical systems show high sensitivity to changes in parameters and influences, which makes only short-term prognoses possible. Here only principle behavioral patterns may be examined. Long-term considerations are permissible only at a very abstract level, at which system changes are components of the model, such as adaptation of a company to the market. However, what may be done is to produce stationary behavior with simulation, in that momentary conditions of the real system are frozen. Then the findings on system behavior may be applied to the momentary system structure.

An example

An example should illustrate the application of the theory of non-linear systems and equivalent methods for logistics systems. Utilizing computer simulation, the processes of a rubber and rubber composite parts manufacturer were analyzed. Basically, the manufacturing process was split into two steps: initial manufacturing of anonymous semi-finished parts and materials and a finishing of parts within market-oriented segments. Within the rubber manufacturing process, the blending unit, where raw rubber blends are produced, represented a bottleneck. The analysis therefore focused on the behavior of the blending unit and its influence on surrounding manufacturing processes.

The blending section produces blends of rubber for seven independent market segments. Due to varying market behavior as well as different manufacturing technologies, changes in demand behaved in a 'chaotic' manner. Demand was determined through independent planners within each market segment. The 'chaotic' dispatching of orders resulted from the low delivery performance of the blending unit, which led planners to dispatch orders early or as express orders. This allowed the chaotic behavior of the blending unit to influence the behavior of the market segments, which then again had a negative feedback on the blending unit.

This amplification within the feedback loops of planning was modeled in the computer simulation system. The simulation model took into consideration all causal dependencies within material and information flow. In addition to these internal dependencies, demand changes due to the customer were modeled with deterministic as well as stochastic elements. Within the simulation study, systems behavior was analyzed in relation to changes of the single control variables of planning. Results of the simulation were examined by means of phase projection analysis. As a measuring variable, the overall weekly order income of the market segments was imbedded in a substitution space following TAKENS.

Investigation revealed that the system's behavior had a main fundamental complexity, which can be easily determined by the structure of the phase projection. Figure 16 shows a configuration of variables which still has clear structures. The assumption of deterministic-chaotic behavior was verified through the Lyapunov exponent. The procedure following Wolf shows convergence. In this state, even small variations of control variables leads to completely different time series. Figure 17 shows an experiment in which the blending capacity was changed by 0.00032 units per week.

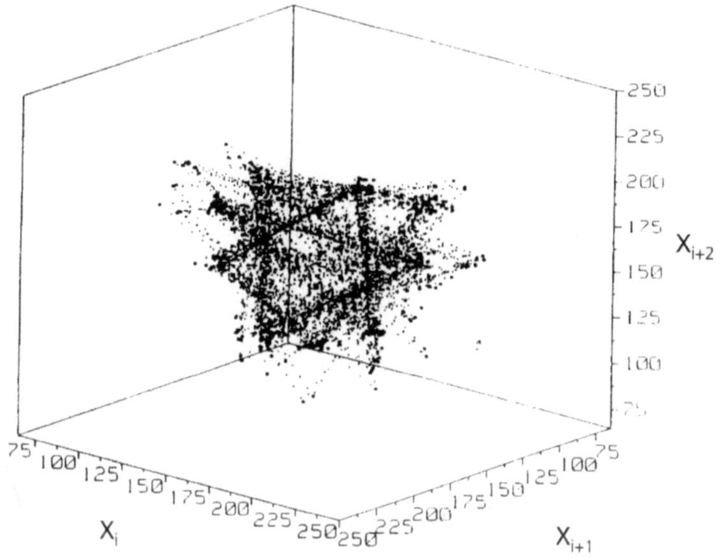

Fig. 16 Phase projection with deterministic-chaotic structure.

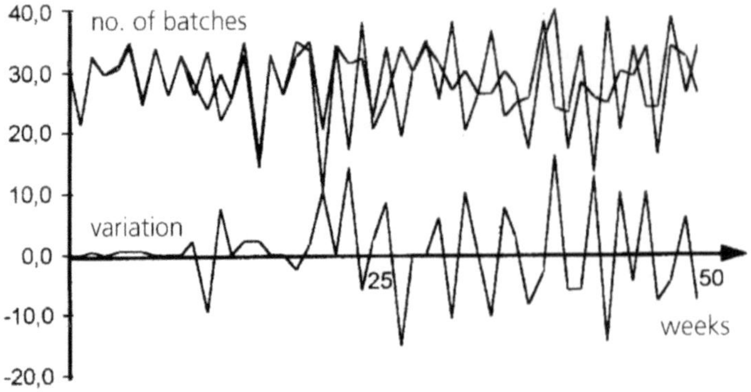

Fig. 17 Sensitive systems dependency on the control variable 'blending capacity'.

Moreover, the experiments showed that changes in planning logic lead to significant changes in the structure of the phase projection. Major random influences lead to stochastic behavior of the system (Figure 18). Under these conditions, no Lyapunov exponent could be determined. The dimensions of the substitution space shift to infinity. But still, the system could be stabilized through a reallocation of controlling competence and an adequate selection of planning information and time scope (Figure 19). By applying different conditions, planning and control became more robust, and the number of express orders was decreased drastically.

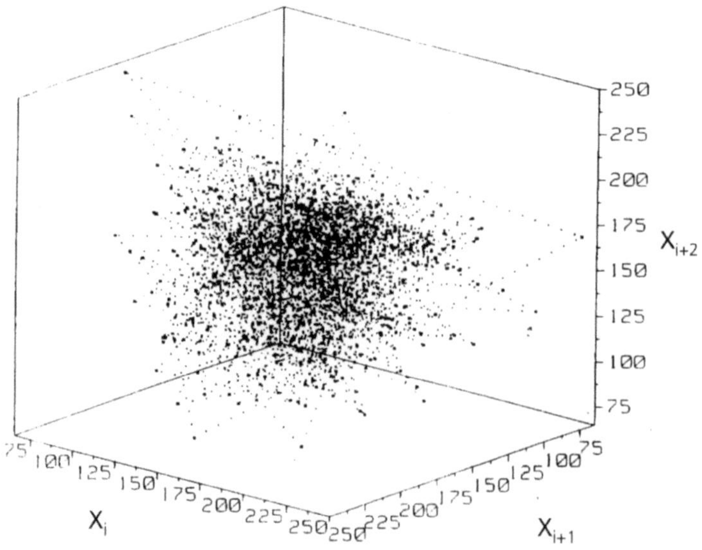

Fig. 18 Phase projection with stochastic structure.

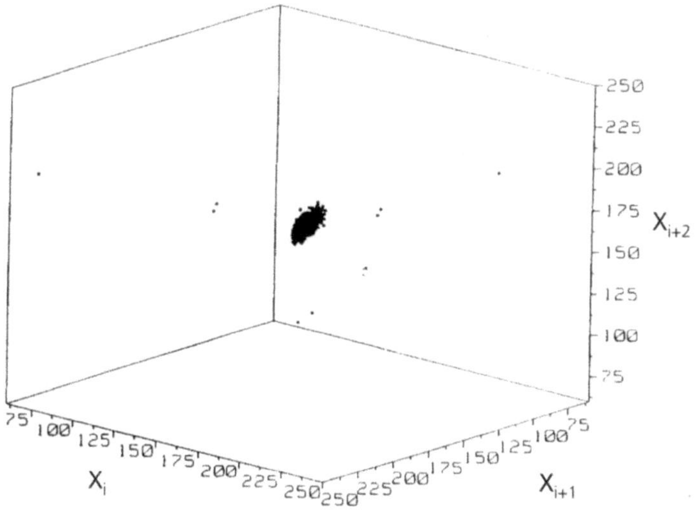

Fig. 19 Phase projection with robust structure.

This brief example indicates some of the means of assessing the behavior of logistics systems using methods of non-linear systems theory. The graphic analysis of the phase projection diagrams already yields first insights into the nature of the system examined.

6. Summary

One characteristic feature of logistics processes is that there is little chance to predict future development and to plan reliably. Describing the processes within logistics systems by means of deterministic models does not represent the reality. For this reason, production planning and control systems, or other computer-oriented approaches based on deterministic models, have not been as successfully applied in industry as proposed. The failure of deterministic approaches is due to the complexity and the dynamics of logistics systems. Stochastic approaches have also failed.

The theory of non-linear systems (chaos theory) provides a new view of complex systems that combines the deterministic and the stochastic world. Chaos theory is based upon a holistic perception of complex systems and seeks to do justice to the fundamentally complex nature of real systems. But applying these scientific findings to logistics systems requires totally new 'thinking' on production and logistics systems. Some of the phenomena which can be observed within logistics systems have already been explained well by this approach. Compared to traditional approaches, chaos theory methods do not mainly analyze the structure of a logistics system, but rather patterns of behavior. In most cases, the behavior of complex logistics systems cannot be described by analyzing and examining the single components of the system. An evaluation of logistics systems is possible only through analysis of the time series of the state variables. The time series can characterize the underlying nature of a system.

If a logistics system shows deterministic-chaotic, fundamental-complex or even stochastic behavior, deterministic planning and control rules will not be successful. Under these conditions, the system structure must be modified before traditional planning and control methods can be applied. In contrast, if a system shows mainly deterministic behavior, deterministic planning and control rules will be successful within the forecast horizon.

The findings of chaos theory enable evaluation of logistics systems even when the original parameters determining the system are unknown. Thus, the methods of chaos theory can lead to a better understanding of logistics systems. This may aid the design of robust logistics systems.

7. References

AHRENS, V., JANSSEN, C. (1994): Analyse von logistischen Produktionsprozessen mit Methoden der Theorie Nichtlinearer Systeme. In: WIENDAHL, H.-P., AHRENS, V. (Eds.): Potentiale der Chaosforschung in den Produktionswissenschaften. Proceedings. Institut für Fabrikanlagen, Universität Hannover.

ASHBY, W.R. (1958): Requisite variety and its implications for the control of complex systems. Cybernetica, 1:2, p. 83-99.

ASHBY, W.R. (1972): Systems and their informational measures. In: KLIR, G.J. (Ed.): Trends in general systems theory. Wiley, New York.

BAGANZ, A. (1994): Vom Nutzen des ganzheitlichen Projektmanagements. Planung + Produktion, 2.

BEER, S. (1959): Cybernetics and management. English Universities Press, London.

BEER, S. (1966): Decision and control: The meaning of operational research and management cybernetics. Wiley, London.

BREUER, R. (1993): Am Rande des Chaos – Einleitung in ein unordentliches Thema. In: BREUER, R. (Eds.): Der Flügelschlag des Schmetterlings: ein neues Weltbild durch die Chaosforschung. Deutsche Verlags-Anstalt, Stuttgart.

BRIGGS, J., PEAT, F.D. (1990): Turbulent mirror: An illustrated guide to chaos, theory and the science of wholeness. Paperback edition. HarperCollins, New York.

BUZUG, T. (1994): Analyse chaotischer Systeme. XBI-Wiss.-Verl., Stuttgart.

CHRYSSOLOURIS, G. (1994): Measuring complexity in manufacturing systems. Working Paper. MIT, Cambridge.

CRAMER, F. (1993): Chaos and order: The complex structure of living systems. VCH Publisher, Weinheim.

CRILLY, A. J., EARNSCHAW, R. A., JONES, H. (1991): Fractals and chaos. Springer, Berlin.

DRESSLER, U. (1994): Über die Steuerung chaotischer Systeme In: WIENDAHL, H.-P., AHRENS, V. (Eds.): Potentiale der Chaosforschung in den Produktionswissenschaften. Proceedings. Institut für Fabrikanlagen, Universität Hannover.

EBELING, W. (1986): Analyse chaotischer Zeitreihen. Verlag Harri Deutsch, Frankfurt am Main.

ECKMANN, J.-P. ET AL. (1986): Liapunov exponents from time series. Phsical Review 34A:6, p. 4971-4979.

EVERSHEIM, W. (1994): Verringerung und Beherrschung der Komplexität stärkt die Wettbewerbsfähigkeit. In: MILBERG, J., REINHART, G. (Eds.): Unsere Stärken stärken – Der Weg zu Wettbewerbsfähigkeit und Standortsicherung. Verlag Moderne Industrie, Landsberg, p. 71-99.

FARMER, J. D. (1982): Information dimension and the probabilistic structure of chaos. Zeitschrift Naturforschung, 37a, p. 1304-1325.

FORRESTER, J.W. (1961): Industrial dynamics. 6th ed. MIT Press, Cambridge.

FORRESTER, J.W. (1968): Industrial dynamics: After the first decade. Management Science, 14:7, p. 398 ff.

GOMEZ, P., MALIK, F., OELLER, K.H. (1975): Systemmethodik: Grundlagen einer Methodik zur Erforschung und Gestaltung komplexer soziotechnischer Systeme. Bern.

GÜNZEL, U. (1993): Entwicklung und Einsatz eines Simulationsverfahrens für operative und strategische Probleme der Produktionsplanung und -steuerung. Hanser, München.

HAKEN, H. (1983): Erfolgsgeheimnisse der Natur: Synergetik, die Lehre vom Zusammenwirken. Deutsche Verlags-Anstalt, Stuttgart.

HAKEN, H. (1991): Zur modernen Chaostheorie. DVS Berichte, 134, p. 1-10.

HAKEN, H. (1993): Synergetics as a Strategy to Cope with Complex Systems. In: HAKEN, H., MIKHAILOV, A. (Eds.): Interdisciplinary Approaches to Nonlinear Complex Systems. Springer Verlag, Berlin.

HAKEN, H. (1995): Synergetics as a strategy to cope with complex systems. In: Managing complexity – Applications of self-organization. Proceedings. SFB 230, Stuttgart.

HAYEK, F.A. VON (1972): Die Theorie komplexer Phänomene. In: Vorträge und Aufsätze, 36. Walter Eucken Institut, Tübingen.

HOMMES, C. H. (1991): Chaotic dynamics in economic models: Some simple case studies. Wolters-Nordhoff, Groningen.

HORNS, A. (1989): Job shop control under influence of chaos phenomena. In: SANDERSON, A.C., DESROCHERS, A.A., VALAVANIS, K. (Eds.): IEEE International Symposium on Intelligent Control 1989. Proceedings. IEEE Computer Society Press, Washington. p. 227-232.

HOULIHAN, J. B. (1987): International supply chain management. International Journal of Physical Distribution and Materials Management, 17:2, p. 51-66.

HÜBLER, A. (1992): Modeling and control of complex systems: Paradigms and applications. Technical Report CCSR-92-12, Center for Complex Systems Research. University of Illinois, Urbana.

JACKSON, E.A. (1990): Perspectives of nonlinear dynamics. Parts 1 & 2. Cambridge University Press, Cambridge.

KAPPLAN, J.L., YORKE, J.A. (1979): Functional differential equations and approximation of fixed points. Lecture Notes in Mathematics, 730, p. 228.

KUNICK, A., STEEB, W.-H. (1986): Chaos in dynamischen Systemen. B.I. Wissenschaftsverlag, Mannheim.

KÜPPERS, B.-O. (1993): Chaos und Geschichte – Läßt sich das Weltgeschehen in Formeln fassen? In: BREUER, R. (Ed.): Der Flügelschlag des Schmetterlings: ein neues Weltbild durch die Chaosforschung. Deutsche Verlags-Anstalt, Stuttgart.

MANSKE, F. (1986): Totalplanung oder Rahmenplanung – Wohin geht die Fertigungssteuerung? Technische Rundschau, 78:43, p. 24-27.
MAYER-KRESS, G. (1992): Chaos and crises in international systems. Technical Report CCSR-92-15, Center for Complex Systems Research. The Beckman Institute, University of Illinois at Urbana-Champaign.
MORFILL, G.E., SCHEINGRABER, H. (1993): Chaos ist überall ... und es funktioniert. Eine neue Weltsicht. Verlag Ullstein, Frankfurt am Main.
MORFILL, G.E., SCHEINGRABER, H. (1994): Die Charakterisierung komplexer Systeme – mit ausgewählten Anwendungen. Working paper. Max-Planck-Institut für extraterrestrische Physik, Garching bei München.
PASK, G. (1978): A conversation theoretic approach to social systems. In: GEYER, R.F., VAN DER ZOUWEN, J. (Eds.): Sociocybernetics. Leiden Bosten, London.
PEITGEN, H.O., JÜRGENS, H., SAUPE, D. (1992): Deterministisches Chaos. Springer, Berlin.
PROBST, G.J. (1981): Kybernetische Gesetzeshypothesen als Basis für Gestaltungs- und Lenkungsregeln im Management. Paul Haupt, Bern.
RASBAND, S.N. (1990): Chaotic dynamics of nonlinear systems. John Wiley & Sons, New York.
RASMUSSEN, D., MOSEKILDE, E. (1988): Bifurcations and chaos in a generic management model. European Journal of Operational Research, 35, p. 80-88.
RICHTER, P.H. (1993): Physik zwischen Chaos und Ordnung – von Pendeln und Planenten. In: BREUER, R. (Ed.): Der Flügelschlag des Schmetterlings: ein neues Weltbild durch die Chaosforschung. Deutsche Verlags-Anstalt, Stuttgart.
ROBERTS, E.B. (1985): Managerial applications of systems dynamics. The MIT Press Cambridge.
RÖPKE, J. (1977): Die Strategie der Innovation. Tübingen.
RUELLE, D. (1980): . Strange attractors. The Mathematical Intelligencer, 2:126-37
RUELLE, D. (1989): Chaotic evolution and strange attractors. Cambridge University Press, Cambridge.
SANO, M., SAWADA, Y. (1985): Measurement of the Lyapunov spectrum from a chaotic time series. Physical Review Letters, 55:10, p. 1082-1085.
SCHERER, E., DOBBERSTEIN, M. (1996): Komplexität in der Produktion beherrschen? IO Management Zeitschrift, 65:3, p. 60-64.
SCHUMAN, M. ET AL. (1994): Trendreport Rationalisierung: Automobilindustrie, Werkzeugmaschinenbau, chemische Industrie. Edition Sigma, Berlin.
SCHÜPBACH, H. (1994): Prozessregulation in rechnerunterstützten Fertigungssystemen. vdf/Teubner, Zürich/Stuttgart.
SCHUSTER, H.-G. (1989): Deterministic chaos: An introduction. VCH, Weinheim.
SHANNON, C.E., WEAVER, W. (1949): The mathematical theory of information. University of Illinois Press, Urbana.
TAKENS, F. (1982): Detecting strange attractors in turbulence. In: Rand, D. A., Young, L.-S. (Eds.): Dynamical Systems and Turbulence. Lecture Notes in Mathematics, Vol. 898. Springer-Verlag, Berlin. p. 366-381.
THOMPSON, J. M. T., STEWART, H. B. (1986): Nonlinear dynamics and chaos. Wiley.
TÖNSHOFF, H.-K., GLÖCKNER, M. (1994): Strukturbedingte chaotische Bestandsschwankungen in der Werkstatt. In: WIENDAHL, H.-P., AHRENS, V. (Eds.): Potentiale der Chaosforschung in den Produktionswissenschaften. Proceedings. Institut für Fabrikanlagen, Universität Hannover.

TOWILL, D.R., DEL VECCHIO, A. (1994): The application of filter theory to the study of supply chain dynamics. In: Production Planning & Control, 5b:1, p. 82-96.
VASTANO, J.A., KOSTELICH, E.J. (1986): Comparison of algorithms for determing Lyapunov exponents from experimental data. In: MAYER-KRESS, G. (Ed.): Dimension and entropies in chaotic systems. Springer, Berlin.
WARNECKE, H.-J. (1992): The Fractal Company – A revolution in corporate culture. Springer, Berlin.
WIENDAHL, H.-P., SCHOLTISSEK, P. (1994): Management and control of complexity in manufacturing. Annals of the CIRP, 43:2, p. 533-540.
WOLF, A. ET AL. (1985): Determining Lyapunov exponents from a time series. Physica, 16D, p. 285-317.
WORG, R. (1993): Deterministisches Chaos. Wege in die nichtlineare Dynamik. BI-Wissenschaft, Mannheim.

Rationality, Culture and Politics of Production

Manfred Moldaschl

1. The dual reality of production, or: Why managers do not meet their goals

The title invites protest. 'We do indeed meet our goals', managers would say, 'just not always to a one hundred percent extent'. Certainly! But the question remains as to whether they achieve their (company, personal) aims, to whatever extent, *because of* or *in spite of* their strategies. The reason why certain goals are not reached, and others in spite of ever more polished management strategies are reached, is the subject of the present contribution.

I wish to examine two phenomena which have engaged me since the beginnings of my social research in the field of industrial work. The first is the phenomenon of the constant rebirth of the *one best way*, the 'key to success' in rational production planning and control which is discussed anew year for year – and its regular failure in production practice. This I experienced and researched in particular in the cases of Computer Integrated Management (CIM, PPC), Artificial Intelligence (AI) and Lean Management (particularly team work and Continuous Improvement Process, CIP) in empirical projects. Some examples will be presented here. The second phenomenon is connected to the first; and it is the enormous *discrepancy between the official claims and the reality* of management concepts.

The main topic of this contribution is less concerned with the reasons for the 'dual reality', and more concerned with the question of why the cycle of euphoria and the 'blues' is repeated *again and again.* What upholds the façade of the movie set, or, why do the discrepancies between model and reality live on, even in 'modern' organizations? The main thesis presented is: Supersedure of the old, hierarchical and Taylorist organization is usually carried out using its very own, inherent methods: strategic management and instrumental rationality. This fundamental contradiction prevents truly new organizations of work. As an alternative perspective of the organization of work, 'cooperative networks' will be discussed in the last section of the paper.

In principle, there are two different interpretations or explanatory approaches: a functionalist which on very rational grounds shows for which company actors the apparent reality is functional, and a more psychological approach, which attempts to explain why the actors are apparently 'blind' to certain realities[1]. Disregarding the theoretical disputes between representatives of these two approaches, my position is that the two should be treated as *complementary* and not as alternatives when analyzing and understanding dual reality in the concrete case. This means that the company be viewed both as a political arena and a cultural world. But before we turn to explanations of dual reality, let us look at a first example from our empirical investigations in the automotive and chemical industries.

Example: The continuous improvement process

The Continuous Improvement Process, CIP, is held to be the incarnation of new approaches to create an 'efficient culture', seeking to link increases in efficiency and social integration. All employees are called upon to apply their experience and their creativity, with no inhibitions, to suggestions for improvement. Monetary prizes, improvements in one's own working conditions and symbolic recognition are offered as incentives in exchange for participation in organizational optimization which transcends one's own work task. The central goal is to 'mine the gold in the heads of employees' and, at the same time, to promote loyalty to the firm. In particular, the basic problem caused by the continuing separation of planning and primary operating tasks – namely, the discrepancy between planning and reality – is to be eliminated. CIP is thus also an approach towards reinstituting the broken flow of feedback from applied experience back to planning, towards *closing the circle of planning and experience*.

If we believe the reports 'from the world of practice' which fill journals and collected volumes, there have only been successes. Oscars for quality are presented on all sides. Studies report increases in productivity up to 20% and the successful forging of links between efficiency increases and employee motivation, which succeeded in 89% of the interviewed companies. Apparently CIP found that which all rationalization concepts had sought in vain: the *magic lamp of the Aladdin Company*. Sesame opens and releases its treasures of experience. Empirical

---

[1] In the 1960s and 1970s, there were a number of studies which fell back upon the work of the sociologist Max Weber. They saw the discrepancy between 'formal' and 'informal organization' as based in the nature of *bureaucracy*, that is, in the power of rules and regulations.

investigations, however, discovered a different reality of massive problems and undesirable side effects.

In the case of the automobile company studied, workers and employees are painting an increasingly negative picture, although for differing reasons. In the opinion of planners, the incentive system has production workers waiting avidly for them to make mistakes. They hardly dare to stop at the machines to discuss possible changes: 'You have to be very careful about what and with whom you speak. Otherwise, there will be an improvement suggestion on the table the next day'. They feel that workers would make sure not to give planners the 'gift' of possible improvement suggestion rewards by giving planners direct feedback. With all this, ultimately a huge improvement-suggestion bureaucracy gets established to check, evaluate, select, report and implement, and the old 'concepts of the enemy' between planning and production become revived instead of eliminated (Figure 1).

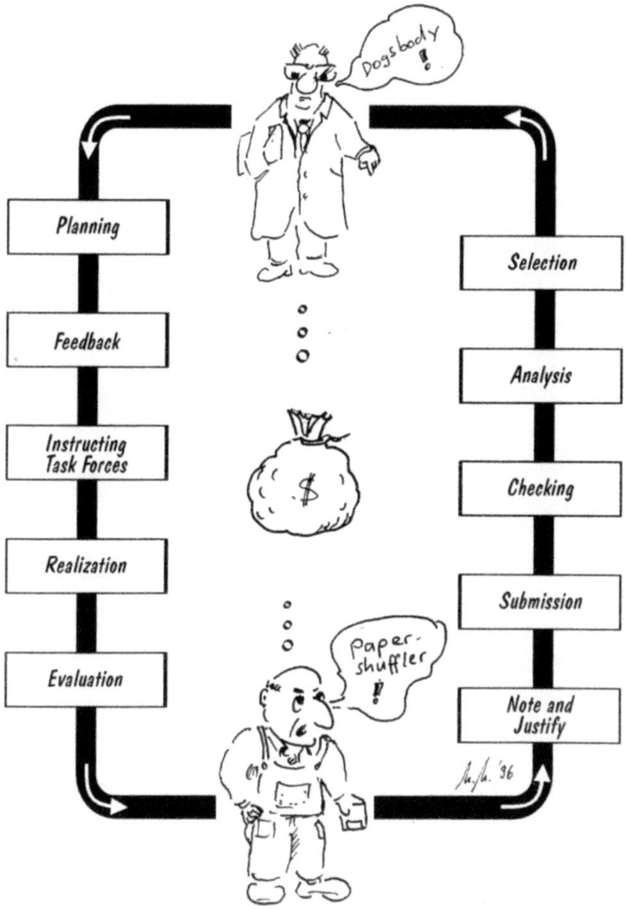

Fig. 1 CIP as a game of power and gain.

But this is about more than money. Where *social distance* between workers and planners was particularly great (as in series production), workers try to make planners look incompetent. 'They let us walk straight into the trap', says a planner, 'and the frustration is terrible'. Planners report that they are forced to protect themselves on all sides: in writing, through witnesses and through coalitions with other experts, who confirm the objective and logical consistency of their decisions. 'Otherwise, we get a threshing'. This incurs not only additional effort and expenditure, but also promotes the revival of the familiar adage that conservatism suppresses imagination. 'We don't want to risk anything', in the words of a planner, and 'God forbid experiments'.

On the other hand, workers view the way improvement suggestions are dealt with as a measure of planners' and management's seriousness of commitment to participation. But to take the improvement of one's own work area into one's own hands is a limited possibility: there is neither enough time, nor resources nor influence to realize the changes workers suggest. The real chance of participation is found to be disappointing. That is why workers sarcastically call CIP the '*concept of imaginary participation*'.

Each of the two groups criticizes equally management's 'cosmeticizing of results'. Because CIP is the thing to have today, a prettier picture is painted: processing and planning expenditures are not included, and the effects of automating and other rationalization measures are reported separately when the numbers have to show the development of productivity and costs since the introduction of CIP. The phenomenon of the stagnation, or decline, of participation and results (Figure 2) are swept under the rug[2].

	1993	1994	1995
Number of proposals	311	102	164
Paid proposals	179	48	4
Cost-savings	345.000,- DEM	117.000,- DEM	22.000,- DEM
Share of paid proposals	57,6 %	47,1 %	2,4 %

Fig. 2 Typical development of improvement proposals and results (engine department of a German automobile factory).

What was intended to provide solutions has become a problem in itself. Workers increasingly restrict themselves to individual CIP participation or withdraw from it completely: 'We work to the rules and avoid thinking'. In the

[2] Behind the scenes, employees find relief in symbolic reinterpretations. The reason the company coined the formula 'CIP^2', workers explain, is because squaring a negative result will give a positive one. The TQM qualification program, which emphasizes behavior norms in its very character, is called the 'reformatory' by the workers.

relationship between worker and planner, the good idea has been transformed into a *cooperation inhibition program* (CIP), into a game for power and gain with built-in brakes to learning, whose rules are set by mutual distrust. 'This whole thing backfired', says a planner in sum. If these experiences were to be discussed openly and worked through in a reflective manner, the actors could gain significant insight into their cooperation problems. But the critique of CIP in planning and production is dominated by *mutual accusations*. Most employees don't question the *system* itself.

2. Approaches to explaining the dual reality

2.1. Functional, or political, explanations

There are obvious rational explanations for the normality of the dual reality. MEYER & ROWAN (1977), focusing on them in their investigation, describe their *function* for management. They have an interest not only in efficiency, but also in securing power. Efficiency requires flexible action and with that, degrees of freedom of action for employees. The securing of power, in contrast, demands control, so that the degrees of freedom are not abused in the service of individual interests. As these two objectives conflict, a precarious balance must always be maintained. According to MEYER & ROWAN, management attempts to resolve the problem by decoupling the strategies to assure power and efficiency. On the one hand, management upholds the claim of extensive plan-ability ('rationality myth'), but informally, it permits certain 'unauthorized behavior'. On the other hand, it tries to move employees – by means of social-integrative and trust-fostering measures – not to utilize their freedoms to interpret and act in ways contrary to the interests of the firm.

> „Still, impracticable regulations are not done away with, because violations are a valuable resource. Superiors can officially insist that rules be followed, but silently tolerate rule-breaking in order to demonstrate their understanding and goodwill towards the rule breakers, who are then obligated to give something in return." (NEUBERGER 1995, p. 276).

This *power play*, of course, has two sides. Employees publicize deviations from impractical or unrealistic regulations and plans only in the case of emergency. They fear to lose their degrees of freedom for action or to be punished for infringing upon the rules. After all, their own power is based on the fact that their experience and proximity to processing allow them to deal with any uncertainty that planning has not been able to conquer, either due to reasons in principle or unjustifiable effort (CROZIER & FRIEDBERG 1977; BURAWOY 1979). Functional explanations here are thus *political* explanations. The term 'political' denotes the issue of: Who acts in what interests and forms an alliance with whom (against whom)?

From this perspective, therefore, there is not only the one rationality, but instead there are different, position-dependent local rationalities; and it seems naive or at least very one-sided to view the company as a 'rational machine' with the manager as its 'strategic controller'. In a similar vein, rational justifications and methods are seen primarily as legitimation for actions led by *vested interests*. Cost-benefit analysis of measures serves, as in the example of CIP, not so much to allow decision-making as to justify actions after the fact. The 'practical constraints' argument is usually complemented at the symbolic level with humanistic slogans ('Our focus is people') or 'we' theories ('We're all in the same boat'). George Orwell, for example, described with his characters „Doublethink" and „Newspeak" in *1984* categories of symbolic management. A more recent version of „Doublethink" is offered by MAZAAKI IMAI (1986, p. 98), who extols the strategic advantages of TQM and CIP as follows:

> „While management's primary concern may be productivity, quality is a concern common to both labor and management. When managent asks labor to increase productivity, labor's response is, 'Why? Doesn't that just mean we have to work harder? What's in it for us?' However, nobody can object when management asks for labor's cooperation by saying, 'Let's talk about quality.'" (IMAI 1986, p. 98)

The success of such strategies, as of any (micro)political or military strategy, depends first upon the extent to which the other party (employee, opponent) sees through the strategy, and second, upon whether he can mobilize his own instruments of power to counteract it. In the CIP example above, symbolic reinterpretation ('concept of imaginary participation') and the strategy of orderly retreat show that employees did not allow themselves to be led by the nose for long.

It is characteristic in this connection that a first draft of a publication we wrote on an investigation of CIP and gave to parties in the firm to read was 'surreptitiously' circulated and left anonymously for managers to see. This was obviously an attempt not only to settle some scores, but also to discretely induce the firm to reflect upon the very mixed feelings and experiences with the new organizational concept.

The fact that with this, discrepancies between objective and result are dealt with only clandestinely, under the surface – and that the critics could not bring themselves to initiate an open struggle over the interpretation of CIP – is surely the result of 'power logics' which no one would call economically rational in the sense of company efficiency. No one wants to be Don Quixote. Successes and failures are dealt with at the personal level – in the form of promotions and career-freezes – and not at the level of the institution. Ways in which differences between model and reality, objectives and results might be worked through in a meaningful manner at the institutional level will be discussed at the end of this paper. At this point, we will leave the strategic interpretation and turn to an examination of the causes which lie within everyday theories, or in other words, which are contained in the implicit, behavior-guiding notions of company actors.

2.2. Psychological explanations of the dual reality

Organization researchers CHRIS ARGYRIS & DONALD SCHÖN, whose theory of organizational learning will be discussed below, make a useful distinction with regard to the way that actors relate to organizational reality:

> „When someone is asked how he would behave under certain circumstances, the answer he usually gives is his *espoused theory* of action for that situation. This is the theory of which he gives allegiance and which, upon request, he communicates to others. However, the theory that actually governs his action is his *theory-in-use*, which may or may not be compatible with his espoused theory; furthermore, the individual may or may not be aware of the incompatibility of the two theories."
>
> (ARGYRIS & SCHÖN 1978, p. 11).

Whether or not the reasons for action – that's the way we choose to interpret the term 'theory' above – can become conscious is, however, important if we wish to understand 'dual reality'. That is why a third 'theory' was mentioned above, namely the ideology in which the actor himself does not believe, but which he puts forward as a pretext in order to veil his actual behavioral motives. However, we find the functional explanation offered by MEYER & ROWAN too one-sided. Therefor, we wish to turn to those theories-in-use which are so deeply rooted, spread and accepted in certain cultures, milieus and worlds of life that they seem to us to be 'natural' and the only option. For these, the historian of science THOMAS H. KUHN coined the term *„paradigm"*.

The paradigm of the controllability of the objective and social world

The main and in a sense most 'reliable' reality-doubling paradigm is that of rationalism, or more precisely, that of *instrumental rationality*. The Age of Enlightenment established the Cartesian model of man's rational mastery over nature (and also human nature) and society through the strength of practical reason.[3] But his intention to free himself of superstition and of absolutist rule became increasingly eclipsed by the side effects of a one-sided rationality making

---

[3] For IMMANUEL KANT, the task of practical reason was to eliminate superstition, unjustified domination, physical and mental lack of freedom and suppression. Making reason so one-sided as to become just a means of reaching any goals at all, independently of social agreement, has been characterized by JÜRGEN HABERMAS as „calculated rational action" (*zweckrationales Handeln*). I use the term 'instrumental rationality' in that sense.

the world bend to its will. The belief in predictability and the conviction that the world of objects may be technically *mastered and controlled* (the 'Laplace demon'), has led many companies into existence-threatening formalization and over-technization. The second component is *utilitarianism,* a view that reduces man and the objective world to their utilizability, combined with the presumption that all human behavior is guided by this principle (homo oeconomicus). Let us look at another example from the world of practice of this conviction of controllability and mastery, which plays an important role particularly within the culture and socialization of the science of engineering.

In a large mechanical engineering firm, a flexible manufacturing system (FMS) was set up for the manufacture of gearboxes. In order to perform all operations completely in the system, twelve mainly different (complementary) machines were integrated. The contracted software company did not succeed, within the twelve-month period agreed upon, to develop a control system that mastered the complex logistics tasks. Two years later, there was still no real progress. At this point, a systems developer from within the company took over completion of the system. Twenty-four months later, the system merely functioned, that is, the system capacity rose to just over 50%.

The reason for the relative and delayed success: the systems developer had moved his workplace right into production, into a small office next to the FMS and it's control system. Through the course of cooperating with systems personnel, he increasingly lowered his expectations of automatic systems control, and instead, implemented more and more intervention options for workers in order to support them in situation-determined control and in the re-scheduling of operations. This was because production proved to be too complex and too dynamic to be represented by algorithms. Context conditions which no one had ever thought about before, such as unexpected interactions of control processes, weather influences or wear-out, regularly threw premises overboard. After the experiences of working with the system over time, one foreman summed up: 'Once the childhood diseases had been overcome, the signs of old age set in'.

The irony in this case was that the systems developer, despite his practical orientation, held on to his theoretical model of predictability, with the goal of complete computer representation of production processes and control of them through planning. In other words, he fulfilled the task *not because of, but in spite of* his guiding concept. Whereas his *theory-in-use* changed through practical application, his *espoused theory* remained totally unchallenged:

'The automatic control system', he concluded, 'worked wonderfully. It is just that the environmental conditions were never right.' In short, the logic is: *The model was good; it was reality that was bad*.[4]

---

[4] As a rule, similar design processes usually take a less favorable course: planners act from a distance, and the organization is adapted to the requirements of the system (as observed in many firms which introduced the SAP system – a PPC system).

The fact that certain points of view are hard to modify even through *experience* is the very stuff of a 'paradigm'. A paradigm is „resistant to disappointment" so to speak, and, as THOMAS KUHN (1962) demonstrated with scientific communities, is not „falsifiable". Once information and engineering scientists, for example, were forced to recognize that their previous models of production control were too inflexible to prove useful in turbulent environments[5], many placed their bets on artificial intelligence (AI). From the 1950s on, AI had claimed that it could better master complex and dynamic tasks than information technology, because it reproduced the functioning of human intelligence. It has yet to give proof of this claim. „That computer programs would in ten years provide the standard model for psychological theories is the only prediction that has been fulfilled" (VOLPERT 1985, 141). The model of man is being adapted to that of the computer.

Researchers within other theory traditions criticized this reductionism and found in it the explanation for the lack of success of its application. DREYFUS & DREYFUS (1986), for example, showed convincingly that the actions of experienced human experts are based, of all things, *not* on the application of rules, which is why even future increases in computer capacity will not be able to supply the algorithms of expertise. The authors developed a 5-level model of expertise, whereby only the actions of the novice are characterized by adherence to rules. The masters at the highest level are distinguished by *intuitive* actions which are marked by sensory experience in situational contexts. It is just this non-formalized thinking with vivid imagination and analogies that characterizes the special nature of this master-level expertise – more so than explicit knowledge and abstract cognition. This criticism is certainly not ignored. At a congress on new approaches in information and control technology where I had presented this view, two software engineers countered: „We're already building expert systems of the 5th generation" (see MOLDASCHL 1992).

This claim may be interpreted in a functional manner: language becomes adapted to market requirements. Discussion which then ensued at the congress, however, placed more weight on the cultural interpretation. The systems developers were completely convinced that every human achievement (which is for them always a cognitive one) could be reproduced by information technology. Even intuition. For them it is always just a question of time (or available computer capacity) until the algorithms are found and the machine can do *anything* better than man. Failure is merely seen as a consequence of a 'not yet' perfect system or model and never as the result of a possibly false paradigm. This is like the driver who has chosen the wrong exit off the highway and hopes, by stepping on the gas, to get close to his destination.

---

[5] In every company we studied, there were numerous anecdotes of the discrepancies between detailed sequences planned with PPC and reality. In two firms, for example, PPC was called *'Pech & Pannen-Service'* (tough luck & breakdown service).

Features of and alternatives to instrumental rationality

If the Cartesian paradigm of predictability proves to be far from reality even at the material level, at least under variable environmental conditions, this is, of course, all the more true for social relations and interactions. Nonetheless, the paradigm is not only entrenched in management practice, but also in the social sciences (economics, sociology, psychology).[6] The essential features of the paradigm can be summarized as follows:

a. The goals of the organization are seen as given and absolute. Therefore, the question of how goals arise is not relevant, and the only interest is in the efficiency with which the goals are achieved. From this it follows that only *the one* rationality is stressed which views any deviations as irrational and illegitimate.

b. Every element of the organic and inorganic world, even people and their relationships, are seen as potential *means* of achieving one's own goals. 'The end justifies the means' is the genuine moral of this *utilitarian* rationality. Man is seen as 'rational egoist', who will cooperate only if it pays to cooperate. Therefor, organization is a necessary means of keeping him in check.

c. The actor interacts with other subjects and objects in a 'calculating' manner and evaluates their characteristics either *egocentrically*, in reference to his own subjective interests alone, or from the *distanced* (or de-centered) perspective[7] of an observer, referring on organizational norms. Putting himself in the place of other people or cultures is hard for him.

d. Because the actor restricts his attention to 'useful' features of the outer world, he overlooks other characteristics and underestimates their own individual logics. He places faith in planning methods and is less apt to anticipate and consider non-intended *side effects*.

This manner of thinking, which involves vastly insufficient reflection upon oneself and the unexpected side effects of one's activity, is the reason for his 'resistance to disappointment' and the reason for the constant supply of new 'promising' methods (at present: reengineering, post-lean, super-lean, etc.). It's the hope to reach the permanently receding horizon. Certainly there are situations in which utilitarian action is appropriate and effective. But in an increasingly interdependent, turbulent and 'politicized' world, it soon reaches its limits, produces fictitious worlds and adaptation problems. Innovation processes or changes involving intensive knowledge and cooperation will not be achieved on such a basis.

---

6 This is not only evidenced in their models ('rational choice' models, for example), but also their, in part out-of-hand, methods of investigation to 'measure' and 'objectify' company conditions in order to derive plans for change. Their results are mostly used by hegemonial groups within an enterprise to legitimate their own strivings.

7 The extent to which a person understands himself as an integral part of a collaborative community is modelled in the cultural-historical activity theory on the levels: egocentrical, decentered and recentered (see RAEITHEL 1997).

Critics of this instrumental rationality have proposed alternatives: The social philosopher JÜRGEN HABERMAS (1981), for one, put forward „communicative rationality", which, instead of rationally-grounded „practical constraints", defines rationality as that to which actors can come to agreement upon through discourse. A similar concept of „social rationality" was presented by the organizational sociologist CHARLES PERROW (1986). And theories of self-organization (e.g. BATESON 1972) help to understand, why „living systems" can be controlled even less than the world of objects. The kind of rationality described in these approaches I would like to unite here under the term *reflexivity* (see Figure 3):

a. Action as negotiation: The social actor assumes that in his field of action, differing, position-dependent rationalities coexist. He does not expect that they might become subordinated to his own or any universally valid rationality. He knows that his acts are embedded in a social field, and that he may only reach his goals if he and others can reach agreement on goals and the permissible means to achieve them.

b. Self-reflexivity: Where the issue in instrumental action is only concerned with the goal – means relation and the effectiveness of the means, the reflective actor expands reflection to include the goal itself, the meaning of goals and their adequacy in the social context. Goals are understood to be produced socially.

c. The type of self-referentiality here is re-centered: the person understands his activity as an integral part of a collaborative activity system. Further, this person knows that his perceptions are subjective (or 'encultured'), and that he makes use of theories, as do others. Putting himself in another's place and observing one's behavior 'through that person', the perspective becomes interpretative, while he accompanies his own actions also with the perspective of an observer.

d. The actor is conscious of the fact that he is embedded within social context and gives particular consideration to the unexpected side effects of his actions, so that he can revise his plans and theories on the basis of feedback. To do this, he must keep them open and adaptable.

Accentuating the social and cultural *embeddedness* of acting this rationality goes beyond the higher-level instrumentalism of 'bounded rationality'. But granted, it does not imply any conception of the organization as a place of grace, empathy or harmonious agreement. On the contrary. The very fact that other rationalities of action are given recognition and one's own is reflected in relation to them is that which allows conflicts to be perceived and dealt with in a constructive way. Power

	Instrumental Rationality	**Reflective Rationality**
Rationality	General and abstract: efficiency ↓ Goals are given, deviations are irrational	Situational and plural: consensus ↓ Valid criteria must be negotiated
Legitimation of Action	Utilitarianism ↓ Maximizing benefits is good and correct	Self-reflexivity ↓ Reflecting on meaning and adequacy of one's own goals in social context
Self-Referentiality	"Egocentrical" or "decentered" perspective ↓ Cooperation for self-interest or to fulfill organizational norms	"Recentered" perspective ↓ Cooperation based on the recognition of interdependance
Learning Mode	Trust in strategic planning ↓ Correction of plans (first order learning) Neglect of side effects	Open, provisional planning ↓ Modification of theories (second order learning) Concentration on side effects

Fig. 3: Types and modes of rationality

and ideology remain as options, if attempts to find forms and media of negotiation, find methods of consensus-building and meaning are not successful – but only as ultima ratio, because the destructive consequences of the use of power will also be reflected upon.

Is there then any alternative whatsoever to instrumental rationalist action for management? After all, is the goal of the organization not set – to increase the capital of the capital owners? The answer to the first question would have to be negative, if with regard to the second there were a scientifically absolutely guaranteed 'one best way'. But that has not been discovered yet, and only hard-core rationalists still believe that a context-independent 'best practice' might exist. What can be made of this distinction in the world of practice? Two examples will serve to demonstrate.

3. Self-organization on demand and feigned company culture

Examination of two themes in the modern literature on management, self-organization and company culture, reveals in many cases that the concepts, recommendations, models and their implementation in practice are still, despite the new vocabulary, firmly embedded in the paradigm of instrumental rationalism. As a consequence, that which should be achieved cannot be achieved, because goals and the path to goals diverge wildly.

Who organizes self-organization?

Be completely spontaneous, and do it now! This is the paradox to be found when self-organization is given as an order. No new orientation, no culture is immune to instrumental rationality. In all of our company case studies on team work (literature quoted in MOLDASCHL & WEBER 1997), use of the term self-organization put a strain on the concept. With one exception, in all cases the principle of self-organization was again misunderstood – or misused? – as a control instrument (Figure 4).

In a chemical firm, Management by Objective (MBO) was introduced into the entire blue-collar area after team work had been instituted. MBO was extolled as a 'participatory concept' according to which management and workers could mutually define achievable goals in the (financial) interests of both parties and negotiate the necessary conditions to reach those goals. In other words, a typically reflective principle was presented. Already with the first goal agreement (for 6 months), however, workers saw that of goal criteria to be discussed, three were predefined, and only the fourth could be chosen. Moreover, management had simply raised the set targets, which workers had judged to be achievable, at their own discretion, and no further negotiation was permitted. In the second agreement, the goals of the first phase were taken as the starting basis and – once

Fig. 4 Self-organization.

again by management alone – increased. In the understanding of management, self-organization meant that workers were to figure out themselves just how they were to achieve the very high goals which had been set. What had functioned for the most part the first time around, ended in the second phase in disillusionment of the workers. Hardly any group made efforts to achieve the goals.

That an interpretation like the one above cannot be dismissed as the typical limited understanding by the practitioner, but rather is also described and recommended by scientists as 'cooperative goal negotiation' is clearly illustrated by an influential professor of industrial management:

> „The task of all executives is to break down the superordinate goals of the organization in the form of goals to be achieved and to follow those goals consistently with their workers. [...] Agreement upon goals takes place through a process in which the superordinate objectives of the enterprise are concretized all the way down to the level of operations as team goals and are accepted by the workers. [...] Goals, once achieved, are then set higher." (WILDEMANN 1995, p.9).

It is often difficult to discern whether this is ideology or paradigm. In any case, this understanding of 'goal agreement' as setting goals, with a built-in increase-effect, is becoming more widespread. However, its unexpected and undesirable consequences, such as withholding of knowledge and refusal to cooperate by employees, usually remain hidden and elude the company's cost-benefit calculations. This example from the world of practice impresses upon us the fact that the instrumental rationalist strategy can indeed function. But where it does, it is at risk of destroying its own prerequisites, which is not reflected upon. It is as HANS MAGNUS ENZENSBERGER aptly described international tourism, with its unceasing search for undiscovered terrain: 'Tourists destroy that which they seek – by finding it'.

How do you make the culture of a company?

The image of man as homo oeconomicus has been much criticized. The insight that man does not only act rationalistically (for pay incentives, etc.) has led to the development of many proposals to complement rationalistic control concepts by means of emotional leadership and integration. Many hold to the Human Relations approach (ROETHLISBERGER & DICKSON 1939) as one of the most important forerunners. There is no doubt, however, that this approach has to do with the „intelligent control of men and situations", as ELTON MAYO put it (see MOLDASCHL & WEBER 1997). The ideal type of the culture of an enterprise is described as the „best practice" components of strategic management in PETERS & WATERMAN'S (1982) best-seller. In comparison to this instrumental understanding of culture, we offer a reflective one:

a. An instrumental rationalist understanding of company culture views it as a complex, ideal control system of management – the software of a control system, so to speak, for the needs of the workers for meaning. As technical and organizational strategies assure efficiency, company culture is seen to produce consensus on company goals and identification with the company, in that it provides binding values, points of view and conventions. In other words, it is assumed that 'culture' can be instrumentalized to serve company goals (organization of culture).

b. In a reflective understanding, company culture is the totality of social relationships (which have grown) and the practices and interpretations that characterize the living organization of an enterprise. It is understood as an emerging order which reflects the real experiences of members of the organization with regard to cooperation and the balance of interests among all involved (but also professional self-definitions acquired external to the company). In this light, culture develops out of practical interaction, and it can change. It cannot, however, be manipulated in any lasting way by management or any other group of actors according to their own designs (organization as culture).

The boom in topics such as ethics and culture and the shift away from the Taylorist paradigm does not in any way mean simply that we have left instrumental rationalism behind. This theory-in-use is very deeply rooted in our cultural environment, as some typical passages from a manual on 'company culture', which actually intend to criticize the paradigm, remind us: „If employees recognize that they are mere means to the goal, they will view the company and their work as mere means to the goal" (LIETZ 1994, 114). „In what goals is energy invested? In the goals of management or in partly deviating, personal goals?" (112). Because LIETZ assumes the latter, he says that management should develop a „catalogue of values", which should be communicated all the way down to the worker" (121). „If higher values are expressed, employees will also accept the fact that a company provides no lasting usefulness to its customers and employees if it does not make enough profit(...). The profit motive becomes a social duty. A conscious lack of contribution to the community with a meaning damages the community and will not be tolerated by the works council or the employees" (116f). The „natural scientific-reductionist image of man" should be replaced with „socioethical competency in management", and „the employee should be treated with respect and trust. This is what makes work meaningful" (119). The advantages, says LIETZ, are *calculable*: „working with each other and for each other" minimizes friction loss, and „considerably less time and fewer employees are needed to complete projects" (120).

Despite LIETZ'S verbalized critique of the instrumental rationalism of management (his espoused theory), we find here all of its fundamental assumptions presented in their ideal forms: the 'homunculus' theory (the person as an individual competitor guided by economic incentives); 'complementation' of the rationalist image of man by the level of 'irrational' emotions as justification for 'emotion management'; simple causal thinking; and the ideology that culture can be strategically manipulated, and consensus can be reached without communication through management and „Newspeak". The best comment to make on all this was formulated by ABRAHAM LINCOLN over 130 years ago: „You may fool all the people some of the time, you may fool some of the people all of the time, but you can't fool all of the people all of the time".

4. Cooperative networks and organizational learning

A critic must also be able to say what could be done better. A recommendation to 'be reflective' would, of course, be hardly any better than the command to 'be spontaneous'! Instead, I will sketch out a way in which reflexivity might become anchored within the institution. To do this, the concepts of *reflexive organizational learning* and *cooperative networks* will be presented.

Simple and reflexive organizational learning

As we have seen, a central problem of instrumental rationalism lies in the fact that it can not process feedback in the case of unexpected and counter-intended consequences of action. But to transform feedback into improvement and change of one's own 'theories' is precisely the essence of learning. How does the empirical knowledge individuals possess enter into the store of knowledge of an organization, or, in other words, how does this knowledge become represented at the superindividual level? How does the organization learn?

Let us return to our initial *CIP example*. Systematic evaluation of experiences made with CIP was undertaken only with reference to criteria which conformed to expectations: the economic reference numbers. Other effects were not studied. Despite the obviously limited success, reaching not even very narrowly defined goals, the system in its current form does not come under question. Measures to resolve problems remain in the context of the original purpose and conception. Time is not found for regular CIP meetings; the processing of improvement suggestions has not been decentralized, employees do not participate in their implementation; employees are not given any feedback on the processing status of suggestions for improvement; and there is a lack of open discussion within the firm of the problems arising. The planners and the employees also do not question the organization of the *system* (allocation of resources, division of labor, etc.), but rather mutually *blame* each other for the problems.

Here there is obviously no *institutional learning* taking place. Instead of solutions, scapegoats are sought – a typical learning block. Management appears to confuse experience knowledge and improvement suggestions with mineral resources, natural ores, which are to be efficiently exploited. This is a further typical block to learning rooted in instrumental rationalist thinking. After all, knowledge is no raw material, but rather is a resource which increases through use. It is *produced* through shared use. In the company studied, however, the cooperative production of knowledge is not the focal point of its CIP understanding. The Kaizen principle of recognizing that errors are valuable resources, safeguarding learning so to speak, is not applied to the organization and management *itself*. The most important component of organizational learning is missing: *feedback*.

Feedback is the key to institutional learning, or the *institutionalizing of reflexivity*. It consists in organizing open communication of experiences with discrepancies between plan and reality, as well as with differences among the perspectives of the various actors. With this, we find once again, here in the area of organizational learning, our two paradigms, the instrumental rationalist and the reflective. We also discover them in a model of organizational learning processes presented by ARGYRIS & SCHÖN (1978). In this model, *simple* 'single-loop-learning' corresponds to the instrumental paradigm. It describes the adaptation of behavior to given conditions and the correct application of rules. The *broadened* 'double-loop-learning' describes the adaptation or change of the rules, assumptions and values (see Figure 5) that underlie behavior. To the extent that this change occurs only in 'espoused theories', and not in 'theories in use', I would still refer to this as instrumental rationalist learning.

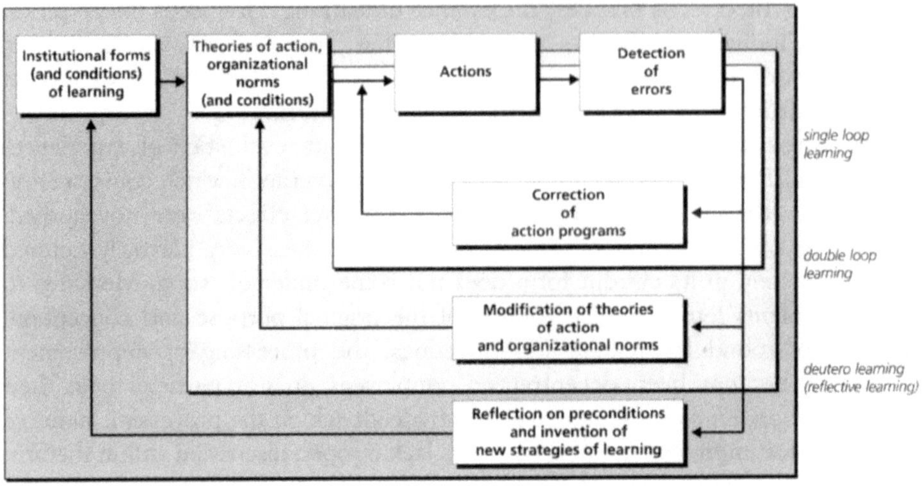

Fig. 5 Levels of reflexivity in institutional learning (in the model of ARGYRIS & SCHÖN 1978).

Typical of *reflective* organizational learning, in contrast, would be „deutero-learning" in the ARGYRIS & SCHÖN model. This consists in making previous learning processes themselves the object of reflection, in order to (further) develop *learning ability* itself. Members of the organization „discover what they did that facilitated or inhibited learning, they invent new strategies for learning, they produce these strategies, and they evaluate and generalize what they have produced" (ARGYRIS & SCHÖN 1978, 27). It should be noted, however, that this model focuses only on the *cognitive* processing of experience and not on the context conditions of behavior as well as their production and change[8].

But how can we support reflective organizational learning? How does one promote reciprocal feedback and the recursive processing of experience? What forms of the division of labor and of cooperation are advantageous? There are many starting points (for example, organizational development, strategic decentralization, etc.). But I would like to look at just one perspective which is very consistent with the principles of reflective rationality, namely cooperative networks.

Cooperative Networks

Cooperative networks may be seen as an alternative to current concepts of team work and project management which also deal with the problems of the division of

[8] For this reason, context *conditions* are set in parentheses. Actually, the loops at all three levels should be doubled in order to adequately represent the cognitive and the structural aspects of reflexivity.

labor and the production of knowledge. Many of these concepts are committed to instrumental rationalist thinking[9] and therefore contain within themselves, in an unprocessed manner, the contradiction of „externally organized self-organization" or „forced autonomy" (MOLDASCHL & WEBER 1997). Also, they mostly overlook the fact that team structures not only dissolve the traditional lines demarcating cooperation, but rather, at the same time, create new ones, which themselves produce undesirable side-effects.

The term network has in recent years experienced an undreamed-of boom, even prior to the 'Net of nets' (Internet). It has spread throughout the most various areas, such as biology and computer science (neuronal networks), linguistics (semantic networks) and psychology (networks of social relationships). In sociology and industrial economics and management, for example, supply networks at the trans-corporate level have become the focus of interest. Inter-company cooperation is described as a form of strategic coordination of activities between market and hierarchy, which allows flexibility – despite specialization – of company units. It is interesting to note that networks, when described at the level of the individual company, however, are always treated as informal structures. And they are usually described as informal structures that are directed against the strategies and interests of management (just as in the famous older study by ROETHLISBERGER & DICKSON 1939). For this reason, there have always been attempts by management to make use of something that they could not eliminate. Earlier, this was *Human Relations*, today it is increasingly *context management*.[10]

But up to now there has been nearly no attempt to apply the network idea to the design of cooperation within a company. How can cooperative networks be formed here? Not at all. In view of the preceding argumentation, this misunderstanding – that at this point in the paper options of instrumentalizing informal social relationships would be presented – should not arise You can let your beard grow, but you can not 'grow' a beard – just as you can not 'emerge' networks. What you can do is create the conditions under which cooperative networks arise, that is, under which they can organize themselves. With this, we come to the characteristics of networks, which are very closely linked to the concept of self-organization (see GRANOVETTER 1985, PERROW 1986, POWELL 1996):

- Networks are *emergent*, that is, not planned; they function, however, because and to the extent that they form in practice and prove worthwhile (coevolution).

9 As an example of a markedly functionalist and hierarchical concept of project organization, the study by CLARK & FUJIMOTO (1991) is recommended reading. Here the further problem becomes apparent that project groups and similar forms of trans-functional cooperation are being conceived as 'parallel organization' which does not question the functional division of the whole.

10 This means that an organizational unit is allowed relative autonomy and certain resources, hierarchy is broken down and only context conditions and results are controlled.

- Links among the elements are not set givens: they organize themselves (*learning*); the structure and history of a network can not be separated.
- The degree of attachment, that is, the dependency among the elements or partners, is slack (*loose coupling*).
- A network has no center: all vital functions are distributed; if a part is missing, the function does not break down (integrality; *error friendliness*).
- The multitude of links provides various, parallel routes of processing on the way to realizing solutions (*redundancy).*
- Networks have blurred edges; that is, it is not always unambiguously clear what belongs to the network and what not (*openness*).

Let us concretize these heuristic principles which serve as a guide to discovering, allowing for and fostering cooperative relationships in the enterprise. What is the difference between networks and traditional organizational forms (Figure 6). Loosely coupled networks are characterized by the relative *autonomy* of their units. Does this automatically preclude unequal distributions of power? In trans-corporate networks this is apparently not the case. But the significant advantages of networks from their *heterarchical* or *polycentric* structures, whereby it is the task of all specialized units to contribute to coordination and innovation. Networks are thus not free of power, but there are, to a large extent, free of hierarchies. For this reason, different forms of 'reward' for achievement are required, such as horizontal careers, which promise greater degrees of freedom for action, resources and income while not being bound to management functions (which actually are rather aggravating to many people who are interested in the content of their work).

The company Semco (SEMLER 1993), for example, has many of the characteristics of a network, which not coincidentally are oriented towards democratic principles. Workers elect their supervisors, engage in improvement considerations with them and, if necessary, vote them out of office. Traditional personnel evaluation is turned around: managers are evaluated by their 'subordinates' twice yearly, as well as vice versa – there is mutual feedback and shared development (*coevolution*). Management itself is not carried out by isolated managers, but rather, as are all other work processes, in teams. SEMLER'S book is a fountain of innumerable further ideas on networks.

Cooperative Networks	(Neo-)classical Organization
Loose coupling of organizational units (OU)	▪ **Tight coupling of organizatinal units (OU)**
- Relative autonomy of OU - Own resources - Exchange of equivalents	- Command and control - Central management of resources - Allocation of resources
Polycentrism, heterarchy	▪ **Hierarchy**
- Numerous direct contacts - Hierarchical structures not excluded - Horizontal carreers - Mutual assessments - Innovation as task of all OU - Professional heterogeneity in OU	- Contacts through representatives - Hierarchical structures obligatory - Vertical careers - Top down assessments - Innovation by R&D and top management - Professional homogeneity
Permeable Demarcations	▪ **Sacred departmental delineations**
- Multiple membership in OU possible - Exchange of personnel between OU - Alternation between expert and management roles	- At most multiple responsibility (in matrix org.) - Alternation of positions only in defined careers - Loss of management tasks as proof of personal failure
Process organization	▪ **Functional or matrix organization**
- Organization of activities in the form of projects - Easy recombination of activities	- Organization in the form of single purpose departments - Modification of organisational structure as indication of failure

Fig. 6 Characteristics of cooperative networks.

In contrast to the uncooperative social interactions on the market guided by personal advantage motives, the bonds among actors in a network grow out of commitment and trust. Partners depend on each other, but in contrast to the hierarchy, there are alternatives to their *loose coupling* and network relationships (*redundancy*). For instance, a production unit does not have to have maintenance services from their own service department if it is not able to cooperate. For this, however, economic autonomy of those in cooperation with regard to their choice of partners would be prerequisite (in profit centers, for example).

Openness or *transparency* of boundaries between the units means that the number of network members and their degree of integration is fuzzy. Exchange in personnel is desirable in order to promote the transfer of knowledge and diverse bonds, even beyond the company. Team work and project groups are not at all taboo. Rather, team work is required by networks, but networks loosen up 'compulsory' team structures by allowing for multiple simultaneous or successive memberships. LIKERT'S (1961) model of multiple group memberships approaches this, but remains fixed in hierarchical thinking. Networks, in contrast, can take on changing arrangements – task distribution, rank order, management – in dependency upon situation and problem. For example, independent music ensembles today, which act as 'small companies' on the art market, engage their own project leader for a production, or a member of the ensemble takes on the role if a piece of his or her own music is to be performed. An important part of activities is thus organized in the form of *projects*, without separating or centralizing these in the form of a parallel organization. MINTZBERG'S concept of „adhocracy" (1994) and MARCH'S „organized anarchies" point in this direction. MARCH (1988) emphasizes the importance of sufficient resources, or „organizational slack", as a prerequisite of cooperative networks, as this reduces the competition for resources and allows for the coexistence of competing objectives and, as a consequence, heightened creativity.

Cooperative networks are particularly well-suited for fields of work which are complex, involve much knowledge, are dynamic and uncertain, where the value of the product exchanged is difficult to measure and where highly qualified actors are accustomed to working independently without direction. POWELL (1996, p. 228ff) speaks of „project markets“ and of „cooperation, which does not erase the independent identity and personality of the cooperating partners“ (p. 241). POWELL sketches out examples from the field of publishing, the film industry, building and construction trades and industrial districts and shows that their exchange relationships and their forms of the cooperative production of knowledge can serve to stimulate ideas for the intra-company network perspective. Research and scientific endeavors are equally predestined candidates for network perspectives.

For example, the research institute where I work shows many of the characteristics of a network. There is no hierarchy, no director. The organizational units are research teams and the service team, which are autonomous and have no internal director. Except for the service team everyone does their own and collective acquisitions. Project responsibility ('project direction' to the outside) rotates according to the project among team members, whereby the more experienced members more frequently take on the internal coordination of the team. Internal coordination of the institute takes place through an elected institute council and through institute-wide constitutional meetings. Research projects with designated and fixed personnel are increasingly giving way to rotating and simultaneous memberships in various teams, whereby external people are sometime integrated. The division of labor and cooperation 'organize themselves' according to certain subject areas, which recur in the different, individual projects. Ambivalence also becomes apparent at times, in the form of discontinuous work processes, the great effort required of individuals to maintain coordination, responsibility for 'everything' and occasional competitive situations with regard to acquisition.

Developments on the labor and sales markets (turbulence, demands for flexibility, the importance of the production of knowledge) are creating increasingly appropriate field conditions for network organization in other sectors of the economy as well. Furthermore, as market-type forms of coordination are imported into the enterprise (strategic business units, profit centers, virtual enterprise), and with the erosion of the traditional work contract, the boundaries between intra- and inter-corporate networks are fading away. Network structures are diffusing, so to speak, from the outside into the enterprise.

Many strategical questions remain open. What kind of exchange relationships – beyond market price mechanisms – can networks utilize in order to regulate their economy of knowledge (consultations, exchange of know-how, personnel and services, budgeting, etc.)? What are most suited to what objectives? What is the role played by power, what means of power can the actors mobilize, for example, in order to achieve their demand for knowledge, or for holding back knowledge? How do conventional forms of organization become networks? What types of experience in trans-corporate networks can be applied to intra-corporate networks? Where do the problems lie (such as limited control over cooperating partners, or the difficult balance between elements of cooperation and competition for resources)?

5. Conclusions

The more we plan, the more vulnerable we are to chance. This idea, once expressed by the playwright FRIEDRICH DÜRRENMATT, describes our starting point. But the 'movie sets' we visited were not classical and Taylorist, but rather those which are oriented towards the most 'modern' concepts of management. We examined the issue of why their façades are often so stable, and how it is that positions can be taken up comfortably on both sides of the façade. Explicit calculating power considerations and the hidden power of a paradigm guiding action – instrumental rationality – were identified as the successful blueprints for the building of the façade. But the 'outside' world pushes its demands through the most beautiful of façades. The rationality ultimately proves to be highly irrational and dangerous for survival.

We examined that understanding of team work, self-organization and company culture as mere instruments of mastery and the attempt to impose them through means of force. Examples showed these attempts to be paradox, at best half-successful, strategies. There is a danger here that cannot be overlooked – not only in the failure to develop the potentials in self-organization, but also in destroying them and discrediting them in the eyes of employees, in spite of mimicry and seeming conformity. Real, 'informal' cooperation fights its way to a large degree by itself, and the relationship structures which arise in this way have more or less network character. If this is so, we conclude, networks should become explicitly allowed, and their development should be supported through appropriate framing conditions. Cooperative networks may be seen as a useful metaphor and applied as the stuff for thought in the following way: they should no more be idealized than team work or other concepts and certainly not stylized as the 'one best way'. We also do not expect them to fulfill management intentions without side-effects. With these premises, however, the network metaphor can yield very constructive stimuli to our thinking on the design of the cooperative production of knowledge and on institutional learning within the enterprise.

6. References

ARGYRIS, C., SCHÖN, D.A. (1978): Organizational learning. A theory of action perspective. Addison Wesley, Reading/Mass.

BATESON, G. (1972): Steps to an ecology of mind. Chandler, San Francisco.

BURAWOY, M. (1979): Manufacturing consent. Chicago: University Press.

CLARK, K.B.; FUJIMOTO, T. (1991): Product development performance. Harvard Business School Press, Boston.

CROZIER, M; FRIEDBERG, E. (1977): L'Acteur et le Système. Edition du Seuil, Paris.

DREYFUS, H.L.; DREYFUS, S.E. (1986): Mind over machine. Free Press, New York.

GRANOVETTER, M. (1982): The strength of weak ties. A network theory revisited. In: MARSDEN, P.V; LIN, N. (Eds.): Structure and network: analysis. Sage, Beverly Hills. p. 105-130.

HABERMAS, J (1981).: Theorie des kommunikativen Handelns. Band 2: Zur Kritik der funktionalistischen Vernunft. Suhrkamp, Frankfurt.

IMAI, M. (1986): Kaizen: The key to Japan's success. McGraw-Hill, New York.

KUHN, T.S. (1962): The structure of scientific revolutions. University Press, Chicago.

LIETZ, J.H. (1994): Von der Zweck-Gemeinschaft zur Sinn-Gemeinschaft. In: KAMISKE, G.F. (Ed.): Die hohe Schule des Qualitätsmanagement. Springer, Berlin. p. 111-131.

LIKERT, R. (1961): New paaterns of management. McGraw-Hill, New York.

MARCH, J.G. (1988): A chronicle of speculations on decision-making in organisations. In: MARCH, J.G. (Ed.): Decisions and organisations. Wiley, Oxford, New York. p. 1-21.

MEYER, J.W.; ROWAN, B. (1977): Institutionalized organizations: formal structure as myth and ceremony. American Journal of Sociology, 83, p. 340-363.

MINTZBERG, H. (1994): The rise and fall of strategic planning. Free Press, New York.

MOLDASCHL, M. (1992): Thinking machines, dreaming engineers? – towards applied expert system technology. In: ALTMANN, N., ET AL. (Eds.): Technology and work in German industry. Routledge, London/New York. p. 92-103.

MOLDASCHL, M.; WEBER, W.G. (1997): The „Three Waves" of industrial group work – Historical reflections on current research on group work. In: The use of social science for social reconstruction. Human Relations, Special issue to the 50th Anniversary of the Tavistock-Institute, London (in print).

NEUBERGER, O. (1995): Mikropolitik. Enke, Stuttgart.

PERROW, C. (1986): Complex organizations. A Critical Essay. 3rd Ed. : McGraw Hill, New York.

PETERS, T.J.; WATERMAN, H. (1982): In the search of excellence: Lessons from America's best-run companies. Harper & Row, New York.

POWELL, W.W. (1996): Weder Markt noch Hierarchie: Netzwerkartige Organisationsformen. In: KENIS, P.; SCHNEIDER,V. (Eds.): Organisation und Netzwerk. Campus, Frankfurt. p. 213-271.

RAEITHEL, A. (1997): On the ethnography of cooperative work. In: ENGESTRÖM, Y.; Middleton, D.A. (Eds.): Communication and cooperation at work. University Press, Cambridge/Mass. (in print).

ROETHLISBERGER, F.J.; DICKSON, W.J. (1939): Management and the worker. Harvard University Press, Cambridge/Mass.

SEMLER, R. (1993): Maverick: The success story behind the world's most unusual workplace. Warner Books, New York.

WILDEMANN, H. (1995): Der Wettbewerb als Unternehmensprinzip. Methoden zur Etablierung einer lernenden Organisation. Blick durch die Wirtschaft, 84, p. 9.

VOLPERT, W. (1985): Zauberlehrlinge. Die gefährliche Liebe zum Computer. Beltz, Weinheim.

Systems and Organizations

Part Two

Models, Systems and Reality: Knowledge Generation and Strategies for Systems Design

Eric Scherer

Most prevalent approaches to the problems of shop floor control tend to focus on the scheduling problem. Thereby an organizational problem in the real world of manufacturing is replaced by a mathematical problem within a modeled world. To enable sustainable and holistic solutions the question remains on how knowledge can be generated for systems design.

1. On the actual situation in industry

The organization of company logistics is crucial to the success of an enterprise. As a rule, the management of in-house logistics tasks is summed up as the function of 'production planning and control' (PPC). PPC comprises, according to the type of enterprise, numerous tasks – from sales and master production planning, materials management and procurement, to material and capacity planning and shop floor control. In the form of response and delivery times, handling costs, and service quality it determines numerous factors which contribute originally to the fulfillment of the customer's demands. But because logistics are also expressed day-to-day through the repetition of more or less routine activities in the company, they are a decisive factor in the quality of work organization as well. PPC in the broader sense, then, represents one of the formative tasks within every production enterprise and determines the degree of integration and operational flexibility of units and groups in a distributed production system.

PPC is typically designed according to a top-down approach which dominates the formal organization of PPC in company practice. It is characterized by the duality of the terms 'planning' and 'control'. While planning is generally localized within a 'central' department which issues orders, master schedules and instructions and oversees their execution, the execution itself is the responsibility of 'decentralized' shop floor units. Because there are always deviations and adaptations involved in execution, this activity is also called *control*, a term which in a sense combines the activities of '*planning adaptations*' and '*execution*'. The term gives expression to the general reactive character of control activities at the level of the shop floor and its – in contrast to the central planning department – more local area of influence. Those persons responsible for control are charged with keeping corrections to an absolute minimum in order to achieve the greatest possible adherence to the total plan.

1.1. The shift from a deterministic to a chaotic conception of production

From a traditional management perspective, shop floor control is the 'front end' of hierarchical production planning. Its task is to execute the schedules 'as given'. The fundamental functional principle is based on a deterministic view of technological and logistical production processes. The assumption thereby is that production processes are predictable to a high degree of precision. Uncertainties in planning, which are caused by unforeseen variations and interferences in real production systems, imprecise state data and simplified – mainly static – algorithmic models, are compensated for by slack times and security factors such as generously calculated transfer and transport times.

However, interferences, such as machine downtimes, personnel absences, materials shortages, lack of tools and so on, lead to continual changes in the situation, and demand, at times, comprehensive adaptation of the schedules originally planned. Production behaves deterministically only to a limit, and in many cases, system behavior is 'chaotic', meaning that even small deviations in the situation can result in subsequent large deviations from the planned state. In operations, this demands re-planning based upon the actual situation in order to allow work to progress. This leads indirectly to change in the original master schedules and the assumptions upon which these were based. Such changes often have effects which transcend the area directly affected, and so they must be coordinated, in the sense of a feedback system, with the upstream planning level and downstream production areas. In many companies, the success of a deterministic control system is only possible because employees at operational level counteract the formal control system. For example, hidden material buffers are created or orders split up without the knowledge of central planning. Thus the informal system grounded in the physical and social reality of the shop floor allows the functioning of the formal system of production planning and control (see WIGHT 1974, p. 6).

Most approaches in industry try to avoid any sophisticated control[1], e.g., through finite scheduling, and focus on a *stable* design of the manufacturing system and the requirements of systems' *control* and *controllability*. Still, despite all efforts, various approaches that attempt to design production, as the system to be controlled, as a determinable system and, for example, build up physical (material) and informational (slack time) buffers as security, do not really succeed in practice. They function only thanks to the uneconomical measures of stabilizing buffers and constraints which de-motivate workers. They lose their original flexibility and often are no longer a match for the cost pressures of the market.

Other approaches, as they are often proposed by research and nowadays by software development, try to compensate for the poor determinability of the system to be controlled by means of increasingly refined technical design of the control system and by improving the validity of scheduling in terms of its precision (see, among others, TAYLOR 1995). Planning variables reproduce the situation as a model on a computer system with increasing precision and with increasingly precise processing of the data. Still, in most cases, the dynamics of production are not tapped to a sufficient degree, and the system runs – with precision – at a lag behind the real events.

To sum up, approaches predominant in research, software development and industrial practice maintain the fiction of a deterministic world in which one cause determines one result and therefore the course of action can be determined. But in most industry the literal 'chaos on the shop floor' in industrial control systems persists (see WIGHT 1974, p. vii, MELNIK ET AL. 1985, p. vii, JURAVICH 1988) and makes it clear: the deterministic control model does not sufficiently grasp the conditions of the real world of production anymore.

1.2. Turning towards design potentials of organization

While for long years design measures to increase company performance comprised manufacturing technologies, automation techniques and computer systems, new concepts – adopted mainly from Japan, such as KANBAN or quality circles – have drawn attention anew to organizational aspects of production. Such organizational measure in general had a much more positive impact on the efficiency of control than any change of the applied scheduling techniques and computational methods. With the shift from technology-oriented thinking to a process- or organizational orientation, there has been an implicit shift in practice from attempts to 'master complexity' to attempts to 'reduce complexity' (see MOLDASCHL 1994, p. 107-

---

1 For a long time, this neglect of scheduling and control was shown by most commercial MRP II software packages, which typically did not feature sophisticated scheduling solutions (c.f. BŁAZEWICZ ET AL. 1996, p. 423).

110). Most of the current philosophies which aim towards organizational design attempt through some simple measures to reduce the complexity of management of an organization and replace it with self-organization. However, some principle problems remain: because it is not possible to simply reduce the complexity of a system, as a rule complexity will just be shifted around. Complexity-reducing solutions which noticeably lighten the management load, such as KANBAN, can not be generally implemented in the majority of companies due to their heterogeneity in technologies, products and markets. Individual companies require special provisions in order to compensate for newly developed heterogeneity. At the same time, the parameters of a completed reorganization continue to be subject to transformation, and even organizations designed according to the most modern principles are forced to change if they are to be equal to new demands. Continuous change of this kind, however, can only be accomplished by a company's own employees, at all levels of the organization, as they in their individual task situations adapt their work to changing conditions.

Implicitly, realization of new organizational approaches in production has shifted the focus of attention to human workers. The employee as the company's most important 'capital' has become the byword in not a few enterprises (see ROLSTADÅS 1995). REICHWALD (1992) speaks of a „rediscovery of human work as the primary factor in production". Training and qualifications measures, help towards self-help and Continuous Improvement Process (CIP) have become reality in many companies and are now directed particularly towards employees at operational levels rather than towards management alone. Close scrutiny, however, reveals a considerable lack of well-thought-out concepts, so that many companies have reached a standstill 'half-way there'. Although the employee is indeed seen as a welcome key to change and new success, the new recipes are only partially conceived and implemented. The rest is rhetoric. And so employees receive training, but their tasks are not further developed, and the basic structures of the organization – despite increased involvement of employees – remain unchanged. The problems which arise out of this may already be observed in many companies and not infrequently crystallize around the issue of the tasks and the position of shop floor control. Employees in production – theoretically equipped with increasing autonomy – remain caught between a massive increase in the quantity of work, constraining instructions from centralist computer systems and the very demanding requirements of machines and technologies operated at the limits of feasibility (see MOLDASCHL 1994, p. 111ff., SCHUMANN ET AL. 1994, p. 659).

1.3. Concepts and lack of concepts

The call for flexibility, agility and efficiency challenges the shop floor with several problems. With an increasing occurrence of changes and dominating customer demand, management of manufacturing processes and the coordination of the multifarious resources, i.e. machines, materials, information, knowledge and humans, becomes a core task. In addition, not only is optimal management of the current situation necessary, but also practices and performance must be continuously improved.

This leads to the issue of adequate design concepts for shop floor control. Upon assessing industry, global leaders, best practices and hidden champions can be found. But in the industrial reality at large, „the shop floor is not under control, it is out of control“ (MELNIK ET AL. 1985, p. vii). Even though various solutions – ranging from organizational strategies like group technology, team work and KAIZEN to technical solutions like MRP II, real time logistics, knowledge-based scheduling and genetic algorithms – do exist, the problem of the inadequate performance of shop floor control continues to remain omnipresent in most industrial enterprises.

1 The situation can be summarized by the following observations:

Most current managerial concepts tend to address the overall enterprise and are usually very general and not focused on manufacturing industry or the shop floor respectively.

2 The shift towards the entrepreneurial function of management tends to leave manufacturing out of the picture and often leads to the impression among practitioners that manufacturing is of no further consequence.

3 While managerial literature deals with organizations, classic industrial engineering research still tends to neglect the organizational issues of shop floor control while simultaneously producing an immense number of highly theoretical and hypothetical solutions every year.

4 In industry, shop floor control is overshadowed by the paradigm of 'command and control' which leaves management to rely on power, rather than any kind of conceptual approach, as a means of ensuring effective control. Therefore, any conceptual approach as a means of altering the command structure meets with opposition. As a result, operators and workers must deal with an increasing pressure which replaces the empty prophecy of self-control (see GOLDMAN, NAGEL & PREISS 1995, p. 98f).

1.4. Need for action: A systems perspective of principles for design

Shop floor control is an object of study by various disciplines from engineering and management to the social sciences. No comprehensive framework or even overview on shop floor control research currently exists. As a result, many problems of the real world remain unsolved, because predominant deterministic approaches – stemming from both research and practice – have significant drawbacks. Due to the fixation on mathematical solutions in traditional research, many potentials present in real world production are not considered – operator experience, motivation and qualifications are not taken into account. The pragmatic approach of stable systems design is based on a paradigm of social control through command and controllability and therefore tries to prevent any change and innovation as such.

Even state-of-the-art textbooks do not focus on the organizational aspects of the scheduling problem and relegate the real world to footnotes (e.g. footnote in PINEDO 1995, p.277). What is missing is an integrative framework for the design and management of shop floor control which combines organizational and technological issues, acknowledges the dynamic nature of manufacturing processes and market behavior and takes into account the sociopolitical impact upon management and workers. A framework is needed that would facilitate the shift from deterministic scheduling models towards agile operations management.

1 The new objectives for systems design can be summarized as follows:

Due to the dynamic nature of production processes, dynamic control is required. The task of shop floor control becomes the motor of the production process.

2 Due to the changing environment in terms of market and technologies, continuous adaptation of the system and its behavior is necessary. Shop floor control forms the foundation of an adaptive system. Therefore, shop floor control can not be restricted to the control of the production system as such, but has to include control of the continuous improvement of the overall system, control and production.

Here it is necessary to acknowledge that no such integrative framework can truly exist for such a heterogeneous and dynamic topic as shop floor control and remain valid over time. Rather than the 'perfect' framework often sought in the academic circle, what seems to be a useful solution in achieving 'holism' is an *open* set of 'views' from varying, interdisciplinary perspectives. This openness also tackles a major drawback of the anticipated 'complete, general and perfect' framework: it is a framework's incompleteness which allows systems truly to evolve. Instead of a complete framework a new thinking in terms of 'systems' thereby can provide the necessary integration. This 'systems perspective' in the case of shop floor control must support the design of a sustainable shop floor control system.

2. Reassessing research and methods for shop floor control

For a long period, research on production and operations management concentrated on the solution of the scheduling problem at shop floor level. Traditional quantitative approaches, as present in Industrial Engineering (IE), Operations Research (OR), and Artificial Intelligence (AI) as well as the underlying principles in most commercial software development, try to achieve an ideal scheduling system by focusing on the scheduling process itself. The modeling techniques applied thereby result in a simplification of the original problem and have a tendency to create and maintain a model world. The solving an organizational problem was thereby reduced to solving a mathematical model. This situation produces a „theory-practice gap“ (KING 1976) in shop floor control research which has been observed for several years (see also MCKAY ET AL. 1988) [2].

2.1. Simplification of reality: On the quality of models used

The majority of control approaches developed in the past decades is concerned with individual aspects of production. At the core, the approaches attempted to describe production in the sense of a quantitative, mathematical model as a sum of functions, while slack and security factors served to compensate for the model's imprecision. Here the implicit underlying assumption was that production is a mechanistic system, and its sociotechnical character was ignored. The goal of approaches working according to the deterministic principle is to predict the behavior of the production system and to apply this knowledge to control. For this reason, the focus of control activities is placed upon planning. Once the plan has been developed, it is the central task of the shop floor to work according to the plan with restrictive means.

In the context of technological advance – since the mid-1970s given its main thrust by rapid progress in computer science – the attempt was and still is to better the insufficient precision by means of increasing refinement of planning methods. The catchwords in recent years have been 'neural nets', 'fuzzy sets', 'genetic algorithms', 'simulated annealing', 'real time logistics' or 'multiple resource planning' (for an overview on quantitative scheduling techniques see BŁAZEWICZ ET AL. 1996, for an review of practical applicability see WIERS 1996 as well as the review by STUBER in this volume).

Attempts to gain a hold on uncertainty by means of increasingly precise mathematical planning models have not led to success in practice (see MANSKE & WOBBE-OHLENBURG 1985, p. 401). On the shop floor, the result has been the development of individual practices to complement the specified, formal control system. These practices take on the form of 'the routine of the exception' in many companies, every day. The reason is that the complexity of the fundamental problem always exceeds the precision of the model. In most cases, planning can not adequately capture the dynamics of production and ends up – with high precision to be sure – limping behind real events. At the same time, the effort required to capture the data needed to reproduce events in such precise

[2] It should be noted, that production research by F.W. TAYLOR and those of his time was deeply routed in industrial practice. While TAYLOR tried to achieve a step towards a „scientific management“ his work is grounded in his industrial experience and provides a fairly comprehensive view of industrial reality (see TAYLOR 1911 & 1919).

models increases to an unrealizable degree[3] and is economically not feasible. The models utilized themselves become so complicated and non-transparent that they very soon overtax the user (KURBEL 1995, p. 267, NN 1995). The user can no longer interpret the model intuitively and is in no position to judge what data, when and with what degree of precision must be entered into the system. The user experiences the system as constraining, to which he or she can no longer react in a flexible manner. The total control system is robbed of its local flexibility entirely.

In practice, precise and deterministic planning approaches can be implemented sensibly only in those specific cases where quasi-deterministic system behavior can be assumed. This would be the case, for example, with the process industry with its high demands for security of the processes and storage of resources, with highly automated, autonomous manufacturing systems or with continuous, highly automated mass production of semiconductors in clean room facilities (see VOET 1994).

As planning methods were refined, the extent of functions was also increasingly expanded, and the 'routine of the exception' mentioned above was integrated into the standard of computer-aided systems without further ado. Subsequently, formal systems have been developed which are themselves very complex in use and are no longer transparent. The human operator, as an intelligent instance of control, is confronted with continually increasing demands. As complexity grows, the human operator becomes overtaxed and the system looses a major source of flexibility.

2.2. Generating knowledge: On the use of models

The second major drawback of most quantitative approaches and particularly of most scheduling research utilizing OR methods can be seen in the role of models. In much research, the goal of building a model is to be able to derive an optimal solution within the model. This results in a predominance of the model itself, while the transformation back to reality is often omitted or not even possible. This is readily observed in the great number of 'assumptions' that traditionally accompany many research papers. As PINENDO (1995) states, „it is not clear how all this knowledge [gained through mathematical models] can be applied to scheduling problems in the real world"(p. 253), since „real-world scheduling problems usually are very different from the mathematical models studied by researchers in academia." (p. 254). PINEDO then lists 10 problems that hinder the transformation of scheduling models into practice[4]. Nevertheless, in the major part of his

[3] Although much hope has been placed in advancing computational capabilities, applications of 'real time logistics' and 'finite scheduling' systems in industry (see GUNN 1994) have not led to the desired success, if manufacturing is the true focus. This can be easily explained. As long as there is a 'real world' outside the computer, the computer will consume time to capture the data necessary for processing.

[4] Many problems inherent in most quantitative scheduling approaches have been discussed. Besides the prevalent critique of static view a major problem lies in the objective functions usually used. They do not capture the close coupling between various variables and do not take into account the fact, that in the real world the objectives themselves change over time and are largely influenced by such subjective phenomena as the balance of power between different interest groups within the enterprise.

textbook, PINEDO describes static optimization methods only to finally admit that scheduling in reality has a „dynamic nature“ (p. 254) and is exposed to a „relatively high frequency of random events“ (p. 257).

What is lacking is the link to reality, an examination of the question as to what models are used for, and a methodology to transfer results back into the real world. As REISMAN & XU (1992, p. 121) put it in an article on knowledge growth in the management sciences, „much of the institutional infrastructure, e.g., the graduate programs, the professional societies and especially the landmark journals have become inbred. They concentrate on research based on established paradigms, work which is essentially logic-deductive rather than grounded in reality.“ Even worse is ANTHONY'S comment in 1965 (p. 90): „Although the usual definition of operations research as 'What operations researchers do' is probably intended to be facetious, there is much truth in it.“ As an example, while the well-known researcher BAKER (1974, p. 2) describes scheduling as „the allocation of resources over time to perform a collection of tasks“, WIGHT (1974, p. 5) – firmly rooted in practice – expresses the same with a different slant: „Priorities [...] means *knowing* what material is needed and when [...]. This is what many people mean when they use the word 'scheduling'.“ This means that in the real world of shop floor scheduling, the focus is not on the generation of an optimal schedule, but rather on determining the demands of the situation.

In an alternative approach, modeling can be seen as a means of generating knowledge about the system itself. That knowledge may then be utilized to determine possible and adequate changes to the system itself. Accordingly, it is not the model or the generation of an specific quantitative solution within the model which is the concern, but rather the process of modeling itself (Figure 1). The model „becomes a guide to the ways in which the system can be improved, [... rather than ...] a working tool for decision making in the system as it exists at the moment“ (BURBIDGE 1978, p. 145). It is within this perspective that OR methods gain relevance again as means of generating 'understanding'.

It is important to emphasize that the model-based approach will always result in a partial model of the 'real world' and thus only can provide a partial solution. It is this circumstance which makes it impossible to create the 'perfect' framework for research. Multiple views *and* individual experiences, i.e., a thorough grounding in reality, are necessary in order to reach a holistic solution. Thus, the present volume presents different views to allow for more than one perspective.

3. Founding a systems perspective

Many predominant problems of the real world of production operations management stay unsolved, as approaches – both in research and practice – have significant drawbacks. The fixation on mathematical solutions omits several potentials present in real world production, e.g., operator experience, motivation

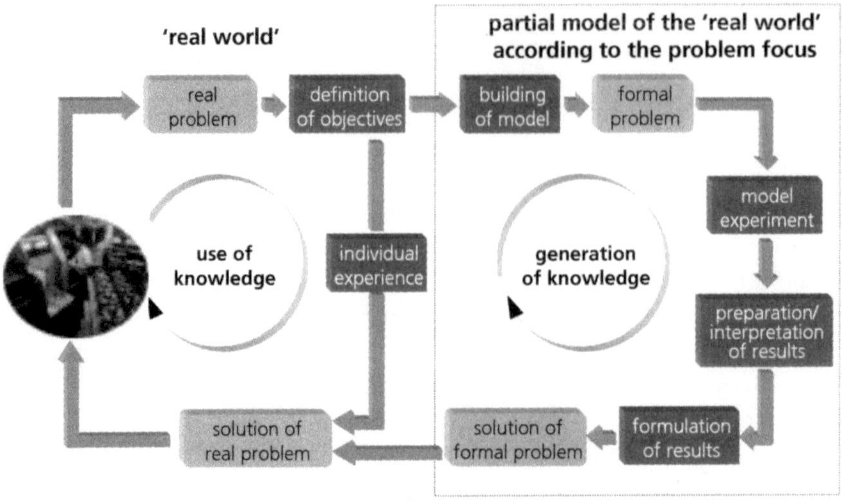

Fig. 1 Model based approach to problem solving (WIENDAHL 1996, p. 9-8).

and qualification. In a systems perspective the problem of control is not a problem of finding an optimal solution, but of establishing and developing a dynamic and flexible organization.

To achieve good solutions in the design of complex systems, a broad analysis of the underlying problem[5], combining various views, and a holistic approach is necessary. In this alternative approach, the design of shop floor scheduling and control can be seen as a task of systems design. Here, principle assumptions about the system and its underlying principles provide a normative foundation for the system designer. In the given context, the shop floor can be described as a system from two perspectives.

- Cybernetic systems theory:
  Production operations as experienced at the level of the shop floor can be described as part of a larger cybernetic system which is highly complex (see the contribution of DOBBERSTEIN in this book). Interactions, feedback and various couplings result in a non-linear system which is easily beyond the boundary of manageable complexity. Therefore systems behavior becomes unpredictable except for a very short-term period. As a result, the system appears to behave in a 'chaotic' manner in the perception of a single observer.

  Cybernetical systems approaches allow to describe and understand the dynamics of behavior of the formal, logic system and its state variables as encountered in the real world.

[5] In a first step, the focus should be placed on 'real life' problems rather than the problems encountered after building the model. The fixation on problem solving within the model can be seen as a fundamental misunderstanding in much IE/OR research.

2. Sociotechnical systems theory:
   Every production system consists of machines and some kind of human involvement. Accordingly, each production system can be seen as a sociotechnical system (TRIST & BAMFORTH 1951, SUSMAN 1976, PASMORE 1988) in which humans and machines interact to produce an optimum result. The task of system design is to optimize both, the social and the technical subsystem. As experience in reality shows, the human controller is often the key to the system's effectiveness and provides the system's ability to function (see DUNKLER ET AL. 1988, MCKAY ET AL. 1989, SANDERSON 1989). Focusing on the design of shop floor control with the sociotechnical system as the framework therefore results automatically in a emphasis on the human role in production.

   Sociotechnical systems approaches allow to describe and understand the courses of action and logic of organizational development of the informal system utilizing patterns of social and human behavior[6].

In the following, corresponding strategies of systems design are described. For effective systems design, the questions raised are:

1. How may the controlled system be designed in such a way that its control be improved?
2. How must the control system be designed?
3. What principles are significant to control itself?

It must be noted that in production the control system and the system controlled can hardly be observed or designed separately. They are very closely linked in function, and in reality they become merged. This effect is reinforced by the sociotechnical character and the open nature of real production, which is subject to environmental influences.

4. Characteristics of cybernetic systems

A central factor in considering production is the complexity of the system, which results from its heterogeneity and interconnectedness as described above. Understanding the complexity of a system and its effects is of elementary importance for its control. As a rule, an observer judges a system to be complex when it can not be described simply. Here we speak of 'subjective' complexity. 'System' complexity, however, is not only dependent upon the complicated nature of a system with numerous elements and relations, but rather also upon its

---

6 It should be noted that in difference to many purely scientific approaches the sociotechnical approach from its very first beginning was directed towards the humanization and quality of working life. But simultaneously it always focused on the aspects of economy.

dynamics – the frequency as well as strength with which the system and environmental factors change over the course of time (see RÖPKE 1977, p. 21). System complexity is thus the result of the meeting of

1 structural complexity and

2 non-linear dynamic behavior.

In principle, complex systems may be viewed at various system levels and from varying perspectives. The complexity of a system is a relative variable and depends upon the position, point in time and perspective of the observer as well as upon the observer's qualifications and experience (see DÖRNER 1992, p. 61). For this reason, in the following we will make a distinction between systems complexity, i.e. the total complexity of a system within its defined boundaries, and relative complexity, i.e. the complexity of a sector of the system from a particular perspective or from a certain aspect under consideration.

On one level, a complex system is comprised of various cells which stand among themselves in an exchange relationship. Each of the cells represents a subsystem which itself has a certain degree of complexity. In the area of production, a cell might be a work team or a foreman's area. The cells are connected to each other by links. Links can be hierarchical-delegated or heterarchical in character. In the case of heterarchy, the importance and direction of the connection arises from the situation. The connections have a transformational character and so take on a certain complexity. The various aspects of complexity in organizations are shown in Figure 2.

In production, structural complexity is reflected both in the structure of the executing system – in the arrangement of machines, material flow, the subdivisions of the shop floor, for example – and in the organization of the control system itself – such as in the form of the distribution of responsibility, of planning and monitoring scope, of applied planning principles, etc. The behavior of complex systems is ultimately dependent upon the interaction among individual subsystems, or cells. Through positive and negative feedback and linked feedback loop systems, structures develop which are highly non-linear[7]. In the case of production, the inherent dynamics are, in addition, overlaid by dynamic influences induced both internally and externally. The dynamic influences are the effects of order changes, interferences and decisions. It is characteristic of dynamic systems that they, in consequence, may exhibit unexpected behavior, such as instability with small interferences, but stability with large interferences, shifts from negative to positive feedback, growth accelerations, and so on (see FORRESTER 1969).

7 A system is non-linear if there is no proportional, reversible function between input and output.

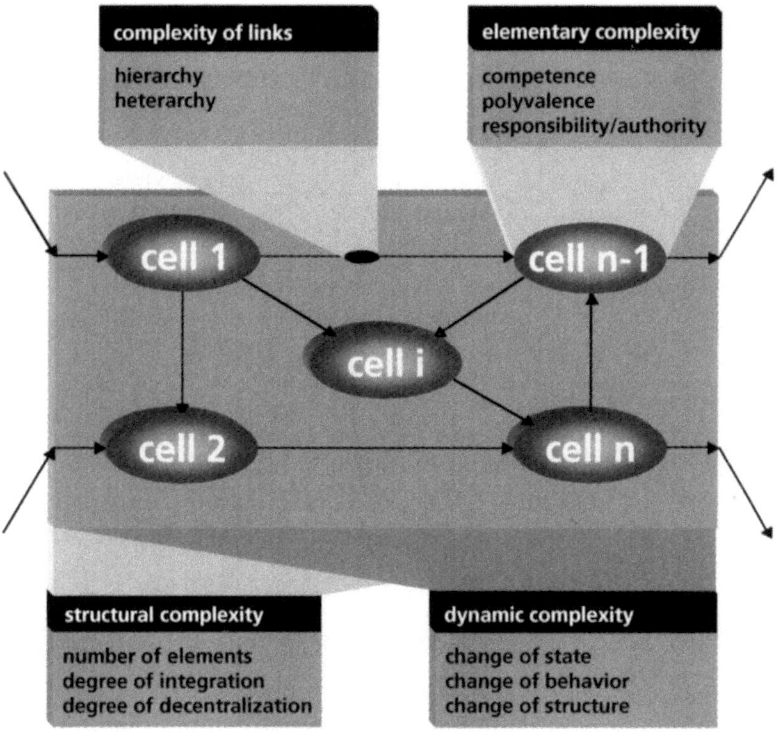

Fig. 2 Complexity in organizations (SCHERER & DOBBERSTEIN 1996).

Because of interconnectedness, feedback loops and high dynamics, the principle of causality no longer holds for complex systems, Events may not be seen as the consequences of single, concrete causes. This state of things has led to the use of the term chaotic to describe complex systems[8]. „This chaos is not hopeless disorder, but rather develops itself according to strict rules. Admittedly, these rules are nor easily recognizable, because all factors involved continue to influence each other mutually“ (SCHULTE & WUNN 1993, p. 36). Events in complex systems may at best be traced back to multicausal relationships, that is, an event is the result of many interconnected causes. „In non-linear systems with recursive process dynamics, knowledge of rules for control is of little help. Where state follows state, that is, where each state is the consequence of that directly preceding, there is predictable system development only in few cases, even if its mechanisms are known, the system is deterministic and there are no interferences“ (KROHN & KÜPPERS 1990, p. 114-115). For this reason, process behavior in open

---

8 Deterministic chaos is the term for the irregular behavior of a non-linear, dynamic system whose temporal development may be described unambiguously through mathematical equations. The equations, however, are not solvable in closed, analytical form. HORNS (1989) demonstrates in his paper demonstrates analytically the development of chaotic behavior due to deterministic planning instructions. AHRENS & JANSSEN 1994 show the non-linear character of production and logistics on the basis of analysis of production data from real production processes (see the contribution of DOBBERSTEIN in this volume).

and complex systems, such as production, is predictable only in the short-term and is of only limited controllability.

With this, the very dynamics of a system hinder an understanding of it. And because production is a multilayered, sociotechnical system, its non-transparency is furthered increased. The attempt to derive generally valid rules for the behavior of even a subsystem leads inevitably to a restriction to single ways of viewing and to abstraction – and thus not to general validity. The demands, increased in this way, on the controller soon lead to errors, which in consequence, result once again in increased system dynamics (see DÖRNER 1992).

> „Complexity means for the context of management that the formal directing organs of an enterprise never have at their disposal sufficient information, sufficient knowledge or sufficient skills to control and design in detail an enterprise which transcends the barrier of complexity" (MALIK 1992, p. 84). Planning and control models based upon monocausal perspectives are thus, due to their principles of effect, inappropriate for complex systems.

At the same time, it must be acknowledged that even an understanding of interrelated nets is of little use due to the dynamics of the system[9]. Therefore, appropriate strategies for design of control are needed.

4.1. Strategies for cybernetic systems control

In principle, two approaches to systems control may be distinguished:

- Avoiding interferences by means of preventive and stability- and equilibrium-oriented systems design.
- Increasing the system's interference tolerance and, with it, the acceptance of errors and instabilities.

In management theory terminology, the stabilization approach is called „*Cybernetics I*" (STAEHLE 1994, p. 41). Avoidance of errors and interferences, and thus the stabilization of the system, is the central goal of systems design. Because errors and interferences can never be entirely prevented in reality, their prevention and stabilization of the system is understood as an independent – that is, separate – task, which is taken on either by an independent organizational unit or by a highly automated „black box" system (see SCHERER 1995). Operations control – at least in the conception of management and the systems designer – is then confronted only conditionally with errors and interferences. Operations is no

9 Approaches which predict the behavior of complex dynamic systems through the aid of simulation also must fail, because in addition to dynamics, there exists also a singularity of individual events. Dynamic simulation can be utilized in explaining the system, but integrating simulation into control in the framework of PPC seems hardly promising.

longer formally responsible for their correction and thus loses its ability to react to errors. The de-coupling of planning and execution increases. Numerous works in the field of production logistics propagate this approach; typical catchwords include, for example, the terms „interference suppression management" (see WILDEMANN 1994) and „zero defect production" (see WESTKÄMPER & JESCHKE 1995).

In contrast, the second approach seeks to increase the system's tolerance for interferences. Interferences and dynamics resulting thereof are no longer excluded from systems design and are also not evaluated negatively per se. The approach called *„Cybernetics II"* accepts interferences, instability and imbalance and views them as normal and as prerequisite to change and evolution of the *entire* system (STAEHLE 1994, p. 41). Responsibility for control, and thus for restoration of stability, lies directly in the hands of the part of the system, i.e. a cell, confronted with the interference. The cell is responsible for control decisions and for instituting measures appropriate to carrying out those decisions[10].

All of this allows to accept production as „beyond control" (WIENDAHL, PRITSCHOW & MILBERG 1993, p. 268). In order to achieve effective control of a production „beyond control", the necessary competencies and authority must be shifted consistently into the area of operations to create individually controlled units which will regulate many interferences and variations directly and locally, thus counterbalancing systems dynamics with their own dynamics. With this, it becomes clear that the dynamics of the system call for an operational network of flexible, local units having powers of decision – and that traditionally-patterned hierarchies are not suitable. In systems theory, however, the complexity of the system controlled demands a correspondingly highly complex control system (ASHBY 1956, p. 202ff)[11]. This can be achieved through decentralization.

This results in a re-orientation of the strategic management of production: systems design is no longer based on the deterministic view of production according to the „Cybernetics I" approach, but rather accepts production as a chaotic system, from the physical level on, in accordance with the „Cybernetics II" approach (Figure 3)[12].

---

10 Using approaches based on „Cybernetics II", the economic efficiency of the entire system can be increased, as costs are incurred only through errors and interferences and as uneconomically high investiture during the design phase of the system can be avoided. Because errors and interferences allow the organization to learn, any costs incurred may be viewed as investiture in an improved organization.

11 ASHBY generalized this recognition as early as 1956, formulated as the „law of requisite variety" (see also ASHBY 1958), whereby the term „variety" came to be used interchangeably with the term „complexity". Since then, the recognition has become well-known as „ASHBY'S Law".

12 I would like to underline here the relativity of my statements. They are always dependent upon the system in question: if a system is relatively stable, further stabilization by means of structuring is certainly effective. But in accordance with our considerations above, we have been taken systems as our starting point which exceed a certain level of complexity.

Principles for deterministic systems...	Principles for chaotic systems...
Complexity must be reduced.	➔ Complexity must be designed.
Detail view.	➔ Holistic view.
Isolation, concentration upon individual elements and aspects.	➔ Holistic perspective under many aspects.
Investigation of the cause of interaction.	➔ Investigation of the results of interaction.
Precise and detailed, but difficult to implement, models.	➔ Models which are not a sufficient knowledge base, but which may be utilized for implementation.
Establishment of defined action procedures for the actors.	➔ Qualifying actors to act in a goal-directed manner.

Fig. 3 Shift in principles of analysis and design of the control of complex systems (following DE ROSNAY 1979, 95f, cited by PROBST 1981, p. 116f).

4.2. Sources of organizational flexibility

Flexibility may be seen as the ability of a system to react to changing demands. Here the system must be able to react in such a way that resulting dynamics do not threaten the stability of the system. At the same time, flexibility of an organization means that it must be able to guarantee the client consistent performance in form and quality despite any internal changes and disturbances. In contrast to the concept of adaptivity, flexibility aims upon the maintenance of the primary structure of the system. Reactions to changing demands lead to adaptive action, which should not, however, require adaptation of the structure of the system.

Flexibility thus represents a potential, or resource, within the organization. If flexibility is required, we must ask how flexibility may become anchored in the system. To do this, two types of flexibility should be distinguished:

1 individual flexibility of a single area, for example of a group,

2 organizational flexibility of the entire system.

Individual flexibility may be promoted at the level of the employee essentially through motivation and qualification. Once we go beyond this definition and view the individual flexibility of a cell within the system, the concept of contingency takes on significance. Contingency describes the ability to react to different and changing facts of a case in varying ways (see WILLKE 1991, p. 18f). With this, the internal complexity of a cell becomes important: in order for the cell to have the ability to act in a flexible fashion, its own internal complexity must – according to ASHBY'S Law (1956, p. 202f) – correspond to external complexity of that sector of the system which places demands on the cell. In dynamic environments particularly, very demands are placed upon employees' qualifications, because individual events occur singularly and stochastically and are therefore not predictable (see WEICK 1990, p. 9). The contingency of a cell in production may be increased by means of improving the polyvalence, i.e. the range of skills, of its workers and by means of designing tasks in a way that promotes their cooperative abilities[13].

Organizational flexibility refers to the entire system and must be ensured through the structuring of the organization. This is achieved by means of slack and redundancy (see STAEHLE 1991). „*Redundancy* means prolixity, repetition or duplication of information and tasks, which assures that the transmission of information and the fulfillment of tasks will be comprehensible or secure. *Slack* refers to the existence of an excess of resources in an organization which exceeds that which is actually necessary" (STAEHLE 1991, p. 314).

In the world of practice, this means that certain competencies must be present at numerous points within an enterprise. Redundancy can thus refer to organizational units, cells or even individuals. Redundancy makes it possible for an organization to react better to individual events such as errors, interferences and special requirements, because it is more probable that the competence to solve the resulting task situation will be right there where the event or corresponding problem arises. Redundancy gains particular significance in production, since the majority of events occur in singular and stochastic fashion, meaning that neither location nor type of event is predictable.

Redundancy stands in contradiction to the concept of hierarchical bureaucracies with their clear, divided-labor responsibilities, but also to the thinking behind some more recent concepts[14], which propagate streamlined design and the elimination of waste. There is an original connection here between the concept of redundancy and the concept of slack. While redundancy refers to operational functions, slack means to have an excess of resources at one's disposal. It is necessary in principle to have an overabundance of certain resources, whether they be authority, personnel or machines, in order to be able to react flexibly to increasing demands without impairment of system behavior towards the customer – in the form of increased delivery times, for example. Conversely, a lack of slack in an organization soon creates stress for individual employees (STAEHLE 1991, p. 320). Thus, slack, like redundancy, not only incurs costs, but actually reduces costs as well.

---

13 This throws a fundamental problem of centrally-controlled organizations into bold relief: because their cells may not act independently, contingency has to be ordered from the 'top', so to speak (STRINA & HARTMANN 1992, p. 171). Whether or not the cell in principle has the necessary contingency at its disposal plays no role at all, as it is already excluded from the control process.

14 'More recent' concepts is a reference to all those approaches which, although they emphasize an attempt to depart from bureaucratization and Taylorism, are based upon mechanistic models, such as Lean Management and Business Process Reengineering.

5. Characteristics of sociotechnical systems

Human resources are a major source of flexibility within a production system. Through his intelligence a human can act independently and creative. A human can provide flexible reactions and adapt his method of working if the overall system suffers situational disturbances, e.g., in case of urgent orders or machine breakdown. Human resources therefore are of key value for an enterprise, not only as bearer of an enterprise's know-how and expertise, but as key to organizational flexibility.

In the perception of work-psychological research, a task performed by an employee forms the nucleus of each sociotechnical system and links individual and organization (see ULICH 1994, 157ff, Fig. 4). Hence a task is not primarily perceived as function or duty to be fulfilled as part of a business process or a technical system but as a number of activities to be performed by an individual. The individual becomes an acting character and adopts certain objectives to lead his own action.

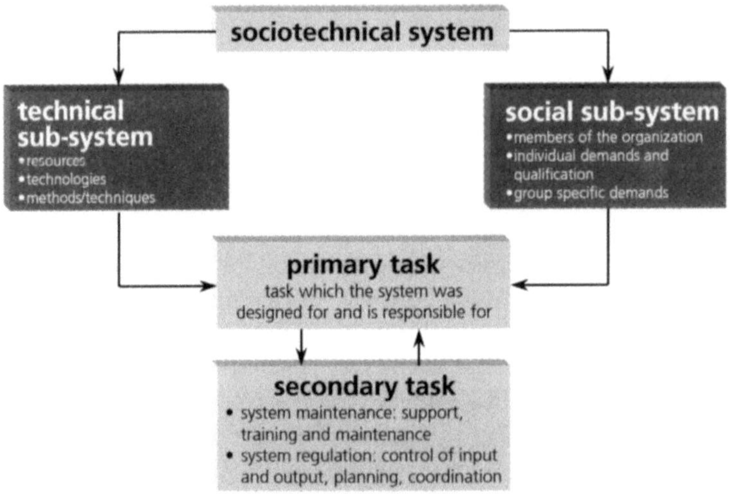

Fig. 4 Definition of a sociotechnical system (ULICH 1994, 155).

Hence to fully utilize these human features, the structure of the decentralized system has to form a positive and demanding working environment for the human actor. Task orientation in sociotechnical systems theory therefore sees the individual human as means of structuring.

In a sociotechnical perspective shop floor control is not viewed as the function of a technical system but as the individual task of every actor within the shop floor organization. In this approach, the principles of social systems theory and organizational development gain relevance, and the human as operator, scheduler and controller attracts attention.

5.1. The role of the actors in the system

The dynamics of the control system arise through the activity of individual cells. The requirement that systems control itself must possess a degree of complexity and dynamics which corresponds to the system being controlled puts the issue of activity within the control system into the foreground. Here persons, as workers in a cell, are the most flexible and, at the same, the only independent resource of the control system, as their intelligence can direct their actions actively and creatively (AGRAWAL 1996, p. 21). It is through human action that targeted control becomes possible at all. Within the framework of the given communications and decision-making structure, systems behavior can adapt flexibly, by means of human actions, to situation-determined phenomena and singular interferences. The person does not only thus contribute to a company's know-how and to mastery of the company's core competencies, but rather, in his role as controller, is also capable of developing and maintaining the system.

Many design approaches view control as a system, which is to be designed from the perspective of management as the traditional source of organizational innovation. This is the only way to understand the much propagated demands for reduction of systems complexity. In reality, complexity is reduced only in the perspective of management which succeeds in transferring complexity to individual operational units and individuals. Total systems complexity will then show a tendency towards increased complexity. The strategy of transferring forces workers in the cells to reorganize themselves, so that the new demands can be met at all. Some very radical reorganizations of companies in recent years have shown that workers are indeed capable of self-organization, even when initial conditions for the development of self-organization are greatly impaired.

This fact demonstrates, on the one hand, that self-organization is possible and is already being out into practice in diverse forms. At the same time, it places a demand on systems design to become aware of the role of actors in the system – and thus aware of the human being as an independent actor – and not leave this role to chance. The role of actors in the system should be designed and promoted. The issue of systems design therefore becomes an issue of double contingency: the human being must be suited to the system, that is, he must bring certain qualifications and flexibility, and at the same time, the system has to be suited to the person, meaning that the system must at once be transparent – to some degree, utilizable and conducive to the development of the person.

5.2. Images of the human

Complexity hinders our understanding of relationships within a system and thus often presents an obstacle to goal-directed action. So the aim of systems design must be to design systems complexity as a whole in such a way that it is utilizable from the perspective of individual workers in the cells. The issue of human strengths and weaknesses thus becomes significant. The basis needs to be a change in the predominant image of man both in industrial management and in the field of software development.

„The human being as a disruptive factor – to be eliminated". This attitude underlies a whole array of management theories and prevails in the theory as well as the practice of PPC. Even today, renown researchers and developers follow the goal of creating the most fully automated technical system possible, with the justification that humans are inadequate and subject to disturbances. For example, SMITH (1995, p. 159), in reference to the development and implementation of artificial intelligence systems in the framework of shop floor control, writes:

> „What about exploiting the expertise of human schedulers? Although we do not discount the experience accumulated by veteran human schedulers, our experience in many manufacturing environments is that schedulers often fall prey to the complexity of interacting constraints and decisions, and tend to adopt myopic 'fire-fighting' tactics (where extinguishing one fire ignites the next)." (SMITH 1995, p. 159).

The author then goes on to propose for the future an autonomous expert system which evades both human checks and influences. TAYLOR (1995, p. 36) suggests, in a basic article on production control titled „Scheduling tomorrow's factories", that human beings be continuously replaced with computers in order to increase effectiveness. MCGREGOR (1960) calls this stand „Theory X": 'The average person has a natural disinclination to work and attempts to avoid it wherever he can'. For this reason, people must be forced, steered, directed and threatened with punishment. MCGREGOR places this in contrast to the more positive „Theory Y": 'For the sake of goals to which the person feels committed, he will subject himself to self-discipline and self-control.'

The extent to which positive and negative images of man are tenable cannot be determined definitively. The issue will remain a matter of personal opinion. But for the world of practice, it seems sensible to reexamine attitudes towards man and so to perhaps reach a more positive image in MCGREGOR'S sense (also see the contribution by SCHÜPBACH in this volume). The main issue to assess is the human role in production. Here the person is both a part of the executing system and the control system. With reference to man's role as a production resource, persons are not comparable with machines; persons are „complementary" to machines (WEIK, GROTE & ZÖLCH 1994). Man and machine complement each other, and together there are necessary to execution of a certain process.

As a part of the control system, the human worker is confronted with the complex structure of a system which is, to the worker, of only limited transparency. Unavoidably, errors in judgment will be made, which are called „cognitive errors" (c.f. DÖRNER 1992, p. 54f). If we compare the human worker to the machine alternative – the computer system which is supposed to replace the human – we find three human characteristics that are important in production:

1. The human being is both a part of the control system and the executing system, so that an additional interface is not required, or not critical in a temporal sense.
2. The human worker is employed throughout the entire system and therefore can make decisions in close proximity to potential events.
3. Human beings are capable of independent, flexible and intelligent action and can gather information independently and individually.

Human action is not only determined by individual qualifications and motivation. It is subject, to a very significant extent, to the system's demands on human employees. If a positive and promoting environment for people and their action is successfully created, then human beings themselves represent a positive variable within the system. At the same time, additional measures can be undertaken to increase a system's transparency in order to reduce the probability of cognitive errors.

5.3. Human potentials

In accordance with the points described above, it is necessary that the positive aspects of the human role be examined, so that human potentials may be utilized as key competencies within the framework of the control system. For this, systems design must be human-oriented, so that human abilities and characteristics may be utilized as well as reinforced. Figure 5 outlines some fundamental human characteristics (see DUNCKEL ET AL. 1993, p. 29ff) and the potentials these characteristics contain with regard to control tasks.

Human characteristics	**Resulting potentials**
human thinking is object-related	▪ reduction of complexity, ▪ fast perception and understanding of situation, ▪ mapping with previous situation and resulting experience.
human behavior is target-oriented	▪ ability to make flexible decisions on alternative working methods to achieve a set target, ▪ ability to determine global impacts on the system, ▪ ability to measure and judge performance and adapt working methods accordingly.
human acting is performed in a social environment	▪ ability to communicate in a personal and flexible way ▪ ability to build up experience, ▪ ability to reflect upon one's own behavior and ability towards self-design.

Fig. 5 Human characteristics and resulting potentials for human-oriented systems design (based on DUNCKEL ET AL. 1993, p. 29ff).

Moreover, humans can learn from their own performance and thereby improve their knowledge and skills. To fully understand the human role in shop floor control, it is necessary to realize human potentials in contrast to technical system potentials in information processing (Figure 6, see also EMERY 1969, p. 149f.).

	Human resources	Technical resources
Information reception		
1. mode	■ mental	■ physical
2. scope	■ large	■ small
3. recognition	■ semantic (form)/ pragmatic (meaning)	■ syntactical
Information processing		
1. cause	■ active	■ passive
2. formulation	■ fuzzy, with debugging	■ exact, no debugging
3. strategy	■ flexible and adaptable	■ fixed
4. mode	■ circumlocutory, redundant	■ concise
5. storage capacity	■ large	■ small to medium
6. prediction	■ large range based on experience	■ short range with narrow scope

Fig. 6 Human and machine characteristics in information processing (adapted from LUCZAK 1983, p. 322).

Humans have the fundamental capacity to understand complex tasks and to adapt their work methods to changing goals and environmental conditions. In the case of shop floor control, humans can react in a flexible way and simply adapt their actions when unexpected interferences and variations, such as machine breakdowns or rush orders, occur. Thanks to the ability to observe the state of the system and the results of actions with regard to goal achievement over longer periods of time, a person if fundamentally capable of changing behavior and also work methods over the course of time. If this learning ability is utilized, it is possible to create an organization which can adapt itself in a continuous process of change. At the same time, the organization's openness to adaptation from the outside, such as when reorganization is prescribed, is increased. In this connection, the relation between complexity and errors becomes relativized: because errors represent learning opportunities, complexity may be seen not only to be a source of errors, but to be a source of learning experiences.

In order to make full use of human potentials, it is not enough to give them verbal recognition and to utilize them in an implicit fashion in the framework of reorganization. Human potentials should be reinforced and promoted in an explicit way. Only then do the characteristics described manifest themselves positively. Thus it is necessary that the human being be the focus of every technical-organizational design measure, and it is necessary that criteria for systems design be oriented to human characteristics.

6. Task orientation and structural design

As mentioned above, the structure of a system represents a significant demand on the actors within a system. The structural design – of both the controlled system and systems control – is crucial to reorganization. If system structure is to be designed in a targeted way, the concept of structure must be examined.

Upon close observation, the structure of any organization reveals itself to be more or less dynamic (c.f. EMERY 1969). Structure itself arises through activity within the system, and thus is based on the actions of the actors in the system. If activity ceases, the organization ceases to exist. As actions are not always the same and also can change form and type on their own through the course of time, we can define the structure of a system precisely only at an exact point in time. Such precise definitions quickly become irrelevant, however, because the structure of the system changes within a very short time. Change is effected through implicit self-structuring as well as through the explicit restructuring measures that are always part of a system and the environment of a system. In this sense, instead of 'structure', we may speak of the permanent 'structuring' of an organization. As WEICK (1990, p. 18) notes, „Structures are both the means and the results of actions". The actions are carried out by the actors within a system in fulfillment of their assigned tasks. In this way, the tasks assigned to actors within a system affect actions, which in turn serve to fulfill individual tasks. The structure of a system, therefore, is not defined exclusively by the formal structural and process organization of an enterprise, but rather results in its origin from the definition of work tasks, whose execution in operations generate the actual, dynamic structure of a system (Figure 7).

In the design of a management and control system for production, the *structure* of the system and the individual work *tasks* represent the most significant design parameters. Structural design is thus both a requirement of and a result of task design. There are no standardized solutions with regard to systems design. But it is possible to derive some rules of systems design. The goal must be to design a system which is flexible, adaptive and robust.

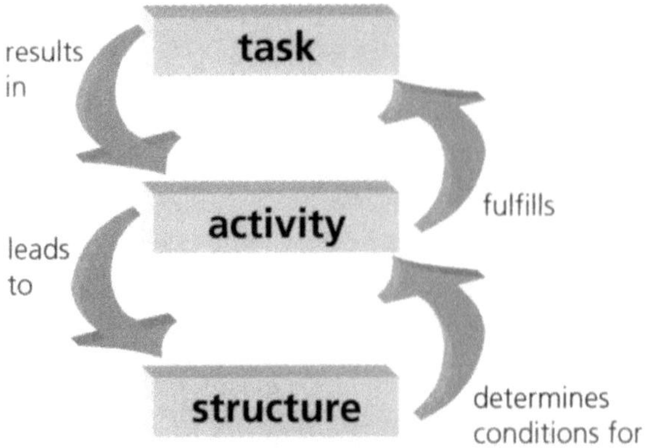

Fig. 7 The relation of task and structure.

Because the work task is the basis for human action, while at the same time the human worker contributes considerably to the adaptivity of a system, a main emphasis must be placed on the task. It is necessary that the system be so designed as to allow use of and promotion of human abilities. System structure and, in particular, system complexity represent the most important demands and restrictions (SCHERER & DOBBERSTEIN 1995). It must also be taken into account that in the framework of reorganization, task design is more amenable to definition and implementation than the comprehensive structure of a system. In turn, task design always has an effect on structure. This structuring effect may take hold very slowly and not always in a predictable form. Task design achieves a certain, yet temporary, system state. This changes with time and can lead to positive or negative effects on the total system.

Work tasks should be designed according to work- and human-oriented criteria (also see the contribution of SCHERER & ZÖLCH in this volume). This allows human potentials to be utilized and, at the same time, a positive work environment is created for people. Humane systems design and economically sound systems design are not mutually incompatible (see KLINGENBERG & KRÄNZLE 1988).

6.1. Design principles

In the following, human-oriented *and* structurally-oriented principles of systems design will be outlined and shown in relation to task design. System structure provides the frame for the entire system, in that it establishes the

1. autonomy and
2. interconnection (linking)

of the cells. The prerequisite for true autonomy of a cell is, in addition to the autonomy to act (sufficient degrees of freedom for decision-making), autarchy. Autarchy is to have the necessary resources at one's disposal. Cells become interconnected via material flow, the exchange of resources and, independently thereof, the exchange of information. The linking of cells determines the effect relations among them. Here information exchange may take place in the form of horizontal or vertical communication, that is, directly between two operational cells or via a hierarchically superordinate cell.

The degree of the autonomy to act and of the complexity to be regulated are, from the perspective of the cell, significant features which determine the degree of effectiveness of control within a cell. They are manifested in the transparency of the work task, in possibilities of intervention and in authorization to intervene, in process understanding and the qualifications of workers.

ULICH (1994, p. 156) makes the following demands upon the structure of a sociotechnical system to ensure task orientation, which are to be met at the level of the organizational unit:

1. Relatively independent organizational unit:
   Is the organizational unit – due to the completeness and independent nature of the task assigned – in a position to itself counterbalance and regulate any interferences and variations at the location of their appearance, so that these do not reverberate in an uncontrolled fashion to other organizational units?

2. Task relatedness within an organizational unit:
   Are the various part-tasks in a work system related in content? Are work-related communication and cooperation required by task fulfillment? Is there a shared understanding of the task?

3. Unity of product and organization:
   To what extent may the work results of an organizational unit be attributed to itself both quantitatively and qualitatively? Can errors or deficits in quality be traced clearly back to the unit? The unity of product and organization allows the unit to identify with its own product.

Through relative independence, task relatedness and the unity of product and organization, it is possible to create teams which may act with clear reference to their surround in the total system.

6.2. The role of autonomy

Teams controlling themselves are only then successful, when the total task does not exceed a degree of complexity which accords with the level of qualifications. „If system interferences become so large or so complex that their mastery within the system overtaxes resources, then it becomes a necessity to shield the group from the interferences“ (ALIOTH 1980, p. 35). The control of these interferences then falls to a superordinate coordinating instance or a superior. A change in the autonomy of a decentralized area and a change in the degree of freedom for action show contrary effects. On the one hand, a minimum level of complexity of the task is required if improvement is to be made. On the other hand, the human worker is limited in his proactive control of complex processes. Based on these considerations, ULICH formulated the hypothesis as early as 1974 that there is an inverse, U-shaped relationship between the degree of effectiveness of work – expressed in the relation between mental and physical expenditure and the performance achieved – and the degree of complexity of a task (Figure 8).

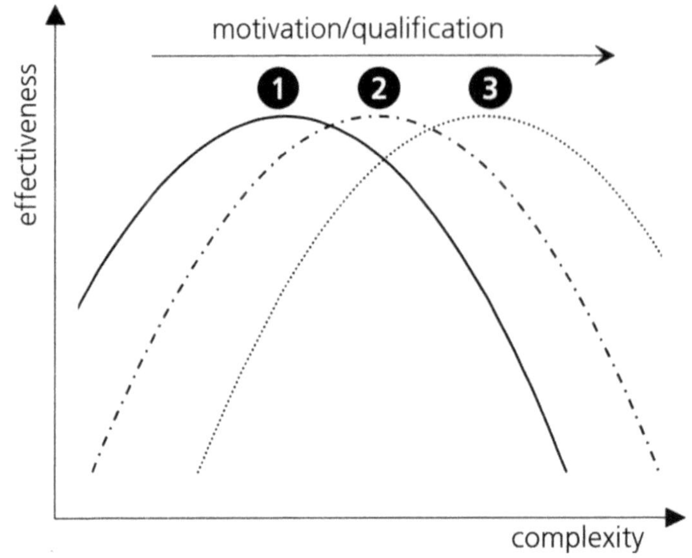

Fig. 8 Probable relationship between degree of complexity of work activity and degree of effectiveness of human work (according to ULICH 1974, p. 193).

To design a system of lasting effectiveness, design must strive towards continuous adaptation of the total system to enduring changes in the surround system. More than primary, temporally limited flexibility is thus required: the entire system must be designed to be adaptive. Here again, the human, as coworker in a cell, is the primary initiator and bearer of possible change. Entrepreneurial thinking leads the cell, through self-organization, towards the goal of creating its own optimal conditions of work. This results in the active modification of structure and tasks from the perspective of operations. Adaptation of the system as a whole is thus achieved via self-organization of the cells. Here the important catalysts are the opportunity to take over further activities with simple environmental conditions, and the need to adapt one's own actions when demands are complex (Figure 9). The extent of change made independently and adaptively is arbitrary and begins with changes in one's own work procedures, with possible spread all the way to change in the goals of the whole system.

System complexity thus represents the essential measuring variable in designing system structure, in that it determines the degree of effectiveness of human work as well as the autonomy strivings of individual cells. Of course, the complexity of a system may not be viewed as something that stays the same in a

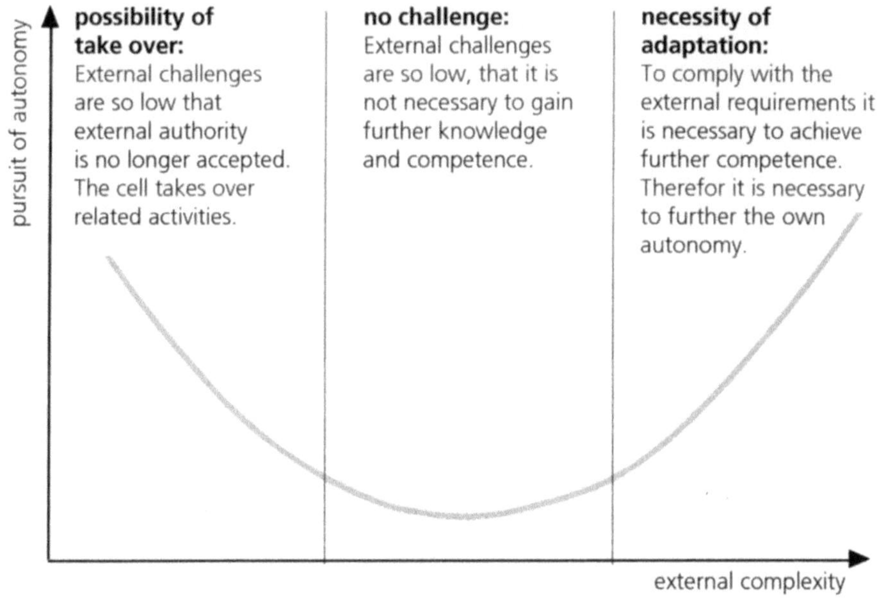

Fig. 9 A cell's strivings towards autonomy in dependency upon environmental influences (SCHERER & DOBBERSTEIN 1996).

lasting way. Instead, complexity is already subject to the structuring process mentioned. The goal of systems design must therefore be the creation of an initial design of the system which allows for and promotes continuous adaptation. There are several conceivable principles here. For one, participation of those involved would foster an understanding of a reorganization project as a process of change from its very start and would allow reorganization efforts to continue in that same vein. Deliberate cutbacks in resources can cause a demand for adaptation and, thus, force self-organization and change. WIENDAHL & AHRENS (1995, p. 4) speak here of „interventionist control" (see also KROHN & KÜPPERS 1990, p. 124-125, as well as the contribution by AHRENS in this volume). Or in another strategy, entrepreneurial action can be promoted by means of an appropriate systems of incentives, by setting up certain market principles among the cells. Here it becomes clear that the shielding from interferences and variations may not go so far that the cells give up their strivings toward autonomy. Shielding must take place only in individual cases. This makes the degree of shielding, in relation to the extent of external demands, a dynamic variable, which can be controlled, so to speak, at the level of management and via the control network.

6.2.1. Integration in distributed systems

While autonomy has to be achieved through a strategy of decoupling the cells, integration depends on a certain coupling of the cells and coordination of the cells' activities. As means to achieve integration the following principles can be helpful:

1 Common objectives
Pursuing a goal is the basic principle enabling activity. A specific goal always depends on the personal interpretation and motivation of an individual (see SCHERER 1996, 33ff). The activity of a single cell has to be integrated into the overall system of objectives of the global production system. This can only be achieved through motivation and incentives. Therefore it is necessary to allow local and even personal objectives in addition the objectives proposed by the higher management.

2 Integrated information systems
Computer-based information offer various means to integrate distributed and decentralized organizations. Thereby it has to be considered, that computer-based systems never will be able to realize a total integration within a sociotechnical system there the human individual still is a major resource. Rather than globally imposing activities and procedures they should support local activity by distributing information throughout the system.

3 Network organization and personal communication
An integrated organization has to picture the dynamic structure of the system and the system's environment. This can be achieved through a network organization of interlinking teams were each team is part of the higher level activities through a representative. This representative forms an 'linking pin' between the different cells and the different levels of management (LIKERT 1961, P. 113).

In respect of integration it has to be considered that integration is the ability of an organization to flexibly act towards a common goal rather than a specific state of an organization of being integrated. Therefore the personal ability to interact and communicate flexibly and efficiently is a major objective for system design.

6.2.2. Self-design and lasting adaptability

In order to guarantee the lasting efficiency of the organization of shop floor control as described, there is a need for the creation of lasting opportunities for organizational development, for learning and for the acquisition of knowledge. To design task that allow learning and knowledge acquisition, the following are required:

1 Ability to *learn*:
Qualifications and motivation

2 Opportunity to gain *experience*:
Interferences and errors

3 Opportunity to apply *knowledge*:
Independent actions

Lasting opportunities to reflect as a group and as an individual and to further develop one's own behavior and performance have to be created. It will be necessary to give work teams the opportunity to examine, question and design their tasks and conditions for execution, as well as their personal and collective goals. This process would include ultimately the question of the meaning of one's work. This leads to a kind of expanded feedback loop (Figure 10), in that not only the direct behavior of a cell, in the form of quality control, is examined, but also the work task, goals and meaning, or necessity, of the cell as an organizational unit within the company (c.f. HENNING & MARKS 1993, p. 86ff.). As individual cells become able to reflect upon the organization from their local perspective, they will contribute considerably to a continuous increase in the flexibility and adaptability of the entire production system, without being prodded to do so from above. Within this layout self-design ensures the system's ability to adapt to a changing situation without changing the overall hierarchical framework.

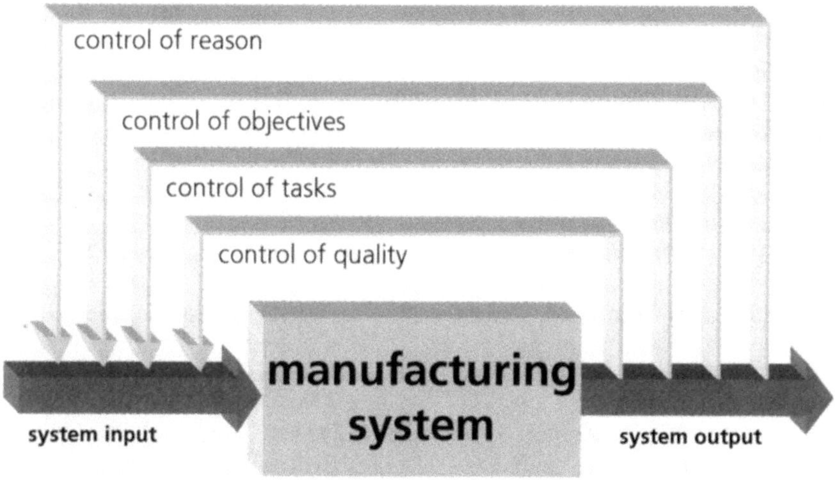

Fig. 10 Expanded control system (based on HENNING & MARKS 1993).

7. Systems design for flexibility, adaptivity and stability

Several principles for systems have been described based on some basic assumptions on how systems and organizations function and what can further human characteristics. The more open a system is the more it has to be able to act ad hoc and independently. To enable optimal behavior and results of shop floor control it is necessary that strategic management in an enterprise rather allows more autonomy at shop floor level than introduces a highly automated and sophisticated computer system for scheduling. Thus achieving flexibility, adaptivity and stability for shop floor control always will result in a change of power and hierarchy in production management. It is this very basic assessment which shows that the process of systems design for shop floor control never can be purely neutral since it always involves humans and their individual will.

8. References

AGGRAWAL, S. (1996): Flexibility management: the ultimate strategy. Industrial Management, 37:6, p. 20-26.

AHRENS, V., JANSSEN, C. (1994): Analyse von logistischen Produktionsprozessen mit Methoden der Theorie Nichtlinearer Systeme. In: WIENDAHL, H.-P., AHRENS, V. (Eds.): Potentiale der Chaosforschung in den Produktionswissenschaften. Proceedings. Institut für Fabrikanlagen, Universität Hannover.

ALIOTH, A. (1980): Entwicklung und Einführung alternativer Arbeitsformen. Verlag Hans Huber, Bern.

ANTHONY, R.N. (1965): Planning and control Systems – A framework for analysis. Graduate School of Business Administration, Harvard University.

ASHBY, W.R. (1956): Introduction to cybernetics. Wiley, New York.

ASHBY, W.R. (1958): Requisite variety and its implications for the control of complex systems. Cybernetica, 1:2, p. 83-99.

BAKER, K.R. (1974): Introduction to sequencing and scheduling. Wiley, New York.

BŁAZEWICZ, J., ECKER, K.H., PESCH, E., SCHMIDT, G., WEGLARZ, J. (1996): Scheduling computer and manufacturing processes. Springer, Berlin.

BURGIDGE, J.L. (1978): The principles of production control. 4th Ed. MacDonald & Evans, Plymouth.

DE ROSNAY, J. (1979): Das Makroskop, Systemdenken als Werkzeug der Ökogesellschaft, Stuttgart.

DÖRNER, D. (1992): Die Logik des Misslingens. Taschenbuchausgabe. Rowohlt, Reinbek bei Hamburg.

DUNCKEL, H., VOLPERT, W., ZÖLCH, M., KREUTNER, U., PLEISS, C., HENNES, K. (1993): Kontrastive Aufgabenanalyse im Büro – Der KABA Leitfaden. Grundlagen und Manual. vdf/Teubner, Zürich/Stuttgart.

DUNKLER, O., MITCHELL, C.M., GOVIDARAJ, T., AMMONS, J.C. (1988): The effectiveness of supervisory control in scheduling of flexible manufacturing systems. IEEE Transactions on Systems, Man, and Cybernetics, 18:2, p. 223-237.

EMERY, J.C. (1969): Organizational planning and control systems. Theory and technology. Arkville Press, London.

FORRESTER, J.W. (1969): Industrial Dynamics. 6th ed. MIT Press, Cambridge.

GOLDMAN, S.L., NAGEL, R.N., PREISS, K. (1995): Agile Competitors and Virtual Organizations. Van Nostrand Reinhold, New York.

GUNN, T.G. (1994): In the age of the real-time enterprise. Oliver Wight Publications, Essex Junction.

HENNING, K., MARKS, S. (1993): Kommunikations- und Organisationsentwicklung. Verlag der Augustinus Buchhandlung, Aachen.

HORNS, A. (1989): Job shop control under influence of chaos phenomena. In: SANDERSON, A.C., DESROCHERS, A.A., VALAVANIS, K. (Eds.): IEEE International Symposium on Intelligent Control 1989. Proceedings. IEEE Computer Society Press, Washington. p. 227-232.

JURAVICH, T. (1988): Chaos on the shop floor: A worker's view of quality, productivity, and management. Temple University Press,

KLINGENBERG, H., KRÄNZLE, H.-P. (1988): Humanisierung bringt Gewinn – Modelle aus der Praxis, Vol. 2: Fertigung und Fertigungssteuerung. Rationalisierungs-Kuratorium der Deutschen Wirtschaft, Eschborn.

KROHN, W., KÜPPERS, G. (1990): Selbstreferenz und Planung. In: NIEDERSEN, U., POHLMANN, L. (Eds.): Selbstorganisation. Duncker & Humblot. Berlin. p. 109-127.

KURBEL, K. (1995): Produktionsplanung und –steuerung. 2nd Ed. Oldenbourg Verlag, München.

LIKERT, R. (1961): New patterns of management. McGraw-Hill, New York.

LUCZAK, H. (1983): Informationstheorethische Arbeitsgestaltung. In: ROHMER, W., RUTENFRANZ, R. (Eds.): Praktische Arbeitsphysiologie. 3rd ed. Thieme, Stuttgart. 321-355.

MALIK, F. (1992): Strategie des Managements komplexer Systeme: ein Beitrag zur Management-Kybernetik evolutionärer Systeme. 4th ed. Verlag Paul Haupt, Bern.

MANSKE, F., WOBBE-OHLENBURG, W. (1985): Fertigungssteuerung im Maschinenbau aus der Sicht von Unternehmensleitung und Werkstattpersonal, Part 1. VDI-Z, 127:11, p. 395-402.

MCGREGOR, D. (1960): The human side of enterprise. McGraw-Hill, New York.

MCKAY, K.N., BUZACOTT, J.A., SAFAYENI, F.R. (1989): The scheduler's knowledge of uncertainty: The missing link. In: BROWNE, J. (Eds.): Knowledge-Based Production Management Systems. Elsevier, Amsterdam. p. 171-189.

MCKAY, K.N., SAFAYENI, F.R., BUZACOTT, J.A. (1988): Job-Shop scheduling theory: What is relevant? Interfaces, 18:4, p. 84-90.

MELNIK, S.A., CARTER, P.L., DILTS, D.M., LYTH, D.M. (1985): Shop Floor Control. Dow Jones-Irwin, Homewood.

MOLDASCHL, M. (1994): „Die werden zur Hyäne" – Erfahrungen und Belastungen in neuen Arbeitsformen. In: MOLDASCHL, M., SCHULTZ-WILD, R. (Eds.): Arbeitsorientierte Rationalisierung: Fertigungsinseln und Gruppenarbeit im Maschinenbau. Campus, München. p. 105-149.

NN (1995): Nicht die Technik, sondern der Mensch bringt PPS zum Leben. Maschinenbau-Nachrichten, 1, p. 49-50.

PASMORE, W. A. (1988): Designing effective organisations. The sociotechnical systems perspective. Wiley, New York.

PINENDO, M. (1995): Scheduling: Theory, Algorithms and Systems. Prenctice Hall, Englewood Cliffs.

PROBST, G.J.B. (1981): Kybernetische Gesetzeshypothesen als Basis für Gestaltungs- und Lenkungsregeln im Management. Verlag Paul Haupt, Bern.

REICHWALD, R. (1992): Die Wiederentdeckung der menschlichen Arbeit als primärer Produktionsfaktor für eine marktnahe Produktion. In: REICHWALD, R. (Eds.): Marktnahe Produktion. Gabler, Wiesbaden. p. 3-18.

ROLSTADÅS, A. (1995): Human capital. Editorial. Production Planning & Control, 6:6, p. 489.

RÖPKE, J. (1977): Die Strategie der Innovation. Mohr, Tübingen.

SANDERSON, P.M. (1989): The human planning and scheduling role in advanced manufacturing systems: an emerging human factors domain. Human Factors, 31:6, p. 635-666.

SCHERER, E. (1995): Complexity and reactivity of decision-making in shop floor control: A human-oriented approach to system design in decentralised, heterogenous production. In: STORCH, R.L. (Ed.): Managing concurrent manufacturing to improve industrial performance. Proceedings of the IFIP WG 5.7 Working Conference, Seattle. p. 472-483.

SCHERER, E., DOBBERSTEIN, M. (1996): Komplexität in der Produktion beherrschen? IO Management Zeitschrift, 65:3, p. 60-64.

SCHULTE, H., WUNN, C. (1993): Sind unsere Manager durch die zunehmende Komplexität überfordert? IO Management Zeitschrift, 63:4, p. 33-37.

SCHUMANN, M., ET AL. (1994): Trendreport Rationalisierung: Automobilindustrie, Werkzeugmaschinenbau, Chemische Industrie. Edition Sigma, Berlin.

SMITH, S.F. (1995): Reactive scheduling systems. In: BROWN, D.E., SCHERER, W.T. (Eds.): Intelligent scheduling systems. Kluwer, Boston. p. 155-192.

STAEHLE, W.H. (1991): Redundanz, Slack und lose Kopplung in Organisationen: Eine Verschwendung von Ressourcen? In: STAEHLE, W.H., SYDOW, J. (Eds.): Managementforschung 1. de Gruyter, Berlin. p. 312-345.

STAEHLE, W.H. (1994): Management – Eine verhaltenswissenschaftliche Perspektive. 7th ed. Verlag Franz Vahlen, München.

STRINA, G., HARTMANN, E.A. (1992): Komplexitätsdimensionierung bei der Gestaltung soziotechnischer Systeme. In: HENNING, K., HARENDT, B. (Eds.): Methodik und Praxis der Komplexitätsbewältigung. Duncker & Humblot, Berlin. 169-191.

SUSMAN, G. (1976): Autonomy at work: a sociotechnical analysis of participative management. Praeger, New York.

TAYLOR, F. W. (1911): The principles of scientific management. Harper, New York.

TAYLOR, F. W. (1919): Shop management. Harper, New York.

TAYLOR, S.G. (1995): Scheduling tomorrow's factories. APICS, 3, p. 32-36.

TRIST, E. L., BAMFORTTH, K. (1951): Some social and psychological consequences for the longwall method of coalgetting. Human Relations, 3, p. 3-38.

ULICH, E. (1974): Neue Formen der Arbeitsstrukturierung. FB/IE, 23, P. 187-196.

ULICH, E. (1994b): Arbeitspsychologie. 3rd ed. vdf/Poeschel, Zürich/Stuttgart.

VOET, M. (1994): Finite capacity scheduling tools: Why and where to implement? In: BENOY, M., DEWILDE, P., VOET, M. (Eds.): Finite scheduling: State-of-the-market. Ernst & Young Belgium, Brussel.

WEICK, K.E. (1990): Technology as equivoque: Sensemaking in New Technologies. In: GOODMAN, P.S., SPROULL, L.S. (Eds.): Technology and Organizations. Joosey-Bass, San Francisco. p. 1-44.

WEIK, S., GROTE, G., ZÖLCH, M. (1994): KOMPASS – Complementary analysis and design of production tasks in sociotechnical systems. In: KIDD, P.T., KARWOWSKI, W. (Eds.): Advances in agile manufacturing. IOS Press, Amsterdam. p. 250-253.

WESTKÄMPER, E., JESCHKE, K. (1995): Massnahmen zur Null-Fehler-Produktion im Maschinenbau. VDI-Z, 137:3/4, p. 84-87.

WIENDAHL, H.P. (1996): Grundlagen der Fabrikplanung. In: EVERSHEIM, W., SCHUH, G. (Eds.): Produktion und Management. Vol. 2. Springer, Berlin. p. 9-1 – 9-31.

WIENDAHL, H.P., AHRENS, V. (1995): Planning and control in self-organized production systems. WGP Annals Production Engineering, II:2.

WIENDAHL, H.P., PRITSCHOW, G., MILBERG, J. (1993): Produktionsregelung – interdisziplinäre Zusammenarbeit führt zu neuen Ansätzen, Teil 1. ZwF, 88:6, p. 265-268.
WIERS, V. (1996): A review of the applicability of OR and AI scheduling techniques in practice. Omega – Int. Journal of Management Science, 55:2, 145-153.
WIGHT, O.W. (1974): Production and inventory management in the computer age. CBI Publishing, Boston.
WILDEMANN, H. (1994): Paradigmenwechsel in der Produktionsplanung und -steuerung. VDI-Z, 136:10, p. 45-48.
WILLKE, H. (1991): Systemtheorie. 3rd ed. Fischer, Stuttgart.

Systems Concepts and Organizational Design: A Basic Review

Julia K. Kuark

In general, systems are considered to consist of *elements*, as well as the *interrelationships* between these elements and the whole. This model can be applied to the system at virtually every level; each of the elements can be considered systems and, in turn, be broken down into a system of subelements. The *boundaries* between the system and its environment can be defined more or less arbitrarily, depending on the question of interest or focus of analysis. The system boundary may or may not be physically observable. The *environment* of a system is understood to be those elements or systems that exist outside the designated boundaries. *Open* systems consist of elements or systems that interact with these outside elements or systems, whereas closed systems do not.

The elements and interrelationships between them and to the whole system make up a *structural framework* in which patterns can be observed. The dependency of the elements and interrelationships with time allows differentiation between *static* and *dynamic* systems. Systems are considered *complex* when „they are comprised of so many elements that the elements can only selectively interrelate with each other. This selection is, at the same time, a reduction of internal complexity“ (LOUIS 1992, p. 10). HIRSIG points out that systems can further be classified according to the determinism of the interrelationships between the elements. „In *deterministic* systems, all the interrelationships between the elements are known or can be inferred at every point in time“ while in *stochastic* systems only part of the interrelationships are known or the „interrelationships can only be inferred with a known factor of uncertainty“ (HIRSIG 1989, p. 2).

1. Theoretical background

1.1. The formal level of system theory

Psychology intensively used the possibilities given by cybernetics for new methodological concepts and experiment techniques as well as for data analysis and presentation as early as 1948 (TIMPE 1989, p. 33; cf. MALIK 1986). *Classic system and control theory* was (and still is) considered to be an exact science in that the formulation of theory is mathematical and objectively verifiable. It is universal enough to be used in a number of disciplines and is particularly applicable to experimental work. It is primarily used as a scientific construct, in accordance with the paradigm that considers the scientist an observer outside the system, making observations without disturbing the system. The researcher is thus able to devise a working model of a particular system according to hypotheses about the influences on that system. Once the system boundaries are defined, the hypothesized variables can be defined and experimentally tested. The hypothesized model can be further developed with each step in the empirical process incorporating the knowledge gained by previous experiments.

This type of model has been widely applied to organizations which are consequently viewed as rational systems. „Structural arrangements within organizations are conceived as tools deliberately designed for the efficient realization of ends [...] for the disciplined performance of participants" (SCOTT 1992, p. 48).

This formal application of systems theory builds on the „black box" assumption about the system elements. The complexity of the system model is thus reduced to the observation of the inputs and outputs to a particular element, and the (mathematical) transformation that occurs over the element can be inferred analytically or experimentally. The inner functioning of the element, how it actually achieves the transformation, is not relevant. With the aid of *feed-forward*, *feed-back* and *supervisory control* loops, very complex system models can be defined and empirically tested.

> „To view an organization as [such] a cybernetic system is to emphasize the importance of the operations, control, and policy centers, and the flows among them. [...] The policy center sets the goals for the system. [...] The cybernetic model places great emphasis on the operational level of the organization – the level at which the production processes of the system are carried out. The analysis of these technical flows – inputs, throughputs, and outputs – is regarded as vital to an understanding of the system; indeed, the control and policy centers are examined chiefly in terms of their impact on these technical flows." (SCOTT 1992, p. 80-81).

The application of formal systems theory in the field of work psychology „demonstrates its power particularly in the analysis of human-machine-systems and their components. [...] System theoretical and engineering analyses of such systems are based on significant similarities with mathematically describable structures. This essentially includes their closed information loops and the fact that the characteristics of the whole system are defined by the single elements" (TIMPE 1989, p. 34). Processes can be simulated with sophisticated algorithms, analyzed and the algorithms further refined. The use of such simulation programs has found widespread use both in academics and general practice.

However, at best, „one must be conscious of the fact that every model can merely be an abstraction of reality, or – more precisely – an attempt of a planner to express his or her image of reality" (HABERFELLNER ET AL. 1992, IX). Although a system theoretical model usually attempts to be as realistic as possible, a tendency towards over-simplification is immanent to the black box approach. TIMPE (1989) observes that „despite the hopefulness invested in the [formal] system theoretical approach, limits exist in the application of it even today. They concern the inclusion of learning processes and the clarification of the connections between psychologically interpretable components and their formal representation [...]" (ibid., 35).

Indeed, the paradigm that this approach works with is a classically mechanical one:

> „The basic conception of a machine is such that it is to be constructed to achieve a consciously planned purpose and that its function, reliability and efficiency are dependent on the corresponding characteristics of its singular components. Further, it is associated with the expectation that each singular component must be constructed according to exact and detailled plans and precisely assembled in a way that has been determined in advance." (MALIK 1986, p. 38).

1.2. The phenomenological level of system theory

In addition to this application of formal systems and control theory on an operative level, the more abstract application of system concepts are possible on a phenomenological level (cf. TIMPE 1989). While the first provides the basis for „scientific observations" and „exact" determining of parameters, a more abstract level offers descriptive and metaphoric advantages, insight on the understanding of *how* a system functions. These two levels of knowledge are not necessarily mutually exclusive, although this may seem to be the case at first glance. Classical theorists and authors of the more modern system approaches are beginning to acknowledge the benefits of both the operational and descriptive levels. The work of SCHÜPBACH (1994) combines both levels; he presents and compares both technical control and psychological action regulation models, embedded in the larger context of the production processes, which must be, in turn, regulated (p. 25 ff.).

The paradigm that rests on a mechanistic view of the world seems to be loosing its monopoly. While the application of these premises have been, and still are, of help in certain special cases, they are undoubtedly much less representative than has been realized in the past. The general transferal of the paradigm to the organizations at large is much more complicated. In order for a mechanistic system to function, „an essential condition must be met: The tuning out of, or control over, unanticipated influences, i.e. the disturbances [...]. This requires on one hand, a severe reduction of the openness of the system to its environment [...] and on the other hand, the openness of the subsystems and elements must also be reduced as much as possible“ (SCHÜPBACH 1994, p. 54). Hence, openness is not actually permitted in this system paradigm. The unsolvable dilemma lies, of course, in the fact that all real systems, particularly social systems, are open, complex and stochastic. The constraints of a laboratory situation as in natural science experiments are simply impossible to set in real life.

At the more abstract level of theoretical application, on the other hand, whole new dimensions have been opened up with the development of more *modern systems theory*. Unlike the „objective“ mathematical principle of the black box model described above, one of basic principles of modern systems theory is that the whole is more than simply the sum of the parts. This has been explored and developed, particularly in the areas of biology and psychological (e.g. gestalt and family) therapy, with respect to living systems.

> „The theory of living systems, that has been developed in the field of the natural sciences in recent decades, that has emerged from the cybernetics of the forties, offers the ideal framework for the scientific formulation of holistic-ecological thinking. Living systems – organisms, social systems or ecosystems – are integrated wholes, whose characteristics cannot be reduced to their smaller elements. Instead of concentrating on basic components, [this] system theory concentrates on basic principles of organization. System characteristics are destroyed when a system is dismembered into single parts, either physically or theoretically. Although in every system single components can be discerned, the whole is always something other than merely the sum of its parts.“(CAPRA 1992, p. 308-309).

Applied in the social sciences, MALIK (1986) speaks of a systemic-evolutionary management concept. The premises of the mechanistic type of management theory „lead almost inevitably to the assumption of a theoretically possible, extensive control of systems. The premises of the systemic-evolutionary management concept destroy these hopes. With its basis, only a form of command that could be called ‘soft’ or ‘fuzzy’ control is possible“ (p. 70). This school of thought admits that there are parameters and processes that are impossible to control and concentrates on understanding and describing the internal workings that enable systems to function.

The sheer complexity of our world apparently evokes a need for reduction of complexity to be able to understand things, but instead of the black box reduction, the reduction here takes place in the level of attention. Instead of concentrating on the details of each element, the more abstract level of the structural patterns are the focus. „The order producing effects of general rules in the sense of the evolutionary concept allows for the extension of general regulatory effects in areas of great complexity; this effect, however, has its price in the form of the unknowns of the details“ (MALIK 1986, p. 46). Trusting the internal evolutionary mechanisms inherent to the system and subsystems to work out the details means accepting and dealing with uncertainty – there is indeed an amount of risk involved. Apprehensions of loss of control are not completely unfounded.

1.3. Organizations as autopoietic systems

While earlier authors broke the ground for the current discussion of modern systems theory in the social sciences (for an overview see SCOTT 1992; BAITSCH 1993), an important cornerstone was set by the theory of autopoiesis for living systems by MATURANA & VARELA (1987). According to BAITSCH (1993, p. 10-11), most of the advocates of the autopoiesis are in agreement that the following eight characteristics distinguish autopoietic systems:

1. Autopoietic systems originate spontaneously, they arise and preserve themselves.
2. Autopoietic systems are operationally closed in the sense of their self-reference and circularity.
3. Autopoietic systems design and organize themselves. This, their autonomy, is not equivalent to independency with respect to outside systems.
4. Autopoietic systems structure their environments themselves. They have the system boundaries at their disposal and control their system identity therewith.
5. Autopoietic systems are complex and can, in turn, comprise complex elements.
6. Autopoietic systems display redundancy with respect to design and leadership powers and activities. A lack of redundancy endangers the conservation of the system.
7. Processes are constantly taking place in autopoietic systems; these are irreversible; an autopoietic system is never balanced.
8. Autopoietic systems change through self- and extraneous observation. Changes are made to the structure, not the organization.

The transference of the theory of autopoiesis to social systems is currently widely discussed, both theoretically and practically applied, by a rapidly growing number of authors (SPRÜNGLI 1992, EXNER, KÖNIGSWIESER & TITSCHER 1992,

BAITSCH 1993). Particularly LUHMANN (1995) has developed the theoretical basis of the autopoietic nature of social systems, while MATURANA & VARELA (1987) themselves are apparently ambivalent toward the transference into the context of the social science. Nonetheless, these ideas can be constructively applied to work organizations with one reservation:

> „Work organizations do not display one central characteristic of autopoietic systems: in the psychological perspective they do not originate spontaneously [...] in a particular set of conditions. Without exception, work organizations are set up by humans and consciously directed toward a goal. [...] However, after having been set up, work organizations show structural developments that are amazingly similar to processes in autopoietic systems. One of the most prominent characteristics of such systems is their potential for regulating system internal distinction and specialization. This possibility of creating, building upon and conserving order within and without is called the capacity of self-organization [...]".
> (BAITSCH 1993, p. 18).

Unlike the more classic definition of organizations as pools of resources such as money, labor, knowledge, etc., their distinguishing characteristic being their formal structure (KIESER 1989, p. 342-343) and their primary purpose being rational goals that have been set in advance, organizations here will be considered to be the context of continuous cyclic self-organization and self-preservation processes. Hence, enterprises, or more specifically production organizations, will be treated as systems open to their environments and inherently dynamic, complex and stochastic. Furthermore, „when we speak of systemic-evolutionary thinking, it is not simply synonymous with thinking in 'complex dynamic systems'. Systemic doesn't only mean consideration of the entire network, but includes the variety of overlapping of levels of perception. [...] 'Evolutionary' doesn't just mean dynamic, but includes the constant transcending of the system structures themselves, which develop out of dialogical processes" (LUTZ 1992, p. 314).

It is important to emphasize that the second characteristic of autopoietic systems presented above, their operational enclosure in terms of their self-reference and circularity, does not mean that they are considered closed to their environment. It can be said that general consensus prevails among the advocates of modern system theory, that with regard to energy, material and information, autopoietic systems are open systems. This also specifically holds true for organizations (KATZ & KAHN 1978, p. 753; MALIK 1986, p. 23; SCHÜPBACH 1994, p. 25). However, the search, choice and interpretations of energy, material and information are selected, or even manipulated, by the system with reference to itself in terms of what it considers relevant. EMERY pointed out early on that „although the dependence of the particular enterprise on its external environment is usually [considered] the most striking aspect, there is inevitably, some interdependence. An enterprise has some possibility for selecting from amongst a range of personnel, resources and technologies offered by its external environment and for developing new markets or transforming old ones" (1959, p. 7; cf. MATURANA & VARELA 1987, p. 95-96). In this way, organizational systems create their own understanding of the world around them, their own perception of reality. Even the economic pressure that enterprises usually use as an omnipresent argument can be interpreted from both sides: „Economic pressure doesn't influence in a deterministic manner, the internal-organizational reference to it is also contingent" (ORTMANN ET AL. 1990, p. 499).

The openness of self-organizing systems is closely coupled with their complexity: „A complex system could not maintain its complexity in a simple environment" (SCOTT 1992, p. 85). This mechanism works through the multiple levels of interactions between the system components and their respective environments: of the system as a whole, of the subsystems or organizational units, as well as that of the individual workers. In addition to the interaction between the complexity of the structure and dynamic relationships and the number of elements in a system, BAITSCH attributes the complexity in autopoietic systems to the presence of elements that are in themselves complex (1993, p. 13) and considers „people in work organizations to be complex elements in complex systems" (1993, p. 43 ff.). Indeed, it „was the people that established the work organization, and through the people's potential autonomous actions and interactions is the organization able to continue to exist" (BAITSCH 1993, p. 30). It is the presence and dynamics of people, perhaps, that make the concepts developed for living systems fundamentally applicable to any social systems, including work organizations.

Moreover, the „alternative conception of [enterprises as] 'open systems' carries the logical implications that such systems may spontaneously reorganize toward states of greater heterogeneity and complexity [...]." (EMERY 1959, p. 3). In this way, open systems draw from their environments. Nevertheless, SCOTT points out that „open systems are subject to what is termed the *law of limited variety*" (1992, p. 85), meaning that a system cannot exhibit more variety than that to which it has been exposed through its environment.

„In general, the central problem of cybernetics is the question how systems of any kind cope with the complexity of their environments, which mainly results from the constant changes and the speed of the changes. The answer lies in the *structure* or *organization* of the systems considered. Organization is the crucial means with which explosively increasing complexity can be overcome" (MALIK 1986).

2. Structure of organizational systems

2.1. Local theory

There are differing views on the issues of organization, structure and identity of systems. MATURANA & VARELA differentiate between the terms 'organization' and 'structure', organization signifying „those relations that must be present in order for something to exist“ (1987, p. 42), giving identity to the system and remaining constant, while the structure is „the components and relations that actually constitute a particular unity and make its organization real“ (1987, p. 47). Hence, structure is the current status of a system and can continually vary. GOMEZ & ZIMMERMANN (1993, p. 112-113) view the development of identity and contextual adaptation as opposite poles of one dimension of organizational structure. LUTZ (1992) considers the structure of social systems to be equivalent to that of personality or identity for an individual, something that evolves over all the changes and roles that a person goes through in the course of a lifetime.

> Lutz describes identity as consisting of a „code“ that prescribes „which observations in the surroundings are taken note of, how and according to which values and priorities impulses are transposed into enrichment of personality, or in actions and statements directed toward the surroundings, and which patterns are significant. It is then the conscious and unconscious perceptive and behavioral patterns, the conscious and unconscious ideas about reality and oneself, as well as the ideas about what good and bad is, what pretty and ugly is, etc.“ (LUTZ 1992, p. 317-318).

The development of the collective self of social systems is similar according to Lutz, the „culture“ of a nation, a family or enterprise being its code. „One could say that the culture, in the sense of the code, is the structure that forms the processes in the system at every moment and that is, in turn, during constant inner and outer conflicts, subject to irreversible changes (LUTZ 1992, p. 318; cf. 'organizational culture' MERKENS 1989). In any case, the *structure* of a system is closely associated with its identity: „through specification of structure, a system individualizes itself“ (LUHMANN 1992, p. 172). It constitutes a form of subjective code of guidance and is capable of change.

A plethora of authors in both the social sciences and engineering disciplines deliberate on similar ideas of subjective frameworks, but again, the terminology is neither precise nor consistently used. Other examples are „reference frame“ (MALIK 1986, p. 21), „codes“ or „programs“ (LUHMANN 1995), „norms and values“ (KATZ & KAHN 1978), „body of thought“ (WEICK 1979) „perspective“ (FLOYD 1989, p. 8) and „organizational character“ (EMERY 1959, p. 44). The term „local theory“ will be used in this context, taken from BAITSCH (1993). He defines *local theory* as being the immaterial gestalt of a work organization, embodying an area of consensus. „The degree of explicitness and consciousness can differ. But in any case, the logic of this reference system is recursive in enlarging or completing the system" (BAITSCH 1993, p. 26). Although it is subject to change, it is relatively stable with respect to time.

When a work organization is set up in the first place, it is already biased by the selective premises incorporated in the already existing individual or collective local theory of the people involved. „On this basis, aiming toward a particular goal, a particular structure for the work organization is conceived or copied and realized" (BAITSCH 1993, p. 20). In this sense, establishing an organization never occurs completely „out of the blue", but always starts with and carries the selective perspective of the originators, a bias. The role that perception plays is crucial because it is always selectively influenced by societal values, norms and expectations. GOLEMAN (1985) refers to this structure as a frame:

> „The frame gives us the context, telling us how to read what is going on. [...] The world offers a vast amount more that we might attend to in any given moment. The frame is highly selective; it directs attention away from all the simultaneous activities that are out of frame [...] These parallel tracks – in frame and out of frame – create a structure in social awareness that duplicates the division within the mind between conscious and unconscious. What is out of frame is also out of consensual awareness, in a sort of collective nether-world." (GOLEMAN 1985, p. 201-202).

Thus, the implicit values and norms of the general paradigm that prevail in the environment (cf. MATURANA & VARELA 1987, p. 95), particularly those that are unconscious, automatically flow into the design of the new organizational system. Since the turn of the century the prevailing paradigm has been, of course, one that incorporates a mechanistic view of the world. The resulting organizations have been predominantly tayloristic – characterized by highly specialized division of labor.

2.2. Expression of the local theory through material and immaterial means

When an organization is set up, it has already reduced the complexity of the task at hand by incorporating its unique local theory, and sets out on a path that has thus been – more or less consciously – chosen. The first decisions that are made already set tracks that pave the way for further development. The organizational and procedural structure – what constitutes a department, which processes belong together and which are separated, how the interfaces between the separate stages of processing are designed, all the way down to how partialized the individual work tasks are – all emanate from the local theory. In this way, the local theory takes on structural form in the organization and can be – must be, in fact, to secure the survival of the organization – transferred on to the other elements of the system. These are, in addition to the organizational structure and procedures themselves, new employees and their assigned tasks.

> „Members of the organization that have been with the company for long periods of time guide a new employee in two ways. First, the new member looks to them as an example of how interactions and interpretations of tasks and processes normally take place. Secondly, they influence the new person directly, introduce him or her in a manner that can have similarities with initiation rituals. [...] The supervisors have a special role; they are agents of the system as a whole, that are qua position responsible for the transference of the local theory. They have the job, in particular, of passing on certain immaterial meanings and to stop the development of others. The context of patterns of interpretation are not only transferred by persons. That which has materialized plays a particularly important role. The actual tasks and the conditions under which they are executed, claim the largest portion, with respect to time, of the interactions of a new employee and builds an intensive structural coupling, whose factuality cannot be denied over a long period of time.“ (BAITSCH 1993, p. 64).

Additionally, the technical tools comprise what is probably the most obvious part of that which has become material evidence of the local theory. In accordance with the attitude towards automation – which tasks will be executed automatically and which manually – irreversible decisions are made as to the characteristics of the tools that are chosen for supporting those tasks. Investments „for the future“ are made and therewith switches set for the following generations. Moreover, the „transference of system immanent meaning that occurs via the material evidence is independent of people. This increases the efficiency, because material evidence is difficult to refute and difficult to discuss“ (BAITSCH 1993, p. 31-32).

The interactions between these three system component types also serve to reinforce and propagate the local theory. The socio-technical system theory (EMERY 1959) emphasizes the inseparable interrelationships between the organizational, technical and human aspects: TRIST'S study broke with the tradition in social scientific analysis that the technology and its requirements are relatively unimportant compared to the social and psychological situation of men at work „by treating the problem as one of relating two systems, themselves part of a more inclusive system.“ (EMERY 1959, p. 9) „So close is the relationship between the various aspects that the social and the psychological can be understood only in the term of the detailed engineering facts and of the way the technological system as a whole behaves in the environment [...].“ (TRIST & BAMFORTH 1951, p. 11, cited in Emery 1959, p. 9).

How tight or loose the interdependencies within and between the organizational, technical and human aspects of the system are has been found to be an important factor for survival of an enterprise.

> EMERY considers the „technological component, in converting inputs into outputs [to] play a major role in determining the self-regulating properties of an enterprise. It functions as one of the major boundary conditions of the social system of the enterprise in thus mediating between the ends of an enterprise and the external environment. Because of this, the materials, machines and territory that go to making up the technological component are usually defined [...] as 'belonging' to an enterprise [...]. They represent as it were, an 'internalized environment'" (EMERY 1959, p. 5).

Thus, the 'tightness' within and between the system components is directly tied to the openness of the system.

Similarly to the axioms of the socio-technical system theory, in WEICK's theory of organizational action the „proportion and meaning of socially shared assumptions about tasks characterize 'the' organization because they indicate the extent and intensity of the real dependencies in the system. [...]." WEICK believes that the chances of survival are greater for loosely coupled systems than those of strongly coupled. Disturbances in processing remain more or less confined to single parts of the organization or need – if they propagate – more time to infect the other parts of the organization." (HOLLING & MÜLLER 1993, p. 62). The key issue is allowing for enough latitude for arising disturbances to be regulated locally (cf. SCHÜPBACH 1994), which requires, of course, a certain degree of openness.

Although the physical system components, through their independent presence as „facts", make a strong contribution to the internal stability of the system, they, at the same time, make the system vulnerable. This double characteristic, called *duality*, also pertains to all system elements and is closely tied to the openness, complexity and uncontrollability of organizational systems.

> „*The need and the capability of people to design and define reality can neither be totally suppressed nor be completely to the service of the work organization*. People always remain a source of deviation and disturbance [...]. Within a work organization, all the members try to influence its design, either in the sense of stabilization or denial. Likewise, the material evidence develops direct and indirect effects. Therefore, in every work organization phenomena occur that are neither planned in advance nor anticipated."
> (BAITSCH 1993, p. 185).

The various system elements are continuously involved in a process of supporting their own meanings and those for each other in an effort to secure the survival of the system, which, of course, requires communicating the interpretations of both the immaterial and material components between the carriers of the local theory. But this transference of meaning itself is of dual nature – not only is the „interpretation of material evidence [...] actually open, i.e. it doesn't induce a particularly fixed meaning" (BAITSCH 1993, p. 32), the people who are responsible for passing on the message are not always reliable, let alone controllable. People, as elements in the system, are themselves complex systems, each bringing in his or her specific perspective. Furthermore, the fact that communication processes between people are not always direct and clear makes the situation even more complicated. Ambiguous contexts allow much room for free translation – misunderstanding and contradictions are apt to arise.

But once again, the dual nature of the system is at work. This inherent openness for interpretation is also essential for the survival of the organizational system because it provides the potential which makes the system capable of change and therefore adaptable in a changing environment.

> „The unreliability of the people describes only one side of the picture, because it also guarantees the input of deviating information: extraneous information flows in through the people. How the work organization deals with foreign, potentially *disturbing* but also potentially *helpful*, information has a determining influence on the development of the system. The ability to reflect puts people in a position to exert influence on the system. The more comprehensively this capability is permitted and utilized, the greater the redundancy of guiding power in the system." (BAITSCH 1993, p. 32-33).

3. Dynamics of organizational systems

3.1. Dynamic nature of structure

In the previous section, the formation of the structure of a work organization was presented. Different types of system components as well as the beginnings of the process of giving them and their interactions meaning were discussed and with that the theme of dynamics already touched on. The structure of the organization not only includes the formal organizational design, but the norms and values, prejudices, rules for behavior, and even the rules about what is perceived and what is ignored. Hence, the term *structure* itself is then understood to be dynamic, not static. It is „all types of relationships that contribute to the rise of certain contexts of order or patterns, independent of whether or not they can be directly perceived and visualized or not" (MALIK 1986, p. 174).

Structure, in this sense, is not only a direct product of the local theory, it also propagates the local theory in that it provides the instructions for the further actions, the self-regulation, of the system. „Structures, then, are premises that serve the regulation of internal processes. They are generalized expectations of behavior, limiting possibilities, making connections between actions (events), making it possible and reproducing it“ (EXNER, KÖNIGSWIESER & TITSCHER 1992, p. 211). Furthermore, by reproducing the local theory, it also influences it – and the duality of the structure comes into play.

> „[...] the procedural and dynamic aspects of the work and organizational structuring as well as of the use of technology play an important role not only as objective and direct determinants and conditions of individual work tasks. They also influence constructs such as the 'work atmosphere', the 'organizational culture' and the 'mentality' towards given or chosen actions, preservation and persistence or change, learning and development, status positions or flexible task orientation [...], thereby in subjective form directly influencing the understanding of the jobs and tasks.“
>
> (SCHÜPBACH 1994, p. 18).

Again, the network of supporting relationships between the different components arises. These interrelationships, characterized by the tightness or looseness of their interdependencies, is called *structural coupling* (cf. MATURANA & VARELA 1987, p. 75 & 193) and have a stabilizing function for the system:

> „The structural coupling [...] implies that the change of one element brings on that of another. In other words: no component can be changed without the others. This yields the further important effect of a growing resistance of the quality of system toward change, all the more so since the material evidence increases. Systemically speaking, this corresponds to the expansion through the development of new components reproductively. [...]. If this dynamic ran on uninterrupted, the system would end up in a state of stagnation and consolidation of structure [...].“
>
> (BAITSCH 1993, p. 25).

However, the elements themselves, as well as the interdepencendies between them, are also of a dual nature. They both enable and limit the survival of the organizational system with the passing of time. „[...] every structural change occurs in a living being necessarily limited by the conservation of its autopoiesis; and those interactions that trigger in it structural changes compatible with that conservation are perturbations, whereas those that do not are destructive interactions“ (MATURANA & VARELA 1987, p. 100). System elements, on the one hand, are compelled to conserve the structure as it is, but they also have to perpetuate the system, which means perceiving and adapting to changes on the other hand. But „the changing of actions (system elements that only have one point of existence) alone does not signify any lasting changes in the state of the system“ (EXNER,

KÖNIGSWIESER & TITSCHER 1992, p. 218). Any non-superficial changes require changes to the structure. „The interventions in a system always have to be structural if they are to be effective“ (MALIK 1986, p. 174). This, in turn, implies changes in the local theory.

This ability to allow the local theory evolve with the system is what makes up the crucial difference between the self-regulation of living and the control of artificial systems.

> „[...] self-organization with the help of control loops also exists in mechanistic systems, as is made obvious by the large number of automatic controls around us. What is the difference between these types of self-organization from those of living systems? A rigid system code. A system code contains all the information that determines how a system processes impulses from the environment [...]. Mechanisitic systems generally have a rigid, extraneously designated code. Living systems have a self-organized code that [...] holds itself constant or [...] is able to develop and enrich itself.“ (LUTZ 1992, p. 315).

Furthermore, unlike „biological or physical systems, social systems don't have a solid structure that exists when it is not functioning“ (HOLLING & MÜLLER 1993, p. 54). While the material components, for the most part technical elements, are visible when not in use, the interpretations that give their neutrality meaning are only present when the people that carry the local theory are present. In this sense, the employees fulfill the function of the „memory“ for the system, the understanding of the code is „stored“ in them. Not only is the local theory „stored“ in the minds of the members of the organization, it is also carefully kept and executed in their patterns of behavior. GOLEMAN points out that a paradigm is the „sum total of shared schemas“ and that it „resides not only in the minds of each member, but also in the interaction between them.“ (1985, p. 171).

3.2. Roles and rules

The idea that roles constitute the formal structure of organizations (KATZ & KAHN 1978, p. 45) aids in describing this invisible, compelling framework.

> „Roles are designated to be specific forms of behavior, that are associated with certain tasks and ultimately traced back to the task requirements. They are standardized patterns of behavior that are demanded of persons within the functional network. Roles are independent of personal wishes or interpersonal duties, that don't have functional meaning. The assuming of roles by the members of the organization takes place through the communication of the expectations of the roles. These are defined as expectations of the functional associations between members of an

> organization with respect to the fulfillment of the roles, tasks or the underlying capabilities.“ (HOLLING & MÜLLER 1993, p. 54).

Thus, the system only „lives“, i.e. functions as a system, with the presence of the people – and only then can the theories of living systems be applied.

With the description of roles and how the actors are to play them, the system specifies the rules of the game. This ordering according to rules is seen by MALIK as an achievement in orientation, crucial in overcoming complexity and hence to survival of the corporation. He notes that the „rules that determine behavior are usually of a negative sort. Instead of postulating what is to be done, they more often postulate what is to be avoided and omitted. These rules have the character of prohibitions and determine the scope of acceptable and safe actions“ (1986, p. 41). In other words, the rules state that everything that is „out of frame“ is off limits, and more often than not, even questioning the boundaries themselves is taboo.

In more detail, three types of rules are described by MALIK, the first only existing as unconsciously but being observed, the second becoming conscious only after long period of unconscious patterning, and the third being of more of a formal nature, being consciously articulated and only being followed by action later on:

> „The first [type of] rules, [...] are actually being observed, but are never articulated in words and possibly cannot be expressed. An example for this first case is the capability of a child, to use grammatical rules long before it is able to understand and articulate them [...]. The second type of rules are those that are indeed articulated and expressed in words, but often only after a long period of time in which they have actually been observed; long after a subjective reality had been constructed, that made it possible to formulate them. The third art of rules are those that are first formulated and only then, usually after a formal ritual of inauguration, are observed.“ (MALIK 1986, p. 44).

Because the rules exist at different levels of consciousness, access to influence on them, making re-shaping possible, must also exist at different levels. This differentiation is important for the discussion of organizational strategies that will be taken up later on.

ORTMANN ET AL. (1990) see these rules of perception and forming as comprising, along with resources, power structures. They describe organizations as „arenas of micropolitical negotiation processes and struggles, in which everyone plays ‘his’ or ‘her’ game, and the whole only functions if the games in the organization are favorably structured and arranged with each other“ (1990, p. 8). They outline two forms of games, „innovation games“ and „routine games“. Routine games are those that are „played“ on a regular basis, those that make up normal daily life, so to speak. Each group in the organization plays their own game with their own modified rules. Innovation games are played on a different level, and usually involve actors of upper management, where „change, improvement, modernizing is their business" (ORTMANN ET AL. 1990, p. 8). This includes, in particular, the introduction of new material components.

People, however, are inevitably unable to identify themselves with the local theory so strongly that they personify and propagate it one to one. Although they normally do strive for the goals defined by the organization – after all, they have agreed to join the ranks – they are also caught up in competition with other members of the system in an effort to achieve and maintain their own identities and advantages for their own conditions of existence. In other words, part of the game is a struggle for power. Power can be defined as „every chance within a social relationship to have one's way, even against resistance [...]. Control means the chance for an order with a specific content to find obedience with given persons" (WEBER 1972, cited in ORTMANN ET AL. 1990, p. 13).

Power not only exists on a individual level, it is the roles that bestow the actors with *structural power*. If one „looks at power as a binary, dyadic relation, where it is always possible to identify an offender and a victim, one emphasizes its *(inter-) personal* aspect. The alternative would be to speak of structural power, which is hidden in technology, norms etc." (NEUBERGER 1989, p. 301). Structural power is hence tied up with the local theory.

The exertion of power is „selection of behavioral premises for another" and is „dependent on system structure in this function [...] systems produce power through selection of certain constellations of alternatives [...]. Power is consequently *structurally dependent selection*" (LUHMANN 1969, cited in ORTMANN ET AL. 1990, p. 20). And, of course, „all micropolitics have to deal with this structurally dependent selection. Vice versa, all organizational redesign and introduction of new technologies always mean structural changes and consequently, changes in the structurally dependent selections that the system produces" (ORTMANN ET AL. 1990, p. 21). Even the selection of what terminology is used is an expression of power. The meanings of the words in the local jargon and the contexts that they are used are also part of the local theory. They serve to identify who is an insider and who is excluded.

Thus, underlying the contest of power is a struggle over who actually has influence on the local theory and in what way. These roles are undoubtedly very powerful. But once again, things are not necessarily what they seem; duality deceives.

> „Power relations in social systems can be regarded an relations of autonomy and dependencies; but no matter how imbalanced they may be in terms of power, actors in subordinate positions are never wholly dependent and are often very adept at converting whatever resources that possess into some degree of control over the conditions of reproduction of the system. In all

social systems there is a *dialectic of control*, such that there are normally continually shifting balances of resources, altering the overall distribution of power."

(GIDDENS 1982, cited in ORTMANN EL AL. 1990, p. 15).

„A member has power within an organization if he or she controls relevant zones of uncertainty for the organization or other members of it: the data processing expert, who can call on his or her expertise and the indispensability of computers; the salesman with exclusive client contacts and knowledge of the market; the member of the board with the urgent contacts at the bank; the supervisor of the corporate hierarchy with his or her power of sanctioning and gratifying subordinates, (who can, in turn, have positions of power because something is expected of them); the maintenance personnel, who – in the extreme case solely – can avoid the collapse of production." (ORTMANN ET AL. 1990, p. 15).

Thus, the structure of the roles only exists because the actors are willing to play them in accordance with the script, the expectations of them, and in doing so they reproduce the structure and close the circle. The fact that „these actions can and must rely on the unequally distributed resources that are made available by this existing power structure explains its perseverance." (ORTMANN ET AL. 1990, p. 12). The structure, by its very existence, and particularly the interactions that are carried out unconsciously, because they are never questioned, is quite inert.

At the same time, though, the actors do have a degree of control over their own actions and therefore potential to influence the system:

„The process of [...] self-conservation is not centrally guided, it is anchored *polycentrically*. [...] The process of steering the system is especially carried by [...] the people, independent of the positions that they have. They embody the redundancy of *guiding potential*. The degree of heterarchical guidance can vary, democratic control is theoretically possible. The amount of competence (in its double meaning, both means and ability) given to the people is the crucial issue." (BAITSCH 1993, p. 30).

The mechanisms of micropolitics help explain the strong resistance to change that project managers or other change agents are often confronted with. „In order to understand the background of resistance, one can understand change processes as exchange processes that go from a given equilibrium through imbalance toward a new equilibrium: The new equilibrium can be worse, the same or better for an individual that the old one" (HANSEL & LOMNITZ 1987, p. 153-154). Thus, the power struggle is triggered, and groups that are confronted with the change, particularly those that fear losses in their advantages can only react with whatever resources are available to them. If the only available tool is opposition, then that is bound to be used full force – either openly or concealed.

3.3. The influence of time

The local theory, then, describes the perceptions, the roles, the resources and the patterns, setting off a dynamic that, while each member in the system has influence, no single person actually has control over. In this way, it utilizes the aspect of time in the transference of meaning inherent to the system, and in doing so creates a history for the period that it has survived. The actions can only be understood in the context of their patterns, it is the patterns that give them meaning, not only of the interactions between the elements at a given point in time, but the interactions over time.

> „Indeed, the whole system is not manifest for the individual and is also not observable as a whole with sense perception. It reveals itself such that single acts make no sense if they weren't elements of a comprehensive pattern. The employee doesn't behave with respect to single cases, as if there was no such thing as past and future. The single case is a constituent in an entire system of events that are related to one another."
>
> (MALIK 1986, p. 50).

The duality and the circularity features of the system mean that the local theories are themselves products of history, they evolve with time. Social consciousness is similar to individual consciousness and GOLEMAN points out that „cognitive development [...] is cumulative; understanding grows out of what has been learned already. We have become who we are, learned what we know, by virtue of the schemas we have acquired along the way. Schemas accrue with time; the schemas we have at a given point are the end product of our particular private history" (1985, p. 75; cf. MARCH & SIMON 1993, p. 28). This idea, in simplified mathematical form, has also found expression in technical control theory in the form of an integral transformation algorithm, where the output variables are a function of the development of the input variables with time (cf. HIRSIG 1989, p. 5.72).

With time, a plethora of new elements – both material and immaterial – are integrated into the system, patterned after the local theory and a subjective reality (cf. MARCH & SIMON 1993, p. 159-160) is built up. „Differences between organizations in leadership, communication [...] and organizational design can often be traced back to their histories of origin" (NICHOLSON 1989, p. 338). Moreover, the tactics that have been used in the past – the patterns – are carefully recorded, put on file and elicited as needed. After all, they have always worked before.

„Interpretations are quasi-historical. Much organizational sense-making consists of writing histories. A history is a kind of cause-effect map that depicts events in sequence. Retained histories can clarify either a tangible present outcome or a future outcome that has to be imagined. History in hand, people who select interpretations for present enactments usually see in the present what they have seen before. In the strategy formulation process in organizations, hindsight is a primary determinant of current planning. Outcomes from previous decisions, especially the mistakes, are disproportionately salient in current thinking. Musing such as 'We should have gone into home computers 5 years ago,' and Digital watches were a loser,' have greater bearing on current interpretations that the current events being interpreted." (WEICK & DAFT 1983, p. 75-76).

The existing perspective of reality supplies stability because it acts as ballast in a turbulent sea and a map for orientation. An illustrative example of the weight that historical development carries is given by the sequence of the uppermost row of letters on a typewriter. The explanation for the sequence QWERTY is simply a historical one:

„It was introduced as a solution to a problem of the early typewriters: the keys often jammed. That problem was supposed to be minimized by separating the keys that often followed one another. A couple of years later general technical advances overcame the problem of the jammed keys, but QWERTY remained. [...]. If, on the other hand, one talks with people about the QWERTY phenomena, they justify it with 'objective criteria' [...]. They say that one thing 'optimizes' and another 'minimizes'. Although these justifications have no rational basis, they depict a process, a social process of the making of myths, which allows us to find justification for primitiveness in every system."

(PAPERT 1982, cited in ULICH 1986, p. 161).

WELTZ denotes the discrepancy between a formal and an informal reality as *double reality*, and warns that: „by neglecting the double reality, a too smooth, streamlined image of rationalization processes could be painted, a rather flattering *overemphasis of rationality in the company's approach*, implying a heroizing of the efficiency of management (1988, p. 102). The need to justify past events, to re-interpret them so that they fit in with our notions about the world, is apparently very strong in us. „Every decision maker will value, after the fact, being the one who knew before – so that he or she can take credit for success, and account for failure as the taking of the smaller evils among the alternatives. The tendency of subsequent fitting of decisions in the current state of information certainly has psychological roots." (LUHMANN 1992, p. 167). It stems from a need for harmony;

discrepancy is difficult to endure. It makes us uncomfortable because it may lead to difficult changes, to unknown states – and so is often repressed, denied, refused or projected.

> „The work on the reconstruction of the 'sense' of social systems carries the inherent tendency to polish history and to orient the streamlined enterprise toward a planable future. If a process leads to such ideas, then the fact that every interpretation is capable of contradiction, that every system has very inconsistent expectations of actions: The workers should be creative but think along the given lines; they should keep the designated rules of conduct and at the same time mobilize their potential; the company must grow or even gain new markets but nonetheless be secure."
> (EXNER, KÖNIGSWIESER & TITSCHER 1992, p. 217-218).

It should be becoming slowly apparent that the recognition of inconsistencies are the first step toward change. The need for harmony, for the tuning of the picture of reality and the local theory, can then become the motivating factor that leads toward a change, solving the tension between the new realization and the old interpretation. Baitsch considers organizational learning to be based in the interaction processes that take place within a work organization or between the organization and its environment, where internal or external discrepancies potentially occur. The pre-requisite for a learning process is „the arisal of an *inconsistency*. Inconsistencies are deviations from or disturbances in the normal interaction process. Development and organizational learning occurs during the process of dissolving the inconsistency. The path of learning processes and organizational development depends on how an organization deals with contradictions" (BAITSCH 1993, p. 33; cf. SCHÜPBACH 1994, p. 12).

As was already pointed out, perception is colored by the local theory, which includes what is perceived as a contradiction and what is not. By choosing not to recognize a discrepancy, a choice has already been made about how to react to it – namely, not doing anything about it. If an inconsistency is recognized, it can still be ignored and consequences avoided, or it can be confronted, dealt with and structural changes made. This process is made even more complicated by the fact that people are the promotors and carriers of it. Structural changes are made through changes in the interpretation of reality through these people, which means changes in the local theory (BAITSCH 1993, p. 35-38).

Furthermore, when changes are introduced to the local theory, earlier renderings will continue to exist as parallel structures of interpretation. Employees might interpret part of a change that occurs, new working tasks or responsibilities for example, in accordance with the previous mindset because they have not yet incorporated the new thoughts. Thus, the same events and introductions of material components to the system can have different meanings for different people at the same time. Organizational and psychological change processes and their parallel courses of development are very complex and go beyond the scope of this work. Attention will instead be turned to the planning and decision processes that occur in the context of organizational systems, i.e. their structure and dynamics, that has been thus far described.

4. Conclusions

In contrast to the mechanistic paradigm, in which rationalists consider themselves in a position to indicate formulas for the successful execution of organizational and technical projects, only more general recommendations are appropriate in the systemic approach. Every proposed change in the organizational system can only be considered in its local context at that particular point in time that it comes into question. However, design suggestions that enhance the polycentric system processes themselves are pertinent. ULICH's postulate can again be adduced here as a basis for organizational redesign – decentralization at the enterprise level, functional integration for organizational units, local regulation in group work structures and qualified work and complete tasks for individuals (ULICH 1994, p. 290ff).

> „One design consequence of the understanding of the concept of limited rationality would be replacing centralized structures of supervision for complex work systems through decentralized structures. More local regulation and decision autonomy at various levels of the organization also have, in general, a greater distribution and diffusion of the demands of problems. [...] Design models such as semiautonomous work groups, quality circles or project teams are some other examples of application, even though they seldom refer to theories of rationality as a basis."
>
> (HOLLING & MÜLLER 1993, p. 60).

The growing turbulence of market demands require, in turn, more and more flexibility from enterprises, organizational units, work systems and individuals. „The more open the subsystems and elements, the more they must be able to act situatively, and the larger their latitude for local regulative behavior (autonomy) has to be" (SCHÜPBACH 1994, p. 49).

A design strategy then, to be effective, must include not only visionary, long term goals from the management perspective, it also has to incorporate the perspectives of the various levels in the organization. „Holistic, consistent concepts for the structural and procedural organization of production, of work design and of the use of technology, that meet the criterion of short term efficiency as well as the longer term capacities of learning and development, must be backed by management as strategic guidelines as well as being implemented locally in the form of concrete design and development projects. Such an approach corresponds with the idea of a *combined strategy*, 'top down' *and* 'bottom up'" (SCHÜPBACH 1994, p. 20; cf. MALIK 1986).

The point that the introduction of new technologies and organizational changes are not merely „logical" problems, they are political issues has been clearly made. Thus, „one of the central issues is who should be involved in the decision making and in what way: management, computer experts and users" (ALGERA, KOOPMAN & VILJBRIEF 1989, p. 88). A further group that cannot be neglected is that of external consultants - „neutral" consultants as well as delegates from the producers of hard- and software. Although mixed results from empirical studies on participation have been reported (ibid. 89), there are several reasons that do speak in favor of participatory project management: the improvement of communication and information (CASANOVA 1986, p. 58) and the quality of decisions, convincing users of the need for a change (ORTMANN ET AL. 1990, p. 71) and stronger motivation and commitment toward the change (BRUNSSON 1989, p. 208). FLOYD goes even further, and speaks of design in dialog in project groups:

> „Dialogical group work cannot spring forth from implicit assumptions, instead it must be the basis for the common task itself. This means that defining and establishing the project, responsibilities, the division and coordi nation of labor, the conventions and standards must be done collectively for the common tasks ... The prerequisite for design in dialog is trust. It can only originate where the interests of the participants are taken into consideration, and demands the willingness to create and maintain a socially supportive atmosphere of the participants, particularly the project leader." (FLOYD 1989, p. 16).

Furthermore, participation process itself is a learning process for all participants. This was reported as one of the main results of the research project 'qualification in work tasks': „While we originally assumed that - through participation - work structures could be designed that ... intrinsically enable qualification inherent to the work, we know today that it is first - sometimes foremost - the *process of the participation* in the design of the work itself that creates the inherent possibilities of qualification" (FREI 1986, p. 18). Hence, the increasing awareness of the other perspectives involved in the design process can be one of the most important ends in itself, particularly for interdisiplinary design teams.

5. References

ALGERA, J.A., KOOPMAN, P.L., VILJBRIEF, H.P.J. (1989): Management strategies in introduction computer-based infromation systems. Applied Psychology, 31:1.

BAITSCH, CH. (1993): Was Bewegt Organisationen? Selbstorganisation aus psychologischer Perspektive. Campus, Frankfurt.

BRUNNSSON, N. (1982): The irrationality of action and action rationality: Decisions, ideologies and organization action. Journal of Management Studies, 19.

CAPRA, F. (1992): Systemisches Denken – das neue Paradigma. In: KÖNIGSWIESER, R., LUTZ, CH. (Eds.): Das systemisch-evolutionäre Management. Orac, Wien. p. 307-310.

CASANOVA, R. (1986): Phasen im Prozess geplanter Veränderung. In: DUELL, W., FREI, F. (Eds.): Arbeit gestalten – Mitarbeiter beteiligen. Campus, Frankfurt. p.-45-59.

EMERY, F.E. (1959): Characteristics of Socio-Technical Systems. Document No. 527. Human Resources Centre, Tavistock Institute, London.

EXNER, A., KÖNIGSWIESER, R., TITSCHER, S. (1992). Unternehmensberatung – systemisch. In: KÖNIGSWIESER, R., LUTZ, CH. (Eds.): Das systemisch-evolutionäre Management. Orac, Wien. p. 204-235.

FLOYD, C. (1989): Software Entwicklung als Realitätskonstruktion. In: LIPPE, W.M. (Ed.): Software Entwicklung: Konzepte, Erfahrungen, Perspektiven. Proceedings der Fachtagung der GI. Springer, Berlin. p. 1-20.

FREI, F. (1986): Zum Forschungsprojekt 'Qualifizierung der Arbeitstätigkeit'. In: DUELL, W., FREI, F. (Eds.): Arbeit gestalten – Mitarbeiter beteiligen. Eine Heuristik qualifizierender Arbeitsgestaltung. Campus, Frankfurt. 12-23.

GIDDENS, A. (1982): Power, the dialectic of control and class structuration. In: GIDDENS, A., MAKENZIE, G.: Social Class and the Division of Labour. Essays in Honor of Ilya Neustadt. Cambridge.

GOLEMAN, D. (1985): Vital lies, simple truths. The psychology of self-deception. Simon & Schuster, New York.

GOMEZ, P., ZIMMERMANN, T. (1993): Unternehmensorganisation: Profile, Dynamik, Methodik. Campus, Frankfurt a.M.

HABERFELLNER, R., NAGEL, P., BECKER, M., BÜCHEL, A. & VON MASSOW, H. (1992): Systems Engineering: Methodik and Praxis. Verlag Industrielle Organisation, Zürich.

HANSEL, J., LOMNITZ, L. (1987): Projektleiter-Praxis. Springer, Berlin.

HIRSIG, R. (1989): Systemtheorie. Eine Einführung für Psychologen. Psychologisches Institut, Universität Zürich.

HOLLING, H., MÜLLER, G.F. (1993): Theorien der Organisationspsychologie. In: SCHULER, H. (Ed.): Organisationspsychologie. Huber, Bern. p. 49-69.

KATZ, D., KAHN, R.L. (1978): The social psychology of organizations. Wiley, New York.

KIESER, A. (1989): Organisationen und Organisationsgestaltung. In: GREIF, S., HOLLING, H., NICHOLSON, N. (Eds.): Arbeits- und Organisationspsychologie. Internationales Handbuch in Schlüsselbegriffen. Psychologie Verlags Union, München. p. 342-349.

LOUIS, E. (1992): Umweltschutz in Schweizer Unternehmen. Unpublished Thesis. Seminar für Angewandte Psychologie, Universität Zürich.

LUHMANN, N. (1969): Klassische Theorie der Macht. Kritik ihrer Prämissen. Zeitschrift für Politik, 16.

LUHMANN, N. (1992): Organisation. In: KÜPPER, W., ORTMANN, G. (Eds.): Mikropolitik: Rationalität, Macht und Spiele in Organisationen. Westdeutscher Verlag, Opladen. p. 165-185.

LUHMANN, N. (1995): Social systems (Writing science). Stanford University Press, Stanford.

LUTZ, C. (1992): Grundzüge des systemisch-evolutionären Weltbildes. In: KÖNIGSWIESER, R., LUTZ, C. (Eds.): Das systemisch-evolutionäre Management. Orac, Wien. p. 310-320.

MALIK, F. (1986): Strategie des Managements komplexer Systeme. Ein Beitrag zur Management-Kybernetik evolutionärer Systeme. Haupt, Bern.

MARCH, J., SIMON, H. (1993): Organizations. Blackwell, Cambridge.
MATURANA, H., VARELA, F. (1987): The Tree of Knowledge. Shambhala, Boston.
MERKENS, H. (1989): Organsiationskultur. In: GREIF, S., HOLLING, H., NICHOLSON, N. (Eds.): Arbeits- und Organisationspsychologie. Internationales Handbuch in Schlüsselbegriffen. Psychologie Verlags Union, München. 365-370.
NEUBERGER, O. (1989): Macht. In: GREIF, S., HOLLING, H., NICHOLSON, N. (Eds.): Arbeits- und Organisationspsychologie. Internationales Handbuch in Schlüsselbegriffen. Psychologie Verlags Union, München. p. 301-304.
NICHOLSON, N. (1989): Organisationaler Wandel. In: GREIF, S., HOLLING, H., NICHOLSON, N. (Eds.): Arbeits- und Organisationspsychologie. Internationales Handbuch in Schlüsselbegriffen. Psychologie Verlags Union, München. p. 336-341.
ORTMANN, G., WINDELER, A., BECKER, A., SCHULZ, H.J. (1990): Computer und Macht in Organisationen. In: Der Minister für Arbeit, Gesundheit und Soziales des Landes Nordrhein-Westfalen (Ed.): Sozialverträgliche Gestaltung. Westdeutscher Verlag, Opladen.
PAPERT, S. (1982): Mindstorms – Kinder, Computer und Neues Lernen. Birkhäuser, Basel.
SCHÜPBACH, H. (1994): Prozessregulation in rechnerunterstützten Fertigungssystemen. Verlag der Fachvereine, Zürich.
SCOTT, W.R. (1992): Organizations: Rational, natural and open systems. 3rd Ed. Prentice Hall, Englewood Cliffs, New Jersey.
SPRÜNGLI, R.K. (1992): Zur Evolution von Entscheidungen: Reflexionen aus der Praxis. In: KÖNIGSWIESER, R., LUTZ, C. (Eds.): Das systemisch-evolutionäre. Orac, Wien. Management. p. 57-68.
TIMPE, K.P. (1989): Systemtheoretische und ingenieurwissenschaftliche Grundlagen. In: GREIF, S., HOLLING, H., NICHOLSON, N. (Eds.): Arbeits- und Organisationspsychologie. Internationales Handbuch in Schlüsselbegriffen. Psychologie Verlags Union, München. p. 336-341.
TRIST, E.L., BAMFORTH, K. (1951). Some social and psychological consequences of the Longwall method of coalgetting. Human Relations, 4, 3-38.
ULICH, E. (1986): Neuorientierung des Managements. In: DUELL, W., FREI, F. (Eds.): Arbeit gestalten – Mitarbeiter beteiligen. Eine Heuristik qualifizierender Arbeitsgestaltung. Campus, Frankfurt. p. 160-169.
WEBER, M. (1972). Wirtschaft und Gesellschaft: Grundriss der verstehenden Soziologie. Studienausgabe, Tübingen.
WEICK, K. (1979): Cognitive processes in organizations. Research in Organizational Behavior, 1, 41-74.
WEICK, K.E., DAFT, R.L. (1983): The effectiveness of interpretation systems. In: CAMERON, K.S., WHETTEN, D.A. (Eds.): Organizational Effectiveness. Academic Press, New York. 71-93.
WELTZ, F. (1988): Die doppelte Wirklichkeit der Unternehmen und ihre Konsequenzen für die Industriesoziologie. Soziale Welt, XXXIX, p. 97-103.

Control Strategies for Complex Manufacturing Systems

Part Three

From Central Planning and Control to Self-Regulation on the Shop Floor

Heinz Schüpbach

The propagation of ever new individual approaches to work and organizational structuring as well as the implementation of new technologies have not solved the crises of shop floor management. Experience has shown that many enterprises have been tempted to address the mere 'symptoms', but not the real weaknesses of shop floor organization. Here we will attempt to outline these weaknesses – rigidity and restrictive utilization of human resources – from a developmental perspective, long-term and cyclical, which is well-founded in work- and organizational psychology. As an alternative, a decentralized, flexible model of 'self-management' on the shop floor will be introduced and discussed. It will be shown that from this, forms of work and organizational structuring may be derived which are not new, but as practice evidences, none the less promising.

1. Shop floor management: Current developments in a work- and organizational psychological view

Those who follow interdisciplinary discussion of the problems of shop floor management in the literature and in practice will have noted agreement, on the one hand, and diverging foci, assessments and future perspectives on the other. From a work- and organizational psychological viewpoint, this state of affairs may be summarized as follows:

1 It is agreed that the tasks of shop floor management may be outlined, in very brief form (see elaboration in SCHÜPBACH 1994), to consist in the part functions of decision, planning, proactive control, and reactive control of operations. These may be viewed under the aspect of information or under the

aspect of their material, objective content. In the forefront stands the question of the organization of 'information management' of the material, objective processes. Traditional models with an orientation towards central and deterministic decision making, planning and control are confronted by decentralized approaches which emphasize flexible self-regulation on the shop floor. While the former continue to be widespread (KIRSCH, STROHM & ULICH 1994), the latter are considered to hold more promise for the future. The unanimous reason given for this opinion is the fact of great change in the competitive circumstances in client and supplier markets.

2 Long lead times, high stock as well as low delivery reliability are held to be internal main symptoms or indicators of the weaknesses of traditional shop floor management. Significant improvement in the area of these variables is therefore seen unanimously as the goal criteria according to which the success potential of 'new' concepts be measured.

3 Several popular 'new' concepts – CIM, Lean Production, Total Quality Management and Business Process Reengineering, for example – claim potential success for themselves. Leaving the very different sorts of presentations aside, it seems to be considered just as important to propagate these approaches as to warn of their one-sidedness, inadequacies and repercussions. This creates uncertainties for enterprises, which then do not dare to make fundamental changes, but rather, limit themselves to partial optimizations and the implementation of new technologies.

4 There has been increasing agreement that new concepts of shop floor management must take aspects of the implementation of technology and personnel and qualification (training) as well as organizational structure into account and view them all in a related whole. In spite of this recognition, however, there is widespread doubt in regard to the reliability of, and skepticism with regard to the ability and willingness of, shop floor workers to work as a team, carry self-responsibility, show their own initiative and exercise task-related self-regulation. All agree that, for this reason, the introduction of semi-autonomous work teams is not at all as far advanced as would be expected according to the number of intentions which have been declared. Examples of enterprises which have made lasting investments in promoting their human resources are still rare.

5 One significant deficit seems to lie in the lack of willingness to expend efforts on careful theoretical foundations. Postulates relating new concepts of shop floor management to recently popular theoretical constructs such as complexity, openness, integration, systems, chaos, fractal, self-organization and so on seem to be readily made, but seldom examined in detail. The lack of solid theoretical foundations is, by the way, not a new phenomenon connected with the rise of new production concepts. The traditional forms and principles of shop floor

management, although developed at the turn of the century under the label of 'scientific management' (TAYLOR 1911), were hardly based on careful theoretical underpinnings, and astonishingly, this state of affairs was seldom criticized. It is, of course, generally known that 'TAYLORISM' has something to do with the mechanistic, centralist, and deterministic as well as excessive specialization and hierarchization. But elucidation of the premises and prerequisites that must be given for the functioning – long-term as well – of systems so conceived are missing. This leads not infrequently to a sweeping dismissal, incomprehensible to the practitioner, of traditional shop floor management, with no reference made to the increasing disappearance of mass or series production, simply structured products, program-oriented production planning, and so on.

6 Finally, it is interesting that there is a stubborn, persisting predominance of structural models. While all new approaches do emphasize their process orientation, up to now what is found is really a shift in focus from models of organizational structure to models of process structure. The predominantly short-term and linear design of these models continues, blocking sight of the long-term, inherently cyclical developmental dynamics of sociotechnical systems. Long-term developmental effects – in particular the way phenomena of a strongly social character, such as the culture of an enterprise and its management, which become set through positive and negative reinforcement – are left out of consideration, as are other longer term developments such as openness to innovation, and the attitudes and motivation of workers in regard to innovation, further training and qualification, team work, and so on. Unambiguously ordered and stable structures – for an order cycle, for example, structures analogous to a canalization system – take preference over more open arrangements with degrees of freedom and capacities for situation-appropriate self-regulation ('order processing') analogous to a 'natural watercourse with compensation basin'. The fact that stability can easily become rigidity and that local self-regulation is a necessary prerequisite of flexibility is often overlooked.

The interest of work and organizational psychology in the discussion on shop floor management is not only theoretical in an objective scientific sense. 'Subjective theories' stand behind shop floor management in every enterprise in the form of the experience and convictions of those involved. One such conviction is the idea that whatever has been held to be correct for a long time cannot suddenly be wrong. Workers who have proven themselves through many years of shop floor management and are proud of their accomplishments do not accept the accusation that they have actually done everything wrong. New management and production concepts have a chance of success in tradition-rich enterprises only if they can show, in a developmental process over the long term, that not human weaknesses, but necessary structural adaptations give the impetus for innovation. Workers must not be seen as victims, but as participants in the innovation process;

qualification measures as investiture and not mere cost factors. If not, these concepts are in danger of being rejected as 'new wine in old bottles'.

2. Traditional organization of shop floor management: Centralized planning and control

2.1. The principle of centralized planning and control

From an organizational perspective, shop floor management has the important role of a link between planning and implementation on the shop floor. For this reason, it requires closer examination. The cybernetic viewpoint (SCHIEMENZ 1979) is most interested in the management function of control on the shop floor (or, more formally expressed, in the interrelation between controller and controlled object). In the traditional model of shop floor control, it is assumed that prior to the start of an operation or a sequence of operations – the throughput of a manufacturing job through all workshops, for example – all required information to secure the availability of work materials and work plans as well as to release the order must be in place (calculability). Further, the material processes then set into motion must run disturbance-free (controllability) so that feedback from process, or process-related, information can be dispensed with. As a totally proactive principle, namely, control is 'blind' to deviations of the process from the planned input – analogous to the flow of water through a closed canalization system. The consequence for organization is that it becomes possible to separate planning and control (information management) completely from process management and to place it up front. Process management in the narrow sense must actually be dispensed with altogether, as the processes are supposed to run a practically inevitable course, with no active intervention, like a row of falling domino pieces (cascade effect). However, this holds only under the condition that the domino pieces have been set up in the exact, desired manner, and remain in place. In other words, the processes are totally calculable and controllable!

2.2. Diagnosis of symptoms: Buffers of time and material to cushion variations and disturbances

Empirical facts alone can answer the question of whether these prerequisites actually exist in production plants. We can assume that, partially at least, there was empirical support at the time of mass and series production beginning in this century – the birth of scientific production management. Evidence is found in the enormous production increases thanks to marked rationalizing and streamlining, higher wages despite sinking prices and sales figures rising over a long period. But conditions have been changing continuously over the past years. These changes became, and become, visible in the form of the unceasing increase in material inventory and time reserves (distribution times) between shop floor and individual work systems. With this, it became possible to ensure a certain degree of planning ability in spite of variations and planning uncertainties. The cost of tying up capital in inventory was passed on to prices, and customers were taxed with long delivery deadlines resulting from very high lead times (and delivery agreements often not met).

But now the tables have definitely turned. Prices are calculated with an ever-smaller margin, short delivery times and high delivery reliability are important competitive arguments. The consequences for the shop floor, namely shorter lead times and low inventory in the place of the highest possible capacity utilization, as well as contract-related disposition for small batches and products according to customer specifications, confront traditional logistics and traditional shop floor management with insurmountable problems (HACKSTEIN 1989; WIENDAHL 1995). Internal analyses within enterprises illustrate the dilemma: 85% of the lead time of a job, or even more, is allotted to waiting times caused by the organization of procedure, and material inventory is usually clearly too high, as shown by STOMMEL & KUNZ already in 1973. But neither waiting times nor material inventory may be cut down indiscriminately, because they represent absolutely necessary buffer zones which prevent that disturbances be passed along to subsequent work systems. It could also be shown that not only frequent rescheduling on the shop floor (for example, accommodation for job-related rush orders), but also failure in organization of supply of materials, tools, equipment, work plans, etc. to the work systems cost the shop floor about 20% of their planned work time in the form of organizationally determined idle time of machines and process engineering systems (compare summary in SCHÜPBACH 1994). More serious, however, is the fact that the wide spread range of idle times – weeks without any significant disturbance can be followed by days of idle time – makes it practically impossible to plan times too marginally. With this, it is clear that central and deterministic shop floor management hinders just that which it should achieve: a *high reliability of planning*. To return to our analogy, the canalization pipes become thicker and thicker, ever new conduction and supply pipes are built on, but still backups, burst pipes and flooding cannot be avoided (the analogy is not chosen by chance – the idea of 'Lean Management' starts here).

2.3. Diagnosis of basic weaknesses: Rigidity and restrictive implementation of human resources

On closer scrutiny, high lead times combined with a large portion of passive, unproductive idle times plus high inventory reveal themselves to be 'symptoms', and not the actual, underlying weaknesses of traditional shop floor management. An interesting question is why it took such a long time for these symptoms to be recognized. The reason is probably a simple one: no one was looking for them! They only become apparent, namely, when one's viewpoint shifts and enlarges from a short-term, narrow perspective of individual work systems or cost centers to consideration of total order processing (from calculating offers to delivery and start-up of the finished products). Instead of focusing on linear connections, one's viewpoint must be concerned with long-term, cyclical reinforcement or positive and negative feedback effects (in the sense of learning and unlearning) in a work organization that has taken shape over years of development – 'the steady drip carves the stone'. As mentioned in the introduction, in this perspective the particularly sensitive phenomena are overwhelmingly social in nature, such as the development of the 'culture' of the enterprise, management, quality and conflict, the openness of workers to innovation and further qualification as well as management's related view of human resources. Humans do not simply react impassively to the factors determining their actions (as implied by the control model), but rather adapt actively to them – even if this adaptation be resignation, disinterest or withdrawal. In the model presented below (Figure 1), such long-term and cyclical developmental processes stand in the foreground.

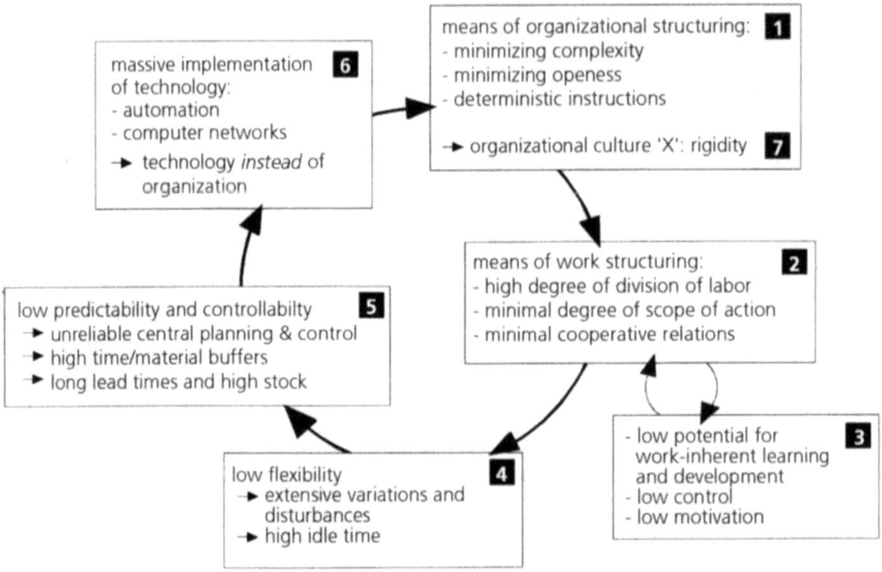

Fig. 1 Cyclical, long-term development model of traditional shop floor control.

The model's premises are the forms and principles described by traditional shop floor management, in particular the principle of central planning and control of the shop floor process. In practice, this results inevitably in the necessity to structure and organize production in such a way that – as established – the processes are calculable and controllable. The means of organizational structuring (1) begin with minimizing the dynamics of individual work processes. These is achieved, firstly, by minimizing complexity, that is, limiting the related net to the necessary hierarchical relations (note that procedures thus become more complicated, more elaborated and awkward and therefore, more susceptible to disturbance – complex and complicated are often seen as equivalent!). Secondly, openness is minimized, in order to exclude disturbance effects and to channel information flow. Too little considered is the fact that the effects of other important information are excluded as well. Thirdly, deterministic instructions and control are meant to ensure controllability of the processes. For the next step, work structuring in the sense of implementing human resources (2), this means: high degree of division of labor and specialization (reduction of horizontal complexity), minimal degrees of freedom of action and minimal cooperative relations and communication. The dominos must be set up in a row and, until the signal is given by control, stand still. Stable structures (analogous to a canalization system) are aids. Activity analyses in metal cutting production reveal that only 2% of operators' work time is filled by their own active planning and only 9% by cooperative/communicative part-tasks (SCHÜPBACH 1994, WEBER 1992). This work situation produces consequences for the worker which have been criticized by the work sciences from the start (3): Little work content and social isolation lead to atrophy of active ability to respond and to atrophy of qualifications. Low potential for work-inherent learning and development lead to loss of control over one's own activity and, thus, to a high degree of psychological stress and the loss of the kind of motivation that would grow naturally from the work itself. MCGREGOR (1960) described these effects in his 'Theory X' as a vicious circle: Humans adapt to the work conditions which limit them and allow only reaction and, with time, they cannot even imagine another mode of work wherein they would be self-sufficient and show their own initiative.

Current pertinent discussion of organizational culture (7) indicates that the repercussions on humans of work and organizational determinants should not merely be understood as an individual problem. Rather, in the course of shaping and reinforcing the relations illustrated in Figure 1 (called 'organizational culture X' to follow MCGREGOR'S 'Theory X'), typical features of organizational culture develop. NEUBERGER (1989) has shown that excessive implementation of traditional organizational principles does not only lead to desired (functional) results, but also to undesirable (dysfunctional) effects (Figure 2).

The effects of work and organizational conditions on the behavior and experience of workers on the one hand, and organizational culture on the other hand, mutually reinforce each other. Practice has shown that if 'bureaucratization' of an organization reaches a certain degree, the dysfunctional effects begin to gain the upper hand. In place of the desired effects, that peculiar frozen rigidity sets in which nips all flexibility of reaction to unforeseen variations and disturbances in the bud. This rigidity expresses itself not only in work and organizational structures, but also in the thinking and actions of workers, in the management style

Organizational principle	Desired effects (functions)	Unintended effects (dysfunctions)
Centralization	Unified management, Speed in decision making	Sluggishness, susceptibility to disturbance, disinterest, resistance
Stabilization	Long-term arrangements, reliability, ensured planning	Rigidity, lack of dynamics and adaptability, exclusion of alternatives
Routinization	Allows self-understood habits	Drop in initiative and creativity, ritualized behavior
Standardization	Norming, formalizing, standardizing	Pressure to conform, schematize, rigidity
Differentiation	Simplifying, separation into comprehensible units, specialization	Loss of meaning, significance, the whole picture, relations
Delimiting	Comprehensibility, ensured controllability	Prevents development, self isolation

Fig. 2 Functional and dysfunctional effects of one-sided implementation of traditional management principles (according to NEUBERGER 1989).

of shop floor management as well as in the culture of the organization, and it stands in sharp contrast to all that is generally meant by flexibility. Here a divide cuts between the claims and the reality of traditional shop floor management (see Figure 1, (4)). It has been shown that the day-to-day real world of the shop floor variations and disturbances, even incalculable and uncontrollable events. When planning and control is separated from execution at the level of the shop floor, variations and disturbances are discovered too late and dealing with them involves laborious procedures. This tends to aggravate their effects and to increase time lost; events intensify rather than being caught and mastered. HANDKE (1986) demonstrates that in production plants where there is strongly centralized maintenance in ending technically determined machine idle time, 22 % of repair time alone is reaction time and 20% is used to diagnose the problem. The intensification of events reduces the already limited calculability and controllability even further (5). The system begins to develop the symptoms described above. In order to at least offset the dynamics of intensifying variations and disturbances in a passive manner, sequences are cushioned with time and materials buffer zones. High lead times and inventory are inevitable concessions made by traditional shop floor management. HIRSCH-KREINSEN & VON BEHR (1988) show that many production plants approach these concessions in a very pragmatic way and experiment on the shop floor with new work and organizational forms. Here the attempt is made to plumb the depths of a compromise which, on the one hand, does not call the principle of central planning and controlling into question and, on the other hand, grants shop floor personnel as much freedom and flexibility as necessary. Experience has shown that such freedom is in fact utilized. Many changes of the schedule are made on the spot by means of spontaneous consultation, missing or faulty work documents, materials, tools and so on are purchased or repaired independently, and small repair jobs are undertaken without calling in formally responsible maintenance personnel. Formal and informal shop floor organization threaten to split even further apart.

Production enterprises do not dispute the fact that the situation is unsatisfactory and needs to be solved. The crucial question facing shop floor management is this: Should the premises of central planning and control of shop floor procedures be upheld, and must, therefore, new ways and means of more consistent execution of central planning be devised? Or should the premises be abandoned and alternative forms of shop floor organization and work design be found? The findings of relevant studies allow us to conclude that the majority of production enterprises, at least in the late 1980s, have tended to follow the first strategy (HIRSCH-KREINSEN & VON BEHR 1988; KIRSCH, STROHM & ULICH 1994). The new ways and means of executing central planning and control have been sought through increased implementation of information, communication and automation technology (6).

The following considerations and expectations have dominated the foreground:

1 The use of computer-aided production planning and control systems (PPC systems) was intended to support central information management. The expectation was that the flood of data would become more accessible, and that by means of implemented planning modules, planning data would be ordered and visible. Here it cannot be denied that the first generation of PPC systems at least, according with traditional organizational structures, were indeed designed for use in strongly centralized planning and controlling structures. That is, they supported sequential data processing 'from front to back', and perhaps in an information technical way, although not in the organizational sense, contributed to integration and flexibility. The goal was to remove humans from the direct flow of information and allow them to influence planning and control happenings only at certain defined interfaces.

2 The implementation of CNC control systems was intended to allow progress in the automation of machine processing, material flow and assembly processes. While CNC technologies were introduced mainly in order to increase productivity, they promised to also increase reliability (as opposed to the reliability of humans) and planning reliability. Here again the goals was to remove humans from direct process management and give them the tasks of supervision and maintenance at the periphery, workpiece inspection parallel to machine processing and workpiece change over between processing cycles. That humans thus increasingly forfeited their process-related qualifications and experience knowledge, was hardly considered (BÖHLE & MILKAU 1988; SCHÜPBACH 1994).

2.4. The implementation of new technologies as a solution?

Behind the term 'Computer Integrated Manufacturing' (CIM) stood the prospect of networking the two types of computer systems, thus allowing realization of general data flow between planning and control and the plant with its machines. It is interesting that here indeed the technical possibilities of flexibilization and integration (which is equivalent to an increase in complexity) were utilized, which the development and implementation of so-called 'flexible manufacturing systems' shows.

What has been the real contribution of the introduction and implementation of these computer technologies to production plants? First, we must establish again that many enterprises – according to their own statements – did not seek new work and organizational structures through implementing computer technologies, but rather wanted to support and strengthen the existing ones (HIRSCH-KREINSEN & VON BEHR 1988; KÖHLER 1990; KIRSCH, STROHM & ULICH 1994). One example can be given in the findings of a study by FRIEDRICHS & GROMOTKA (1992). They showed that the expenditure deemed necessary by the product makers for customer-specific conception and the introduction of manufacturing management systems as well as related training was underestimated by two thirds of the interested users. Moreover, closely tied effects regarding organization were underestimated by over 90% of the users! FRIEDRICHS & GROMOTKA (1992, p. 112) emphasize as follows:

> „Particularly critical is the finding that only 9% of the users have an accurate idea of the organizational repercussions of a system of manufacturing management techniques. Either no consideration is given to this aspect, or examination of the issue is left to the departments responsible. In many cases, the attempt is made to force data processing into existing structures, whereby many of the advantageous features of the systems are not utilized." FRIEDRICHS & GROMOTKA (1992, p. 112)

With regard to the goals connected with implementing new technologies, a representative study of Swiss capital goods production firms conducted by the Work and Organizational Psychology Unit (IfAP) at the Federal Institute of Technology (ETH) in Zürich confirms that the new technologies were procured with the aim of achieving improvements in the aforementioned 'symptoms' (Figure 3). The sobering fact is that a total of only a third of the studied enterprises state that they had, in fact, achieved this goal (KIRSCH, STROHM & ULICH 1994).

But the companies studied did report that the neglect of questions of qualification and work and organizational structuring resulted in a whole series of subsequent problems (KIRSCH, STROHM & ULICH 1994, cf. NUBER, KÖHLER & SCHULTZ-WILD 1989).

Company goal	Goal considered 'very important'	Goal achieved
Increase in delivery reliability	68%	29%
Decrease in lead times	65%	27%
Reduction of material inventory	48%	14%
Improved capacity utilization	41%	31%
Improved calculation bases	40%	41%
Increase in internal flexibility	39%	28%

Fig. 3 'Very important goals' and their achievement through the use of computer-aided production planning and control systems (KIRSCH, STROHM & ULICH 1994).

The findings suggest the conclusion that the implementation of new technologies does not replace lasting measures for structuring work and organization or for qualification, but rather at best can support such measures effectively. The question raised at the end of our previous section can now be seen to leave only one option: we must turn away from the premises of traditional shop floor management (from the concept of the central planning and controllability of shop floor processes) and seek flexibility-oriented alternatives.

3. Alternative approaches: Decentralized, flexible 'self-management' on the shop floor

Escape from the vicious circle of traditional shop floor control consists apparently in breaking away from the developmental logics and dynamics presented in Figure 1. In their place, work and organizational structures should be sought which aim towards flexibility and, in doing so, place emphasis on lasting measures to foster human potential. It is not a matter of simply abandoning traditional shop floor management for its opposite, extreme decentralization. Issues of coordination, that is, communication and cooperation in the sense of vertical and horizontal, technical and social networking would too easily disappear from the field of vision. Rather, it is necessary that an optimal balance be found between central planning and coordination in the sense of establishing framework instructions (for example, time 'frames' as opposed to time limits) on the one hand, and decentralized finite scheduling and material planning as well as self-regulation during implementation on the other. Finding this balance is the real 'art' of shop floor management (MALIK 1989). In Figure 4, in analogy to Figure 1, the cyclical developmental logics and dynamics of the path from traditional shop floor management to 'self management' is illustrated. The initial, or starting, link in this model is made up of the 'symptoms' of traditional shop floor management – high lead times and inventory ((5) in Figure 1) – and the goal values – significant reduction of these lead times and inventory through a high reliability of planning ((1) in Figure 4) in accordance with – and this may sound paradoxical – an increase in central coordination ability and the reliability of shop floor procedures. The key here is that no attempt is made to variations and disturbances, but rather, the solution is sought through development of the ability of shop floor workers to react quickly and flexibly to those variations and disturbances which inevitably occur.

In the area of organizational structuring (2), the issues of the complexity, openness and establishing of rules and instructions is raised once again. This time, however, the goal is their optimization, and not minimization. Optimizing complexity means that a measure of horizontal and vertical integration and networking, or communication and cooperation, must be found which would make synchronous and fast consultation, along the shortest routes, possible. This would allow up-to-date comparison of information between central planning and what is happening locally on the shop floor. For traditionally structured plants, this would mean generally that complexity would be increased through thicker nets of relationships and the abolishment of 'territorial borders'. (This would actually result in a decrease in complicatedness, as the information flow would be less laborious and less susceptible to disturbances.) Optimizing openness means that input information is not entered into the work system absolutely selectively. This would mean tolerating the fact that some redundant, ambiguous and even 'disturbing' information would be entered into the work system, but trust would be placed in the workers' ability to filter out the information relevant to the situation

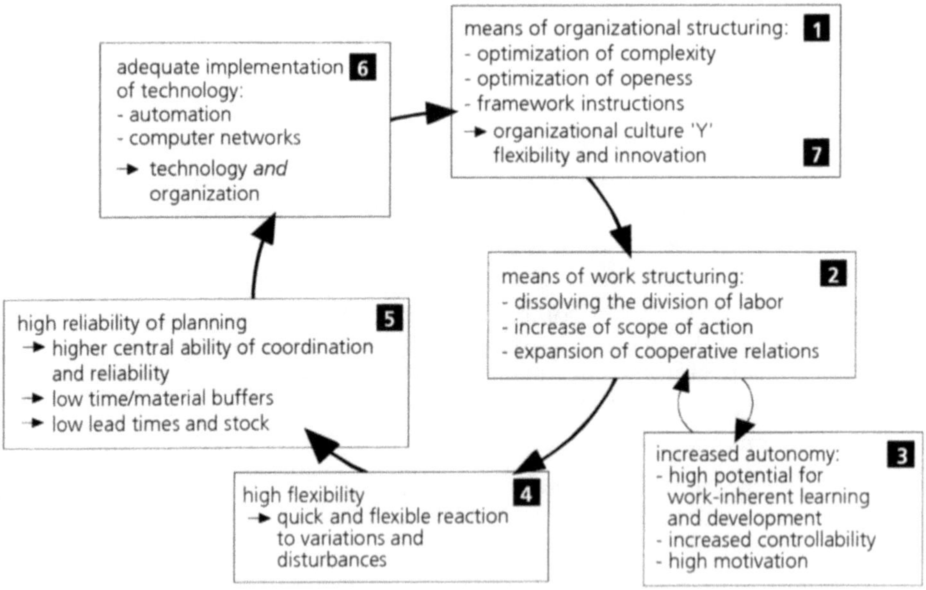

Fig. 4 Cyclical, long-term development model of 'self management' on the shop floor.

and to process this information appropriately. The term 'openness' also contains a psychological dimension: 'open' humans follow the events taking place around them and are willing to entertain and test an interesting impetus or innovation. Herein lies a great potential within the production plant which must not be overlooked with regard to innovation and further qualifications. An important key term here is 'framework instructions'. It means that central planning and coordination confines itself to significant instructions which are clear and binding (such as due dates), but also can be formulated as simple framework conditions (for example, time frames). This represents the crux of decentralization. Decentralization, from the viewpoint of management, means delegation of decision, planning, and control authority; from the angle of operative work systems, it means reintegration of such authority and, thus, the conception of complete tasks and whole activities (ULICH 1994). In concrete form, this means (3): dissolving the division of labor, both vertically, or in terms of information (reintegration of 'thinking' tasks), and horizontally, or in terms of operations (for example, complete manufacturing of parts in manufacturing teams); bound to this, there is expansion of the limits of scope to act and of cooperation and communication nets. All this leads to an increase of autonomy (4), which, however, must indispensably be accompanied and supported by extensive measures to foster qualifications and individual and social competencies. The measures must not only be initiated promptly, but must also be applied long-term (for a period of years) and continuously. After all, the attempt is being made to turn away from thought and behavior patterns as well as from organizational and management cultures which settled in over years of development and now must be replaced with other, lasting, patterns! The goal, and as a rule the result, of all these organizational and work structuring and qualification measures is increased flexibility (5). Variations and disturbances can now be responded to, or even prevented, swiftly and on location – thanks to mutual information exchange. What is crucial is that due to prompter clearing of disturbances, the wide spread of machine down times is reduced: Disturbances still occur, but now clearing may take two hours instead of four. This is what increases – we return to the starting link of the model (1) – the reliability of planning, as a consequence of central coordination and reliability. Time and material buffer zones can be reduced, lead times become shorter, and inventory decreases.

The theoretical examination shows clearly where the main difference between the traditional and alternative approach to shop floor management lies: while the traditional approach is based, in the area of operations, on the conception of *planning* a – passive – order processing, the alternative approach focuses upon *active* order *processing* itself. In the place of stable (or even rigid) compulsory structures there is active 'self management', or active self-regulation. The management principle of control on the shop floor is replaced with reactive regulations as well as continual monitoring of the processes and internal conditions (psychologically, through steady orientation and alertness). Shop floor personnel receives all necessary (but only these) framework instructions from central planning and coordination, compares these out with up-to-the-minute information on current conditions, initiates the required measures within the framework of the available degrees of freedom for decision-making, planning and activity (ULICH 1994) and coordinates with superordinate as well as previous and following work systems.

Empirical validation of such alternative approaches must be found in the results of single case studies; representative before/after studies are not available. THEERKORN (1988) does report on an investigation on production conducted by Felten & Guillaume accompanying the transition from traditional shop floor manufacturing to flexible team structures, the results of which are summarized in Figure 5. In the context of our considerations here, the most impressive finding is the increase in planning reliability of 40%! With this, and along with decentralization and reintegration measures, inventory was reduced by 30% and lead times by 60%. The drastic reduction in scrap rate is just one of the many further criteria which were fulfilled.

Criteria	Change
Inventory	-30%
Lead times	-60%
Scrap rate	-72%
Reliability of planning	+40%

Fig. 5 Results of converting a production shop to flexible team structures (THEERKORN 1988).

Confirmation was made of the positive effects of conversion upon evaluation of the work situation and the behavior of workers:

> „Team work and interesting tasks met with widespread approval, which was reflected also in objective factors such as low absentee rates, marked willingness to learn, high readiness to achieve and good work results."
> THEERKORN (1988).

The implementation of new technologies, in particular computer aid (Figure 5 (6)), does not stand in opposition to the model presented. On the contrary: computer implementation, if designed right, could very successfully support the forms of processing production, or shop floor, operations discussed above. However, a totally different logic of computer implementation would be required. Two main demands must be met with regard to the implementation of PPC systems. The first concerns the direction of information flow. PPC systems of the newer generation are more and more capable of supporting horizontal integration of databases and data flow, that is, of making possible comparison and updating of databases of all production areas along the logistics chain – from sales to procurement and stock keeping, to construction, plant management and production to the distribution of goods and service. This recommends these systems for implementation in flexible production, even in small production plants. In accordance with this understanding, by the way, the term 'PPC system' is increasingly being replaced with the more apt term of 'logistics system' (SCHÖNSLEBEN 1995). The second demand concerns the relationship of these systems to the shop floor. Data editing and manipulation should serve the goal of putting at the disposal of the shop floor the information needed for finite scheduling. In addition, the information system should serve as a means of communication between 'lower' and 'upper' levels for planning purposes. This type of utilization, now becoming increasingly feasible, may be one of the reasons why there have been fewer efforts as of late to develop self-sufficient manufacturing management systems.

3.1. An 'organization chart' of flexible and holistic work and organizational structures

One aspect with regard to decentralized organizational structures has hardly been touched upon in the preceding discussion: the issue of the total internal coordination of work systems having a high degree of self-regulation and holistic task allocation. In Figure 6, the attempt is made to illustrate an organizational structure which takes this aspect into account. In several significant areas, this structure differs from traditional organization structures.

The main feature of the organizational structure presented here in the form of an 'organization chart' is that each element, or each work system, possesses decision, planning, operating and monitoring tasks in the sense of holistic tasks according to ULICH (1994). Input to each work system consists of superordinate goals and framework instructions (controlling variables). These define the parameters within which individual goals become defined and established, planning is made and measures and controls carried out or introduced. In the language of control theory, each work system forms within itself an autonomously controlled system with its own controller and controlled object, which themselves determine controlling values within the given variable range. The production plant as a whole becomes in this way a 'system of linked feedback control loops' (SCHIEMENZ 1979; see also BUBB 1981).

The element at the center of the figure is production management, as the highest hierarchical level. One of production management's tasks is the formulation of strategic goals in reference to the production environment and markets on the basis of marketing and sales data (input). These flow into marketing and sales as superordinate goals and framework instructions (output). It should be noted at this point that the arrows between the work systems are intentionally two-way and represent not just information flow, but also (and

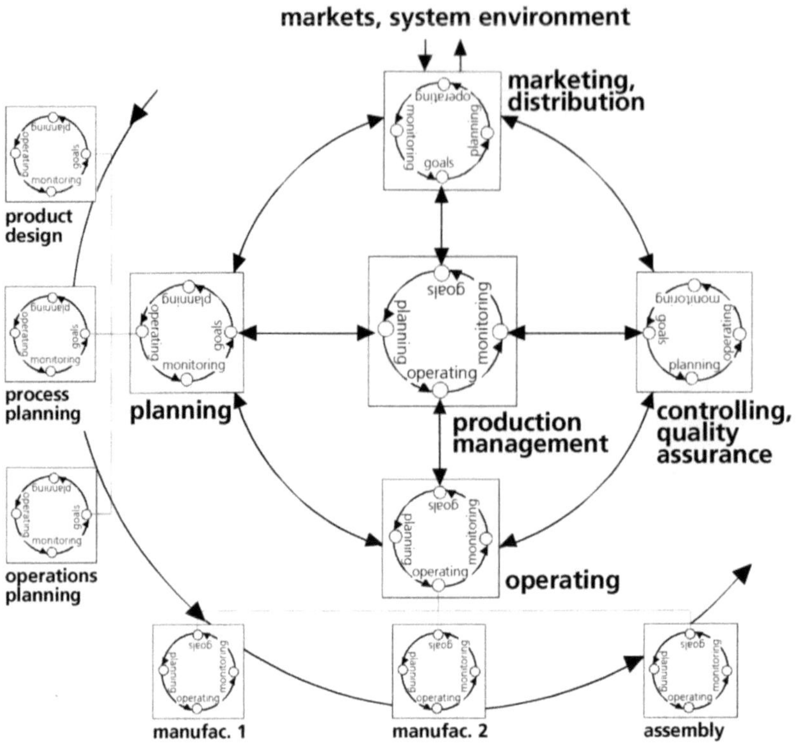

Fig. 6 An 'organization chart' of flexible and holistic work and organizational structures (SCHÜPBACH 1994).

importantly) cooperative and communicative relationships. Another task of production management is, together with planning departments, the strategic planning of production (for example, securing required capacity, constructive adaptation of the product, and so on). Once again, the results will consist in superordinate goals and framework instructions, this time directed to the planning departments. Analogous relationships exist between production management and operating areas (manufacturing and assembly) as well as quality assurance.

The circular arrangement of work systems surrounding production management represents, in the inner circle, 'departments' on the one hand and 'middle', planning management on the other. The work systems in the outer ring belong to the operations level and include, for example, planning teams and manufacturing or assembly teams. The rings, from inner to outer, thus form the structural organization and the arrangement of work systems in the circles and show the cyclical process organization of the production plant. Between hierarchical levels, the relationship is always identical: the superordinate level sets upper goals and determines the framework requirements, the parameters within which self-regulation is possible and necessary.

3.2. Design approaches

The designs with which these organizational structures may be instituted in production practice are multifarious and, in part, were discovered years ago. Noteworthy are the sociotechnical approach in conjunction with semi-autonomous work teams (EMERY 1959; see also ALIOTH 1987 and ULICH 1994), the 'new forms of job design' (ULICH, GROSKURTH & BRUGGEMANN 1973), group technology (see BRÖDNER 1990, 99ff), the production island concept[1] (AWF 1984) and the concept of overlapping work teams (LIKERT 1961).

In the LIKERT model (1961; Figure 7), the connection ("linking") is particularly evident. What is crucial here, as in the other design approaches noted, is that the fundamental organizational unit of the plant is no longer seen in the human-machine system, but in semi-autonomous teams, whose work means have been united and with holistic tasks in the sense described above, such as complete manufacturing of a parts family. The main advantage lies in the fact the 'organization pyramid' not only becomes flatter, but rather also – and most importantly from a process organization perspective – narrower. Namely, task integration does not in the main pertain to middle hierarchical levels, as is usually the case, but rather the lowest level. This means that many interfaces at the horizontal level of operative order processing are no longer necessary and fall away. And not only is the flow of information from 'upper' to 'lower' levels, but, and significantly, the flow between 'front' and 'back' levels, simplified and reinforced.

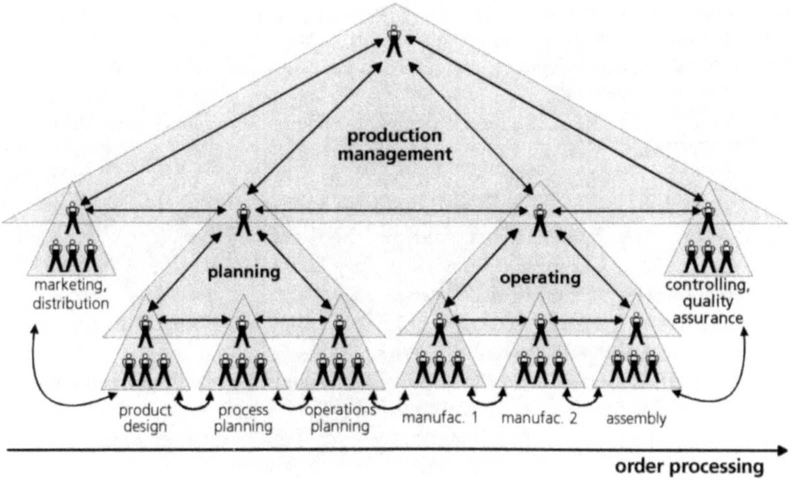

Fig. 7 The production plant as a system of overlapping work teams (elaboration based on LIKERT 1961, p. 105).

1 The concept of the 'production island' was presented as a combination of group technology and team-based work organization. It was proposed in 1984 and strongly promoted by the German 'Ausschuss für wirtschaftliche Fertigung AWF' (Committee for Economic Manufacturing), which is a institution formed by German manufacturing industries and various private and public research institutions.

4. References

ALIOTH, A. (1987): Selbststeuerungskonzepte. In: KIESER, A., REBER, G., WUNDERER, R. (Eds.): Handwörterbuch der Führung. Poeschel, Stuttgart. p. 1824-1834.

AWF (1984): Flexible Fertigungsorganisation am Beispiel von Fertigungsinseln. Ausschuss für Wirtschaftliche Fertigung e.V AWF, Eschborn.

BÖHLE, F., MILKAU, B. (1988): Vom handrad zum bildschirm. Campus, Frankfurt.

BRÖDNER, (1990): The shape of future technology: The anthropocentric alternative. Springer, Berlin.

BUBB, H. (1981). Analyse der Systemdynamik. In: SCHMIDTKE, H. (Ed.): Lehrbuch der Ergonomie, 2nd Ed.; Carl Hanser Verlag, München. p. 286-339.

EMERY, F.E. (1959): Characteristics of socio-technical systems. Document no. 527. Tavistock Institute for Human Relations, London.

FRIEDRICHS, P., GROMOTKA, W. (1992): Fertigungsleittechnik – Marktübersicht und Erfahrungen der Anbieter. VDI-Z, 134:10, p. 112-134.

HACKSTEIN, R. (1989): Produktionsplanung und -steuerung PPS. Ein Handbuch für die betriebliche Praxis. 2nd Ed., VDI-Verlag, Düsseldorf.

HANDKE, G. (1986): Diagnose- und Überwachungsverfahren für die automatisierte Fertigung. In: MIC (Ed.): Überwachung und Sicherung automatisierter Fertigungsprozesse. Tagungsband zum Praktikerseminar. MIC, Stuttgart.

HIRSCH-KREINSEN, H., VON BEHR, M. (1988): Implementation rechnerintegrierter Systeme und Gestaltung der Arbeitsorganisation. In: INSTITUT FÜR SOZIALWISSENSCHAFTLICHE FORSCHUNG (ISF) (Ed.): Arbeitsorganisation bei rechnerintegrierter Produktion. Forschungsbericht KfK-PFT 137. Kernforschungszentrum, Karlsruhe. p. 25-42.

KIRSCH, C., STROHM, O., ULICH, E. (1994): Design of computer aided manufacturing systems: work psychological concepts and empirical findings. In: KIDD, P.T., KARWOWSKI, W. (Eds.): Advances in Agile Manufacturing. IOS Press, Amsterdam. p. 214-217.

KÖHLER, C. (1990): Der elektronische Leitstand – Befehlsempfänger der PPS oder Partner der Werkstatt? VDI-Z, 132:3, p. 14-19.

LIKERT, R. (1961): New patterns of management. McGraw-Hill, New York.

MALIK, F. (1989): Strategie des Managements komplexer Systeme. 3rd Ed., Haupt, Bern.

MCGREGOR, D. (1960): The human side of enterprise. McGraw-Hill, New York.

NEUBERGER, O. (1989): Organisationstheorien. In: ROTH, E. (Ed.): Organisationspsychologie. Hogrefe, Göttingen, p. 205-250.

NUBER, C., KÖHLER, C., SCHULTZ-WILD, R. (1989): Technisch und sozial noch ganz am Anfang. Technische Rundschau, 81:8, p. 30-43.

SCHIEMENZ, B. (1979): Kybernetik. In: KERN, W. (Ed.): Handwörterbuch der Produktionswirtschaft. Poeschel, Stuttgart. p. 1022-1028.

SCHÖNSLEBEN, (1995): Logistik-Standardsoftware – Thesen und Standpunkte. IO Management-Zeitschrift, 64:12, p. 78-82.

SCHÜPBACH, H. (1994): Prozessregulation in rechnerunterstützten Fertigungssystemen. vdf, Zürich/Teubner, Stuttgart.

STOMMEL, H.J., KUNZ, D. (1973): Untersuchungen über Durchlaufzeiten in Betrieben der metallverarbeitenden Industrie mit Einzel- und Kleinserienfertigung. Westdeutscher Verlag, Opladen.

TAYLOR, F. W. (1911): The principles of scientific management. Harper, New York.

THEERKORN, U. (1988): Kleinbetriebe unter einem Dach: Produzieren nach dem Inselprinzip. In: KLINGENBERG, H., KRÄNZLE, H.-(Eds.): Humanisierung bringt Gewinn, Bd. 2: Fertigung und Fertigungssteuerung. RKW, Eschborn. p. 25-52.

ULICH, E. (1994): Arbeitspsychologie. 3rd Ed., vdf, Zürich.

ULICH, E., GROSKURTH, P., BRUGGEMANN, A. (1973): Neue Formen der Arbeitsgestaltung. Europäische Verlagsanstalt, Frankfurt a.M.

WEBER, W. (1992): Eine psychologische Typisierung und Bewertung des Arbeitshandelns in unterschiedlichen CNC-Fertigungsstrukturen. Zeitschrift für Arbeitswissenschaft, 46:2, p. 92-99.

WIENDAHL, H.-P. (1995): Load-oriented manufacturing control. Springer, Berlin.

System Models and Concepts for Distributed Production Planning and Control

Volker Ahrens

1. Foundations of established concepts of planning and control

Production planning and control (PPC) is to a large extent the domain of engineers. This is particularly true for production and production-related tasks which have to do with efficient and effective utilization of technical resources. Such tasks include, for example, scheduling, capacity planning, order release and order monitoring.

One of the most important features of the problem-solving competence of engineers is their use of methods and tools based on simple and clear models. Engineers learn in their earliest college courses that a T beam in steel construction may also be viewed as a beam on two supports which, if cleverly defined, allows simple calculation of the forces working upon it. In this way, engineers are stamped with a pragmatic approach to mainly natural and scientific phenomena.

This has an effect on the way in which they conduct the planning and control of production systems. This is shown by the model of technical systems (Figure 1), used in widespread fashion in all technical disciplines, which also serves as the theoretical basis of most of today's well-established PPC methods and tools. The model stems from systems engineering methodology, whose foundations were developed in the early 1950's mainly at the Bell Telephone Laboratories in the United States (see HALL 1962). In the model, a system is described as a sort of

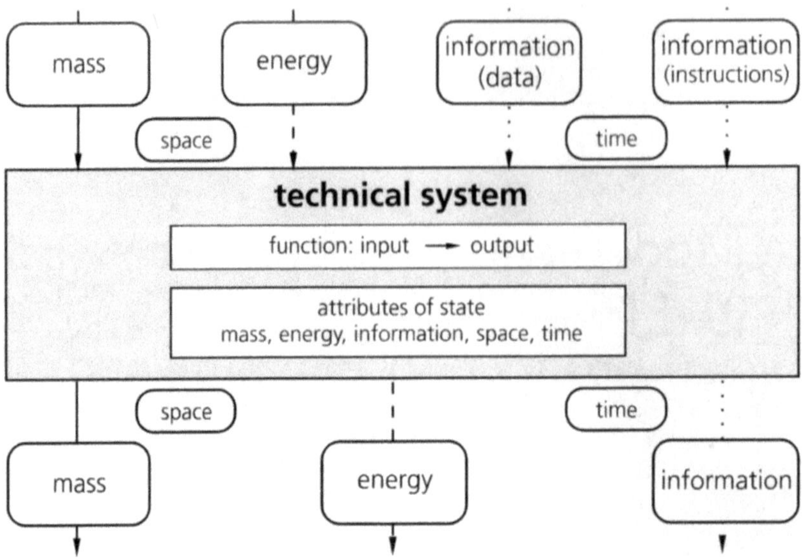

Fig. 1 Principle model of the technical system.

Black Box, about which we need know little except that, according to a usually quite simple function, input variables are transformed into output variables, and the system state itself may become modified. As flow variables, usually a distinction is made among mass, energy and information.

A systems concept such as this can be developed into a model of a production enterprise, as Figure 2 shows. The basic function is described as the „production of industrial products for sale" (WIENDAHL 1989, p. 9). Realization is achieved by the three subsystems of procurement, production and distribution, which themselves comprise a number of subsystems. Finally, there are systems of control, design, leadership and control, which should function to coordinate systems-internal fulfillment of function with input variables from the environment.

On the basis of a model of this type, individual functions may be formulated, as a comparison, as computer programs. The higher the performance of computers became over recent years, the more that production scientists were hopeful that, with their help, all planning and control tasks might be automated. The magic words were 'Computer Integrated Manufacturing' (CIM), and the goal was a factory empty of humans.

This development, as hardly any approach before it, ended in disaster. In the world of practice, one spoke of CIM 'ruins'. Hundreds of millions of US dollars were invested in computers and programs which never really worked. Hordes of planning and control experts made the pilgrimage to the temples of the CIM palaces to gain further training – to no practical gain. No one likes to speak of this any longer. CIM is as dead as a doornail.

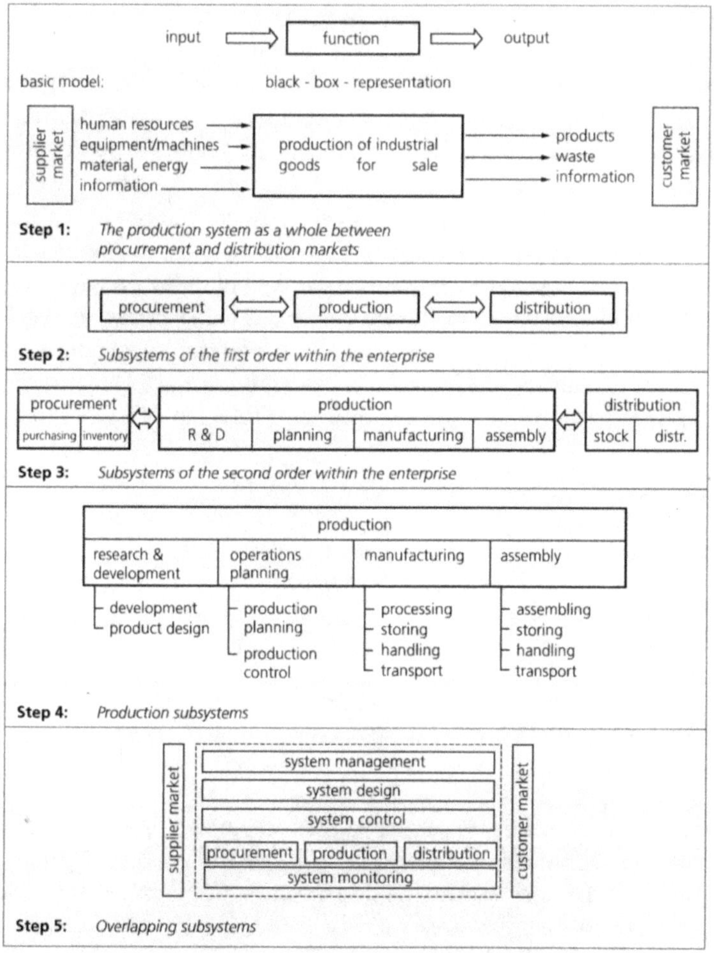

Fig. 2 Model of the enterprise based on the technical systems model (WIENDAHL 1989).

2. New organizational concepts in the factory

Actually, the dead should rest in peace. But this is not possible here, because the some old patterns of thinking live on! It is hidden even in those approaches which at first glance appear to view the task of the planning and control of complex production systems very differently. An example is the proposal, from Germany – with WARNECKE (1992) as its proponent – that companies becomes organized in fractals. In this approach, due to the unfortunate experiences with the model described, the choice was apparently made to simply do away with any model-based foundation. Instead, a vague metaphor is used to explain just what a fractal company is. The attempt is to apply fractal geometry as developed by the American mathematician MANDELBROT (1977). Fractals are geometric shapes having a broken rational dimension. They are characterized by self-similarity: sections of a structure resemble the total structure. WARNECKE attempts to use principles such as this as a foundation for the design and mastering of complex production systems.

This is not an isolated example. It is an ever more frequent phenomenon that engineers turn to unfocused *metaphors* such as this and dispense with formal *models* of the production systems they mean to control. A great deal of money is being put towards the international Holonic Manufacturing Systems (HMS) initiative, which deals with the 'Holonic Factory'. In America, there is a preference for the term 'Agile Factory', and in Germany as well as in Japan, in addition to the idea of the 'Fractal Company', proposals have been put forward for 'Holographic' and 'Bionic Manufacturing'.

The present contribution will show that when model formation is sacrificed, considerable gains in knowledge are lost. Moreover, it is not necessary to limit oneself to blurry metaphors. In recent years, basic research in systems theory has brought forth models which actually do approach the conceptions of production scientists. What remains to be done is to apply the models to concrete production-technical practice. But before this is demonstrated, it will be shown, once again using the example of the Fractal Company, that the classical conception of the functioning of complex production systems persists even in new proposals.

2.1. The goals of an enterprise

The self-similarity of fractal companies is supposed to be expressed in the similarity of goals of all subsystems acting within it (divisions and departments, work teams, individuals, etc.). This is to be achieved by the transmission of enterprise goals, as established by top management, from upper levels in the hierarchy down to the lowest levels. In this connection, WARNECKE speaks of a genetic mechanism such as that used in object-oriented programming to derive categories. „The goal system arrived at through coordination with the level immediately superior to it is supplements by each fractal according to considerations of expediency, for the closer the task is to the manufacturing process, the more specific the goal system must be." (1992, p. 165). However, WARNECKE himself goes on to admit to the limits of such conjunction with the superordinate level: „The unholy and partly grotesque experiences with 'self-commitment' in planned economies are adequate warning." (1992, p. 170). Furthermore, he holds that direct, pre-set goals are in principle possible and explains that „it will not be possible to dispense entirely with unidirectional specification and checking of goals." (1992, p. 172). With this, WARNECKE makes a full turn-around from an excursion into new concepts of self-control back into established hierarchical practice.

Conceptions like these are increasingly criticized – not the least because they contradict day-to-day, practical experience. SPRENGER (1996, p. 84) explains, for example, that „it may well be that the head of the enterprise is permanently concerned with goals. However, for the main body of a company, goals do not play a role. Instead, it is the 'how' which is experienced, the atmosphere, anything that belongs to the path. The time spent by workers in goal involvement is infinitesimally small. This is not a bad idea: those who are stuck at the goal level do not get anywhere. When I look at the amount of energy invested by management in goal-setting and goal 'implementation', and how little attention is given to the practical 'how', then doubt is cast on the whole procedure." (1996, p. 88). MEYER (1994, p. 69) has pointed out that the top-down concept of goal operationalization leaves, to a great extent, the operative level out of the picture. REICHWALD raises the similar objection that conceptions of the role of goals in enterprises today concern too exclusively the goals of management and neglect to consider the goals of workers (1979, p. 531). The most fundamental criticism raises issues of ideology. In this vein, HUNDT & LIEBAU sum up: „It must be understood as a sign of cynicism that the reality that science desires to explain, and the social reality into which it intervenes, is reduced to the reality and practice of goals which are the dominant ones. Here science is biased: it is the weapon used to defend existing power structures." (1972, p. 235).

2.2. Modern goal semantics in the light of the established systems model

The conception supported by WARNECKE of the breaking down of subordinate corporate goals into goals of the actors on the scene, while it is based upon an imprecise metaphor, can be described in terms of the model of technical systems as shown in Figure 1. For it presupposes that superior levels know everything they feel they need to know about lower levels (or at least can find out), and that they have complete influence and control over these subordinate levels in all desired ways.

If this is the starting assumption, there is no reason not to hold to the classical form of hierarchical steering and control of complex production systems. Similarly, there are no objections to the maintenance of established conceptions of PPC systems, which are, after all, reflections of the hierarchical model of organization. It thus comes as no surprise that in many classically organized firms, it is stated again and again that organization had actually always been fractal even though other terms had been used. In adaptations of this rather conservative position, it is easy to observe the progressive variant, according to which established structures are simply renamed, without any real change having occurred. 'Fractal' companies become contradictions within themselves, in that the implementation of established PPC systems continues, even though these were actually designed for centrally-oriented structures.

3. The transition from causal to hermeneutic methods

The underlying reason for such an understanding of the functioning of complex production systems was put forward in our introduction and has to do with the natural scientific slant of engineers' thinking. In the natural sciences, the causal laws explaining the appearance of observed events are sought. But production systems are not systems determined only by natural laws, because in addition to technology, there are humans guiding the events.

Human behavior cannot be explained in causal fashion, cannot be described as simple events and thus simple integrated into technical processes. Human behavior, in contrast to physical or chemical processes, is motivated. There is meaning in human behavior. It follows that in addition to technical issues, production experts must take social and even psychological aspects into account as they seek ways to most efficiently and effectively utilize the technical resources of a production company. This requires a change from causal to hermeneutic thinking and methods, as MAX WEBER (1979) demanded of sociology decades ago. Within production sciences as well, explanations of events must be complemented by the attempt to understand the motives of those who act in the production process and have their own individual understanding of the meaning of the process.

3.1. A model of sociotechnical systems

A change from causal to hermeneutic methods does not require a complete turn-around of the former foundations of production technology. Rather, the hermeneutic method complements these foundations, in that it, for example, incorporates the established model of technical systems (Figure 1) in an extended and expanded model.

As a starting point for the development of a formal model to represent sociotechnical systems, let us take up a proposal made by the production scientist ROPOHL (1979), who also studied the ethics of technology. Further aspects, which stem from sociological systems research, will complement his proposal, whereby the work of sociologist and social scientist LUHMANN (1995) will be emphasized. This is not unproblematic, as engineers often express considerable reserve with regard to the social sciences. ROPOHL himself is very aggressive on this point: „As this systems-theory amateurism annoyingly tarnishes also the reputation of the most serious systems theory in the scientific public, we must very thoroughly and explicitly lay out our own position, so that we will not be subject to attack by critics who are familiar with systems theory only in the guise of these social-philosophical distortions" (1979, p. 49). This devastating criticism is directed to LUHMANN (1995) among others. But LUHMANN's theory of self-referential and autopoietic systems has had a lasting and stimulating effect on modern systems research. In particular, those points in theory which lead ROPOHL to become enmeshed in self-contradiction are explained insightfully by LUHMANN. For this reason, ROPOHL's reservations will be ignored in the following.

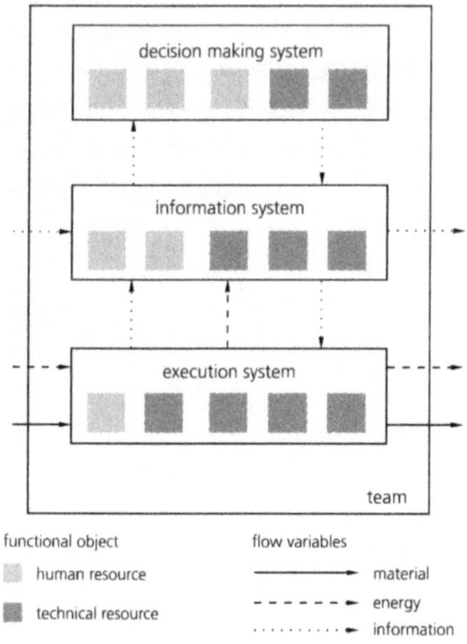

Fig. 3 Model of a sociotechnical action system.

Figure 3 shows the extended model. The subject of an action – in this case a single person – is modeled as a system of action. In its outlook, the model is similar to the classical model of functioning in Figure 1. Just as there, mass, energy and information are processed. Within the action system, three part-systems are differentiated which are responsible for certain partial aspects of action. In the execution system, all material and energetic processes take place, while data and instructions are processed by the information system. Finally, the decision making system, on the basis of information on current state and possible states, puts forward the goal situation.

Here it becomes clear how the classical functional model (Figure 1) is incorporated into the elaborated model. Both in the execution system and in the information system, certain part functions may be delegated to the technical system, that is to say they can be automated. In contrast, the decision system may not be automated. Decisions remain always bound with humans, because it is here that the motives for action exist; it is here that actions are given meaning. Another peculiarity of the decision system is of critical significance: this system has no direct connection to the environment, but rather communicates with the environment only indirectly via the information and execution systems. These two systems function as a kind of filter. For example, if a sender system transmits information to a receiving system, which is meant to use this information in its decision system to decide upon a particular action, the sender cannot, first of all, be certain that it will achieve its intervention goal. Secondly, the receiving system cannot be certain that it has understood the intervention goal of the sender system as intended. This makes clear that SPRENGER (1996) is correct when he says that understanding is improbable. With this a link is forged to the work of one of Germany's most important contemporary philosophers, GADAMER of the HEIDEGGER school. It is not enough, says GADAMER, to explain the world. We must attempt to understand it. Understanding, however, requires patience, time, and the inclination. For understanding is an art – an art because it is hard to learn to understand.

The advantage in the systems theory model of sociotechnical action systems consists in the fact that social science-based knowledge may be translated into the language of engineers. Through this, it can also be expressed in this language what is meant when we say that sociotechnical systems are not open, as assumed by the model of technical systems (Figure 1), but in principle closed. Certainly flow variables such as energy, material and information are transformed, but sociotechnical systems decide to a great extent themselves on the meaning that will be given to these variables. This is called self-reference.

Such considerations take their starting assumptions from psychology and neurophysiology (ROTH 1986, p. 168). It was found that there is no unambiguous relationship between stimuli from the environment and organic systems with their control system, the brain. *Gestalt* theory places particular emphasis on this insight and assumes that only images of reality are acquired, while actual reality remains hidden. Perception actually occurs in specific sensory regions of the brain and not in the sensory organs themselves. The only task of sensory organs is to transform the multitude of varying physical-chemical events such as light quanta, sound waves, olfactory molecules, vibrations and so on into a single class of events, namely electric nerve potentials. Through this, environmental signals lose their specificity. The potentials do not show that which is coded within them. Their origin cannot also not be reconstructed. If we consider that electrical nerve potentials may also be produced by inner processes (thinking), we must conclude that reality is essentially created by the system itself. „Perception is [...] the assignment of meaning to in-and-of-themselves meaningless neuronal processes; it is construction and interpretation.“ (ROTH 1986, p. 170).

Further investigations in other scientific areas have in the meantime indicated that these general laws of organic systems may be generalized (LUHMANN 1995). For example, social systems such as groups or societies can be demonstrated to show self-referent behavior (DRESS, HENDRICHS & KÜPPERS 1986). And as social aspects play an important role in sociotechnical systems, we proposed that production sciences also should make the transition to the theory of self-referential systems (AHRENS 1996a).

For this purpose, ROPOHL's model, in its basic structure, can be well utilized, although it will require some further concretization. While ROPOHL (1979, p. 131) does dispense with a connection between the decision system (he calls it the goal-setting system), there are contradictions in his representation with regard to the openness of the action system. He writes, „An action system is in principle open and only for reasons of simplification in modeling to be viewed as relatively isolated" (1979, p. 110). „Here it will be [...] pointed out that even goals can be given to the action system by the environment; action systems act on the basis of either given goals or self-set goals" (1979, p. 115). But later on in the same discourse, ROPOHL says that the goal-setting system „[...] is generated internally within the system as a guideline for action" (1979, p. 131). From this it is not clear whether goals are given external or internal to the system, whereby as a whole, ROPOHL leaves the impression that he prefers the idea of system openness. This shifting between autonomy, which cannot be ignored, and openness, which exists of course, at least with regard to information, energy and material, is the result of the widespread difficulty with the concept of goal, which we discussed above and which evidently ROPOHL also was not able to clarify unambiguously.

Clarification of the issue of goals can be achieved when the self-referential nature of sociotechnical action systems is taken into account. If expectations requiring decisions by the system come from the outside, the degree of the resulting potential conflict will be largely determined by whether the actors experience subsequent limitations on their freedom to act as help and support in reaching their own goals or as a hindrance which they must reject (TITSCHER & KÖNINGSWIESER 1985, p. 149). Following SPRENGER, we can say that external perceptual stimuli serve only to trigger inner experience, and that external input represents an opportunity for insight and nothing more. It is at each individual's discretion to make use of the opportunity or not (1996, p. 113). Some approaches to organizational psychology go so far as to say that goal and purpose statements serve only to self-represent an enterprise to the outside world or to their own less powerful members. Purpose statements are seen to have a certain mythical character, and to be formulated only after the fact, thus not functioning as causes of action whatsoever (TÜRK 1989, p. 25; KÜPPER & ORTMANN 1988). This is seen not only with regard to the concept of a company as a goal-oriented system, but rather also with regard to the concept of an enterprise as a rational phenomenon. TÜRK explains in this connection that decisions are seldom reached in a rational manner, and that frequently even, companies only function at all because (and in so far as) they divert from purely rational procedures (TÜRK 1989, p. 36).

The sociotechnical model also makes this clear, particularly when we apply the cybernetic principle of recursion to higher system levels such as work teams, the enterprise or society. Figure 4 applies the principle to the work team. It becomes evident that the functions in the three subsystems are carried out by personnel action systems, whereby these of course, as according to Figure 3, may act in connection with the technical system. Further, the work team model shows that there are subsystems even within the decision-making system. In a group, this is group management, which is supposed to lead group behavior as a whole.

However, it is here that the autonomy aspect comes into play, as group management is faced with the difficulty that team members in the information and execution systems have their own decision-making system. They are thus just as self-referential as any other behavioral system. It follows that the opportunity for direct management and control is quite limited. Each individual personnel system organizes itself to a great extent, even then when it is embedded within a repressive hierarchy. For this reason, the behavior of the group as visible to the outside is not only determined by decisions made by group managers, but also, and to a significant degree by, the self-organizing behavior of team members themselves. As a consequence, finally, the vertical structures functional model of sociotechnical systems cannot serve simultaneously as a model of management and direction.

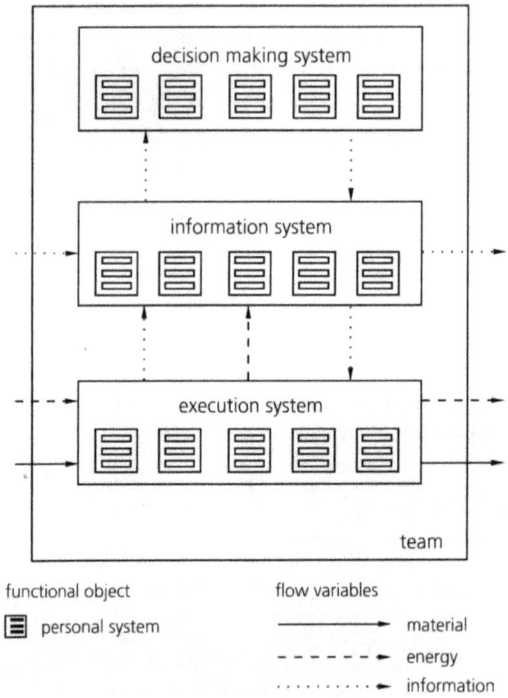

Fig. 4 Team or organizational unit as action system.

3.2. The transition from hierarchical to non-hierarchical management

In his search for answers to the issue of how the closed nature of sociotechnical systems is manifested, LUHMANN (1995) went further. Taking his starting point from the works of MATURANA & VARELA (1980), LUHMANN developed a sociological theory of autopoiesis, which describes systems not only as self-referential, but also as structurally determined, self-producing and self-maintaining (AHRENS 1996a). With this, it might be falsely assumed that systems are autistic in that they are almost completely unconnected with their environments. However, if thought through in a consistent manner, this would make any interaction among systems and any management impossible.

To make use of this model in understanding how sociotechnical systems may be managed and controlled, there is aid to be found in a model by the American sociologist TALCOTT PARSONS (1969). In this way, a connection can also be made to the work of LUHMANN's followers, such as WILLKE (1992), who have made similar connections (MÜNCH 1996, p. 33).

WILLKE, for example, prefers to speak of relativizing programs (1992, p. 179) rather than the goal programs discussed in Figure 2 in connection with goal agreements between management and subordinates. The divergent conceptions of the subsystems in an enterprise are not to be reduced to the perspective of a single, individual system (such as the authoritatively-set goals of management). Rather, the differentiated, individual logic and operative autonomy of each of the coordinated actors should be maintained and, in their conditions and consequences, become tuned and adjusted to one another – in other words, relativized. Such programs fulfill the function of decentral context management, whereby a minimal degree of common orientation is assured, while the shared context, however, can no longer be guaranteed through instructions from a central unit such as the top of the hierarchy (1992, p. 185). Another example of a proposal of context management stems from WEICK, who suggests replacing the traditional close link between systems with a loose coupling of systems. „Loose coupling occurs either when two systems have few variables in common or when the common variables are weak compared to other variables that influence the system." (1979, p. 111). It is in this manner that the stability of a system can be increased, as disturbances and variation no longer simply spread throughout the system as happens when subsystems are closely linked.

The concrete manifestations of context management, or loose coupling, in production systems will be explored on the basis of PARSONS' model.

3.2.1. The structure of modern societies as a model for manufacturing systems

PARSONS deals with the issue of the formation of order in society. Thus, his original model was constructed for this systems level. In the following, the cybernetic principle of recursion will be used to show how non-hierarchical management can be implemented at the company level. For, on the societal level, the process of modernization which has taken place in Western cultures over the last centuries has replaced the original hierarchical order of the Middle Ages and Early Renaissance with non-hierarchical order. Experience has now shown that the elements of this new order – democracy, market economics, liberalism and free press in science, art, religion and philosophy – have assured historical stability. Upon this background, the following considerations are based upon the hypothesis that a similar process of modernization within societal institutions, to which industrial enterprises belong, can lead to similar progress.

The four spheres of social action and their interpretation

To demonstrate this, PARSONS' model is examined in its original form, which refers to the societal level. Figure 5 shows his AGIL schema, according to which four spheres of action in societies may be distinguished – communication, society, economy and politics – and which we will view as self-referential and operatively closed in their ideal forms. This will become evident in the light of their specific characteristics and their 'language'. Further, each sphere of action offers certain services to the other spheres. Exchange media are available for this purpose. Thus the four areas can mutually influence each other – an aspect which in LUHMANN's work is inadequately represented.

In the field of economy, the dominant language is the language of money. Only that which can be expressed in that language is understood and perceived to be relevant to action. The main feature of this societal sub-area is its consequentialist ethics, which seeks, in its utilitarian approach, to secure the greatest happiness of the greatest number of people. This can be seen in the fact that the pure (free) market always divides its participants into winners and losers but does not provide protection mechanisms for losers. It is through this that the field of economic of all four areas has the greatest dynamics. In the ensemble of the four spheres, economy thus has the function of *adaptation* to changing conditions ('A' system).

In contrast, in societal communities, moral ethics are the guiding principles. Justice and solidarity prohibit any exceptions to the rules. If a member of society breaks a rule, this is viewed as an attack upon the social order which must be answered with punishment (repressive sanctions). In the economic field, in contrast, offenders are merely required to make restitution (restitutive sanctions). However, the application of principles is limited to only the members of a societal group (moral particularism) such as families and clans. Outside the group, the

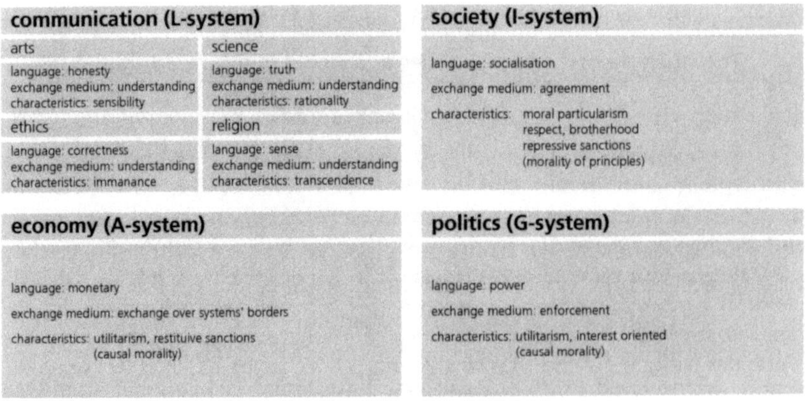

Fig. 5 Elements of PARSONS' model of social systems (PARSONS 1969).

principles lose their binding power. Through the categorization of behavior patterns, there is in this area the smallest number of behavioral options, so that it has a controlling effect on all other areas. This results in an *integration* ('I' system) of all actions within a firm order.

Within politics, the main issues are the obtaining and maintenance of power. With this the possibility comes into being to introduce and collectively enforce binding decisions and policies. In order to be capable of action, the basis, as in the field of economics, can only be one of utilitarian morals, which are oriented towards certain interests. In the total ensemble of spheres of action, this system has its primary function in the establishment and realization of *goals* (goal attainment: 'G' system).

The last area comprises all aspects of societal communication. Art is that which is true and targets the senses. Science, on the other hand, is based on rationality and seeks truth. Ethics deals with issues of right and wrong within a society. While religion limits itself to the inner world (immanence), it transcends it at the same time to ask about the meaning of existence (transcendence). In these ways, fields of action are generalized and maintained as latent *structures (latent pattern maintenance*, 'L' system). The sum of all behavioral options is given a general framework.

The way in which services are exchanged among the spheres of action, each with their own laws, is illustrated in Figure 6. In social communities ('I' system), for example, business, according to principles of community, may initially be conducted along communistic lines. Within families and clans solidarity is reflected in fair sharing of resources among all members. However, beyond the boundaries of the clan, the principle no longer holds. In order to allow for friendly trade of goods, services and rights with other groups, help is needed from the economic area (A). This area provides the neutral means of exchange.

Turned around, the economic area (A) is dependent upon aid from the community (I). For if the peaceful acquisition of goods, services and rights based on the principle of „money in exchange for goods“ would be in question, all economic order would be canceled out. Principles such as „life in exchange for goods“ have no place here. With this in mind, one can say that society, in the economic area, supplies the necessary business ethic, while it, in the other direction, contains the possibility to lend a certain economic objectivity to communal life and thus peacefully resolve conflict within anonymous societies.

An important point here is that the individuality of the governing laws and principles of the four spheres of action are not lost through the exchange of services. PARSONS calls the manner in which this happens 'interpenetration'. For example, if the economic partsystem (A) is offered a service by another partsystem, it will first be translated into the language of economics, that is, the language of money. Only if and when translation is accomplished is the 'foreign' service usable in the economic sphere of action.

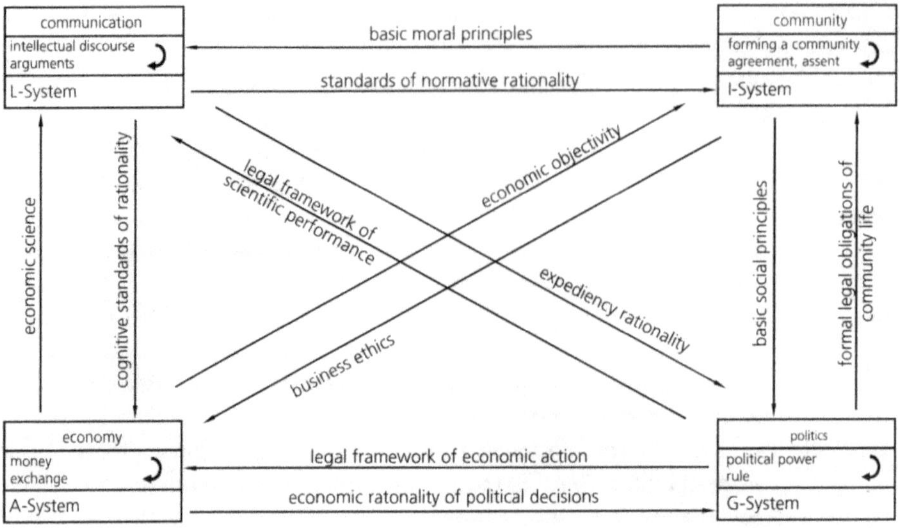

Fig. 6 Interpenetration between autonomous spheres of action (MÜNCH 1996).

Sociotechnical systems in the force field of spheres of action

On the basis of the AGIL schema, the details of the motives of the sociotechnical system governing action as well as the embeddedness of this system within the surrounding action system can now be grasped. We can ascertain how certain aspects of an action are initially determined by surrounding systems at the same level of the system hierarchy. In addition, we may determine how it is that the supersystems to these surrounding systems influence the system examined and how this system itself, in relation to the other systems, unfolds its effects (MÜNCH 1980, p. 32). In a sense, the AGIL schema represents a kind of force field, within which a certain action aspect of reality takes on a certain position. The closer an aspect comes to one of the partsystems, the more it will tend to obey the laws governing that system. The closer the aspect to the center of all the systems, the more it will be determined by an overlay of all such laws.

In the following, we will first turn our attention to the super-societal level in order to investigate the conditions which challenge societies, their institutions and individuals. From these, theoretically-grounded measures towards adaptation of structures and processes may be derived.

3.2.2. Globalization in the light of the preceding considerations

Intensification of the exchange of goods, services and rights transcending all societal boundaries has, at the supersocietal level, resulted in the dominance of the economic sector (A). Internationally, no superordinate order has become established which, through the guiding influence of the other three partsystems, could counteract the dynamic effect of the economic partsystem. The United Nations supplies the international community of nations with a political instance

(G), but this has yet to achieve the necessary power to place economic activity within a legal framework to the extent required. And finally, the area of communications (L) is far from able to present convincing arguments to the other partsystems as to why, and in what form, the dominance of the economic sector at the international level should be tempered by opposing forces such as global human rights. In the force field set up by the AGIL schema, the super-societal system is thus in a position where it is increasingly directly influenced by the A system. The super-societal system has thus in the main a dynamic effect, forcing other systems, which are under stronger control, to adapt.

This affects, on the societal level, all societal systems which have any significant role in world economy. And indeed, the adaptation of political (G), communal (I) and communicative (L) action to economic requirements (A) is clearly evident. In Germany, ever more politicians distinguish themselves mainly as economists, even in parties which previously were dedicated to social goals such as solidarity and justice. Funding for basic research is being cut back for shortsighted reasons of cost, and the community accepts the measures with clear gestures of resignation. This phenomenon increases as the economy demands that we dispense with the growth of affluence and social security.

3.2.3. An approach to modernizing classical production organization

It is not the societies of the Western world which must make fundamental changes in order to keep pace with the globalization of economy, although at this level there is also clearly a need for reform in sub-areas. There are three reasons for this.

First, societies are the wrong addressees with regard to resolving the problems which arise through globalization. These may only be solved at the level at which the causes can be located, and this is the international level. Market dynamics, which develop freely, must become tempered by a world order, in ways similar to measures at national levels. Here, solutions which modern societies have found with the elements discussed above could serve as models.

Secondly, the free play of the forces in the international market is neither fate nor a natural event which societies must obey unconditionally. It results, as does any act, from motives, to which concrete acting subjects attribute meaning. Thus, the motives can be researched, and one can behave accordingly through one's own sovereign actions.

If the dynamics produced by the economic sector (A) are tempered, it follows without doubt that this will happen at the cost of purely economic success. This is the price that must be paid for placing more weight upon other kinds of success of functioning communities, such as advances in communication (L), solidarity and justice (I) and political governability. At the next lower systems level, the level of profit-oriented institutions, it is, however, very tempting to make profit by shifting the social power balance towards the economic partsystem (A). But this does not serve either the institution or society in the long run. For such a shift destroys the basis of affluence and progress upon which both sides owe their existence.

However, it is not the institutions' task to maintain the societal balance of powers. This has to happen at the level of society. Here political insight is required to recognize that this is the only level at which the modernization process founded in the idea of the Enlightenment (KANT 1984) has already taken place, while both the superordinate and the subordinate levels still have this process ahead of them.

This leads to the third aspect, which has to do with societal institutions. Just as at the super-societal level, a modern order must be developed at this level to replace the centralist control model of medieval and early modern origins (feudalism, absolutism, Ordo thinking and so on) with a non-hierarchical model. A thorough treatment of this aspect would be beyond the scope of this paper, and we will examine only that area which plays the leading role in profit-oriented institutions (business enterprises): the economic area (A). The question concerns the type of order which should be created in order to make enterprises equal in strength to the dynamics of their environment. A further question is how management (G) might exercise influence and control within this sphere of action.

Figure 7 presents the AGIL schema for the institutional systems level. In the area of communication and, here particularly, in the rational area (L), concrete research and development take the place of general knowledge. The economic sector (A), at this level, neglects economics to place emphasis upon business management. The community (I) in an enterprise is made up of its staff, and political power (G) derives from property (ownership) conditions and lies in the hands of enterprise management.

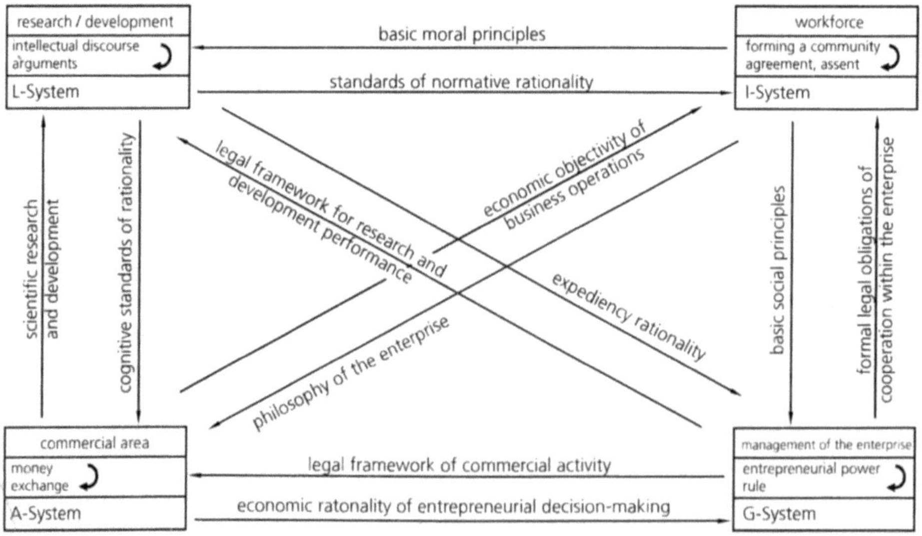

Fig. 7 Non-hierarchical management model for profit-oriented organizations.

With this as a basis, we will examine in a first step how it is that management may have a controlling influence over the economic area (A), if this area in principle is closed and thus not particularly accessible to extreme control measures. For this initial step, we will leave aside the question of how, within the economic sphere of action, order can be established which allows for self-determined behavior without resulting chaos.

The first control function: External control through direct instructions

Three control functions may be distinguished which may be exercised by management (G). The first control function consists in direct instructions and orders. Although this may at first glance look deceptively like the classical control concept, this is not so. In connection with our discussion of the AGIL schema, we noted that each sphere of action offers to all other spheres certain services which are needed and which the spheres cannot provide for themselves. And control which results from power is just such a service. Control becomes necessary in a particular sphere of action whenever a decision can neither be economically or morally justified or based upon communicative extracts (scientific, aesthetic, ethical or religious). In such cases, direct instructions from management are required by the spheres of action not based on power. A prerequisite for the success of such intervention, however, is that external control be explicit and make no attempt to appear as if were self-determined control, as is the case with so-called goal 'agreements' between managers and subordinates.

The second control function: Control of the action context

The second control function applies to those cases where it becomes apparent that direct intervention into the autonomous inner life of the economic sphere is not successful, even though necessary for the reason described above. KROHN & KÜPPERS (1990) propose three approaches here, which are founded upon the following basic cybernetic models:

- cybernetics of observed systems (external control)
- cybernetics of observing systems (internal control) and
- cybernetics of reciprocal observation (this presupposes symmetry between internal and external standpoint, recognizing at the same time that this is impossible)

Interventionist control

According to the concept of intervening control, the system to be managed is viewed as a Black Box which is controlled by an observer external to the box. In a practical sense, the observer will be at least somewhat familiar with the system. Still, the observer cannot know much about the actual effects of his control measures. The main point is that the self-organized system may realize new values through the perception of external changes. The external observer, therefore, can attempt to stimulate the system, by means of targeted changes to its environment, to move into system states which are compatible to his goals. It is crucial to recognize and understand the connection between the system's dynamics and the conditions and circumstances in its environment which stabilize these dynamics. If new conditions are set up in the environment, instability in the system can be produced in a targeted way, causing the system to enter into a new state.

A similar strategy is already implemented in psychology – in 'systems therapy'. Just as in our application, the therapist assumes that the system is self-referential and autonomous. The goal is thus to cause a family to overcome its difficulties itself. The therapist's only task is to find that sensitive point for intervention and to express it in the form of new conditions.

This strategy, however, does not really resolve the problem of self-reference. The problem, which the observer sees as positioned externally even though he must be viewed as part of the system, remains. In systems therapy, the therapist and the group together form the therapy system. Through reciprocal observation, they learn from each other and grow to become a common group in which the therapist becomes ever more a part. Finally, what occurs is not intervention, but rather self-management. The systems therapy model does in fact view the process in this way. Accordingly, the attempt is made to deal with the resulting problem by adding one more instance of observation: a supervisor observes the entire process. This forces the therapist to maintain two points of reference. Within the therapy system, the therapist is the actual observer, and in turn, he is observed by the supervisor.

If this strategy of interpenetrating is transposed to the control of the economic sphere of action in the enterprise, environmental conditions may be altered, for example, through the establishment of target values and deadlines. This corresponds in principle to what was said in regard to the first control task above. But here, in addition, there is no expectation that targets and deadlines be precisely met. It is assumed that the system to be controlled will deal with the new conditions in its own way and come to its own results. These results are then checked in terms of their compatibility with the actual goals of the intervention and, if deviating too greatly, corrected by means of the setting of new conditions. The strength of this approach is its own weakness. Actual reorganization within the system is left to the system itself, and control concentrates on identifying the possibilities for effective intervention in external conditions, the context.

Experimental control

Experimental planning offers an alternative which is located between classical control and a pure evolutionary principle of trial-and-error. Experimental planning is possible, because it stems from a social description and observes self-reorganization. In this respect the observer is actually within the system.

What is required are models and methods to produce the desired effects (statistical experimental design, for example), and usable data. Experiments have the disadvantages of theoretical expectations and the controlled risk of failure, and they represent the type of action which is rational action under partial uncertainty. Success, of experiments that succeed and of experiments that fail, consists in the generation of experience. However, the experience gained with the model must still be transposed to reality.

There are difficulties with this approach as well. If the control instance admits to uncertainty and to taking measures in order to augment its lack of knowledge, it is vulnerable to critique. The willingness of an individual, work team or enterprise to take part in an experiment will be limited if negative consequences cannot be ruled out. One reaction of the control instance might be to send in experts, who in their classical function, make uncertain knowledge appear to be solid. However, once again, this will only in the end strengthen the power of the real facts. The actual goal of experimental planning will be missed, because controlled learning under the condition of uncertainty is replaced by irrevocable decisions.

One answer here is the implementation of special techniques such as planning games or computer-aided simulation. In a special experimental environment under realistic conditions, experiments can be made without the risk of negative consequences in the real world. There is always the danger that results may be achieved which cannot be generalized or applied to reality. For one thing, the modeling required for simulation involves an inevitable reduction of actual complexity, and secondly, the safety of the simulation situation often leads to high risk behaviors unlikely to be observed in reality.

Transformational control

The concept of transformational control follows upon an extension of observation (supervision) as presented in the concept of intervention above. The goal is symmetry between external and internal control, which results from intersystem perception. The problem here is that the formation of symmetry can be only partially successful, as no frame of reference of a supersystem is available.

Structural coupling, hyper cyclicity, systems interference and synreferentiality are conceptual attempts to structure system – system relationships under the difficult condition of autonomy and to achieve more than merely changed environmental conditions with the hope that this alone will lead to the control goal. In this connection, KROHN & KÜPPERS (1990) link the terms autonomy (operational closed-ness) and informational openness. Their starting point is that there can also be non-recursive communication between the system and the environment, or between systems. Where there is this type of communication, however, there is the major problem of translation (transformation). Planner and executor cannot communicate in this case, so that it will be necessary that the executors can interpret the plan in their own context. There is no generally valid metasystem available that would allow agreement on the content of the communication. The use of everyday speech might be a candidate here, but this is as a rule not powerful enough to deal with very difficult instances of communication.

The solution is to be found in organizational specialization in transformation problems. This must occur in hybrid situations, which constitute both an arena for confrontation among varying interests as well as for mutual planning. Within the institutions, plans may be tested in terms of their translatability. Actual translation can then be performed systematically. Practical approaches of this type can already be found in some enterprises today. An example is the so-called 'quality circle'.

The third control function: Establishing the ground rules of self-organization

The third control function, finally, consists in establishing the framework of action of the economic partsystem. This framework sets down the ground rules which all actors within the economic area must be able to follow. This requires explicit and voluntary agreement, based on insight and understanding. A concrete proposal for the design of such a framework has been presented by AHRENS (1996a, 1996b). The proposal is based upon rules of the market, which should be operationalized within the enterprise. With this, the same dynamics can be reconstructed internally that are the foundation of market rules external to the organization.

The market within the enterprise

The idea is simple: customer orders represent, within the enterprise, demands making a claim to the company's own resources. Both sides have specific goals which contradict each other. For example, an order should be processed at the lowest possible cost, but from the perspective of resources, the highest possible price for the product is desirable. There is also competition among the suppliers themselves, as well as among the demands. Orders compete in terms of use of resources, and resources are in competition for the most attractive orders. In spite of these conflicts, both sides must cooperate with all others if goals are to be achieved. Cooperation only arises when all partners agree to the rules and conditions of their own free will. This cannot be taken for granted, and as a rule, the conditions have to be negotiated by all involved.

Origins of the idea of the market within the enterprise

In fact, this idea is not as new as one might suppose. In 1937, COASE, in his now legendary essay on „The Nature of the Firm", gave first impetus to considerations of this kind. If the idea has as yet not been implemented, then there must exist crucial opposing arguments which would have to be disproved.

One of the central questions raised by COASE is the issue of why there are firms at all: „If a workman moves from department Y to department X, he does not go because of a change in relative prices, but because he is ordered to do so. [...] Within a firm, (the) markets transactions are eliminated and in place of the complicated market structure with exchange transactions is substituted the entrepreneur-coordinator, who directs production. [...] Yet, having regard to the fact that if production is regulated by price movements, production could be carried out without any organization at all, well might we ask, why is there any organization?" (ibid., 19). In his answer, COASE points out that there are circumstances under which the price mechanism causes higher costs than planned business proceedings: „The main reason why it is profitable to establish a firm is that it would seem to be that there is a cost of using the price mechanism". From this issue transaction-cost economics arose as an economic discipline, whose core hypothesis is that economic institutions, markets and enterprises, were created to minimize transaction costs which stem from their use (TIROLE 1988).

One of the most important questions addresses the source of transaction costs. In regard to the market, COASE (1937) explains that „the most obvious cost of 'organizing' production through the price mechanism is that of discovering what the relevant prices are. [...] The costs of negotiating and concluding a separate contract for each exchange transaction which takes place on a market must also be taken into account". If costs may be spared by eliminating the market mechanism, why is production not taken over by one single giant firm? According to COASE, planned business proceeding also incurs costs. The most obvious costs are salaries for management personnel. Considerably higher as a rule, though, are the costs which are incurred through poor management. The larger and more dominating an area is, the less possible it is for a central office to gain and process enough secure information to use production factors in an optimum manner. And finally, there are losses through the friction that arises because all members of an organization have and utilize their own sources of power. Central commands and instructions are put into practice locally in a limited way, and the effort put into the development and realization of operative opposing strategies should not be underestimated.

Empirical observation of the behavior of firms today would indicate that transaction costs have shifted to the advantage of market location. Under the banner of the catchword 'outsourcing' (PICOT 1992), firms have begun to outsource whole areas, making subject to the market what was previously organized by planning economics. This often starts with the company cafeteria, but affects services closer to production, such as data processing and even departments involved in direct production, as well. Even an approximate look would suggest that there are two main reasons for the phenomena: those in practice experience the pressure of problems which are theoretically explainable, and great options open up.

The pressure of the problems results from the fact that enterprises having a high degree of integration must master a high degree of internal complexity. Management's resources are thus bound to internal affairs and are no longer available to deal with either external complexity or internal core processes. Further, there are the management mistakes probable in a firm of high internal complexity, and finally, flexibility decreases while reaction time increases when one's own resources are used to promote performance.

Options lie mainly within the area of data processing and information technology. Here development has taken off in such a high degree that the effects of problem-induced activity can be superseded. Figure 8 outlines the development of information technology in comparison to the development of methods and techniques of the company-internal planning and control of resources and processing. Clear parallels can be discerned. The starting point of the development of modern computers were the central mainframe computers. At about the same time, the first programs to support company organization were written. An example is the System R/2 by SAP, which among other things, handles production planning and control. It would, however, be shortsighted to explain the central orientation of such software systems in terms of the possibilities of the existing hardware alone. Rather, it can be stated – even without comprehensive proof – that both developments had the same origin: the Taylor understanding of the organization as a functionally structured and hierarchically controlled system. This perspective was long predominant, although there were early movements away from monolithic systems, and the distribution of local terminals made the first steps toward decentralization. But even the spread of personal computers, which went along with the rapid increase in the performance of the small computer, was marked by the functional view. But there was a change in the basic conception of the functioning of complex organizations. In addition to the functional perspective, there was an increasing orientation towards process. This development is reflected in the development and spread of high performance networks with which today's

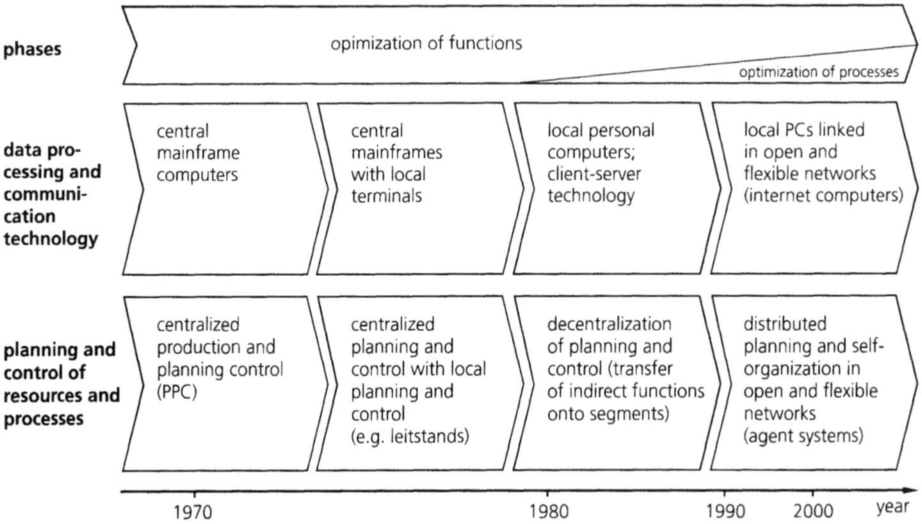

Fig. 8 Parallel development of information technology and production planning and control systems.

heterogeneous computers are connected. In parallel fashion, so-called indirect areas in organizations, mainly planning and control functions, are becoming decentralized. The exchange required by superordinate coordination can be achieved with the aid of standardized methods and tools, such as normed product data in the framework of the international STEP initiative (Standard for the Exchange of Product Model Data). For the transmission of business data, EDIFACT (Electronic Data exchange for administration, commerce and transportation) makes available similar standards.

The emblem of current advance in technology are so-called client server systems. It is characterized by a combination of central and decentral components, whereby as a rule data storage is performed by the central server, while data processing is taken over by local computers at work stations (clients). This technology is now no longer only used by programs designed to aid company functions, as is the case with PPC systems, but rather also by programs aiding company communication (processes). An example are workflow management systems. However, this does not exhaust the potential for distributing company tasks. Experience with the Internet has now shown clearly that even such systems can function that do without central coordination and central control. Upon this background, there are attempts already to distribute functions almost completely and to organize coordination decentrally as well. NEUMEIER-MACKERT, a manager with IBM in Germany, says about this: „There has been talk of client servers for years. Great promises were and are being made to firms. For IBM, this technology is an intermediate step – from a host-centered world [...] to a net-centered world, where applications at the time of execution are set into motion by the user on the basis of standards“ (1996, p. 324). This indicates that the final step of the development outlined in Figure 8 of data processing and communications technology will in all probability be soon realized. From this perspective, the project sketched out in this paper can be understood as a contribution to preparations for the last steps in the decentralization of the planning and control of resources and processes.

3.2.4. Operationalization on the basis of new data processing and communications technologies

Computer science now supplies the basis for the creation of true distributed planning and control tools. To these belongs first and foremost distributed artificial intelligence (DAI) and agent theory, which is directly compatible with the theory outlined here. Agents in DAI are understood as programs which act autonomously and, where necessary, organize work sharing with other agent programs. In order to coordinate agents, various models are utilized. Some examples are shared memory methods, where agents tack their messages to a kind of bulletin board, so to speak, and shared environment methods, whereby agents tap into the activities of other agents only as they perceive changes in the shared environment caused by other agents. But coordination based upon the market has also already been conceived. Vanguards of this development were in the main DAVIS & SMITH (1983), with their contract net approach. And PARUNAK (1987) was one of the first to apply market-type coordination mechanisms.

Within the framework of a research project funded by the German Research Community (DFG), we demonstrated in model experiments that on the basis of such methods and tools, true distributed production planning and control can be realized (AHRENS 1997). However, our concept of the agent was more extensive than the typical concept in computer science. The reason for this is that the necessity to take social and psychological aspects into account, considerations which have formed the core of this paper, prohibit a purely technical viewpoint. As agents, therefore, we characterize the people who are involved in the internal market of a firm. These persons have at their disposal agent programs which can execute certain functions, such as negotiations with other agents (programs), automatically. But human agents control the implementation of these programs, just as they control what is done with the results that the computer agents ascertain.

Upon this background, it was shown in many simulation games that a good basis for the realization of the human-centered approach is a preconception and understanding of market-based rules, which can be assumed for all people. Further, it was a most interesting experience to discover that comprehensive planning at the highest aggregate level is in fact unnecessary. Instead, agents react quickly to new situations, such as a new customer order or interference caused by a machine, and moreover, through dispensing with long-term and tedious planning, they were highly flexible.

4. References

AHRENS, V. (1996a): New organization concepts for distributed production systems. Kybernetes – The International Journal of Systems & Cybernetics, 25:2, p. 24-39.

AHRENS, V. (1996b): Planning and control in self-organized production systems. In: SCHWEITZER, F. (Ed.): Self-organization of complex structures – From individual to collective dynamics. Volume II: Biological and ecological dynamics, socio-economic processes, urban structure formation and traffic dynamics. Gordon and Breach Publication, London.

AHRENS, V. (1997): Mit Agentensystemen zur Beherrschung flexibler Prozeßketten – Workflow-Management und Intranet-Anwendungen in der Produktion. 10. VDI-Jahrestagung Produktionslogistik – Komplexe Prozesse beherrschen. Köln.

COASE, R.H. (1937): The Nature of the Firm. Economica, New Series, 4:11, p. 386-405.

DAVIS, R., SMITH, R.G. (1983): Negotiation as a metaphor for distributed problem solving. Artificial Intelligence, 20:1. p. 63-110.

DRESS, A., HENDRICHS, H., KÜPPERS, G. (1986): Selbstorganisation – Die Entstehung von Ordnung in Natur und Gesellschaft. Piper, München.

HALL, A.D. (1962): Systems engineering. Van Nostrand, Princeton.

HUNDT, S., LIEBAU, E.: (1972): Zum Verhältnis von Theorie und Praxis – Gegen ein beschränktes Selbstverständnis der Betriebswirtschaftslehre als 'Unternehmenswissenschaft'. In: DLUGOS, G., EBERLEIN, G., STEINMANN, H. (Eds.): Wissenschaftstheorie und Betriebswirtschaftslehre – Eine methodologische Kontroverse. Düsseldorf.

KANT, I. (1984): Beantwortung der Frage: Was ist Aufklärung?. In: BAHR, E. (1994): Was ist Aufklärung. Reclam, Stuttgart.

KROHN, W., KÜPPERS, G. (1990): Selbstreferenz und Planung. In: NIEDERSEN, U., POHLMANN, L. (Eds.): Selbstorganisation. Duncker & Humblot. Berlin. p. 109-127.

KÜPPER, W., ORTMANN, G. (Eds.): (1988): Mikropolitik – Rationalität, Macht und Spiele in Organisationen. Westdeutscher Verlag, Opladen.

LUHMANN, N. (1995): Social systems (Writing science). Stanford University Press, Stanford.

MANDELBROT, B. (1977): The Fractal Geometry of Nature. Freeman, New York.

MATURANA, H.R., VARELA, F.J. (1980): Autopoiesis and cognition: the realization of the living. Reidel, Dordrecht.

MEYER, M. (1994): Ziele in Organisationen – Funktionen und Äquivalente von Zielentscheidungen. Gabler, Wiesbaden.

MÜNCH, R. (1980): Talcott Parsons und die Theorie des Handelns II: Die Kontinuität der Entwicklung. Soziale Welt, 31:1.

MÜNCH, R. (1996): Risikopolitik. Suhrkamp, Frankfurt am Main.

NEUMEIER-MACKERT, I. (1996): Wir wollen das Versprechen einlösen, daß Client-Server-Anwendungen mehr Vorteile für das Unternehmen bringen. AV, 33:5.

PARSONS, T. (1969): Politics and Social Structure. Free Press, New York.

PARUNAK, H.V.D. (1987): Manufacturing experience with the contract net. In: HUHNS, M. (Ed.): Distributed Artificial Intelligence, Volume 1. Pitman Publishing/Morgan Kaufman, San Mateo. p. 285-310.

PICOT, A. (1992): Marktorientierte Gestaltung der Leistungstiefe. In: REICHWALD, R. (Ed.): Marktnahe Produktion. Gabler, Wiesbaden.

REICHWALD, R. (1979): Zur empirischen betriebswirtschaftlichen Zielforschung. Zeitschrift für Betriebswirtschaft, 49:6. p. 528-535.

ROPOHL, G. (1979): Eine Systemtheorie der Technik – Grundlegung der Allgemeinen Theorie. Carl Hanser, München.

ROTH, G. (1986): Selbstorganisation – Selbsterhaltung – Selbstreferentialität: Prinzipien der Organisation der Lebewesen und ihre Folgen für die Beziehung zwischen Organismus und Umwelt. In: DRESS, A., HENDRICHS, H., KÜPPERS, G. (Eds): Selbstorganisation – Die Entstehung von Ordnung in Natur und Gesellschaft. Piper, München.

SPRENGER, R.K. (1996): Das Prinzip Selbstverantwortung – Wege zur Motivation. Campus, Frankfurt am Main.

TIROLE, J. (1988): The Theory of Industrial Organization. MIT Press, Cambridge.

TITSCHER, S., KÖNIGSWIESER, R. (1985): Entscheidungen in Unternehmen – Zur Theorie und Praxis des Umganges mit Krisen wechselseitiger Abhängigkeit. Wien.

TÜRK, K. (1989): Neuere Entwicklungen in der Organisationsforschung – Ein Trend Report. Stuttgart.

WARNECKE, H.-J. (1992): The Fractal Company – A revolution in corporate culture. Springer, Berlin.

WEBER, M. (1979): Economy and society: An outline of interpretive sociology. University of California Press, Berkley.

WEICK, K.E. (1979): The social psychology of organizing. 2nd Ed. Addision-Wesley, Reading Ma.

WIENDAHL, H.-P. (1989): Betriebsorganisation für Ingenieure. 3rd Ed. Carl Hanser, München.

WILLKE, H. (1992): Ironie des Staates – Grundlinien einer Staatstheorie polyzentrischer Gesellschaft. Suhrkamp, Frankfurt am Main.

Planning, Scheduling and Control: Aspects of Decision Making

Part Four

Approaches to Shop Floor Scheduling – A Critical Review

Franz Stuber

The problem of the computerization of conflicting objectives of shop floor management serves as an example to illustrate the reasons for the continuing discrepancy between scientific attempts at automated planning and control performed by software functions and their realization in available software systems. The investigation examines optimization attempts in the area of operations research, research on artificial intelligence and more recent, so-called 'nature analogous' methods. Findings provide the basis for a rejection of the notion that production management be automated in the most extensive manner. In conclusion, some theses regarding reorientation in research and development are formulated.

1. Introduction

In many industrial engineering approaches, attempts were and are still made to solve the problems of operative production and shop floor management by means of refining and automating scheduling and control techniques. At the same time, such approaches meet with little response from both users and producers of planning and control systems, those whose functioning the approaches intend to improve. This state of affairs is the subject of controversy within the scientific community (see, for example, MCKAY ET AL. 1988; CORBETT & WASSENHOVE 1996), and discussion becomes more virulent, because many enterprises faced with modern production concepts withdraw from over-designed planning and control methods in MRP systems and prefer to place their stakes on more flexible decision-making within production (see HEEG & HIRSCH 1995; SCHERER ET AL.

1996). In the following, therefore, a critical analysis of well-known methods will throw light upon the difficulties confronting automation approaches.

2. Conflicting objectives of shop floor management

Shop floor management comprises a multidimensional spectrum of disparate goals and objectives. Within this spectrum, there are decisions to be made which pertain to technology, business management and also to the social field. In shop floor management, the production methods, schedules and quantities determined by strategic production management are subject to the determination, distribution and control of processes and resources.

2.1. Degrees of freedom and restrictions

From the perspective of individual processes, choices are made among available alternative processing methods, input materials, means of work and persons responsible for the tasks. For example, differing quantities allow for, according to workpiece geometry, different ordering of the steps within the process. Further, with regard to choice of resources, there is a specific range of variation which prevails. Figure 1 shows an example where diverse means of work are appropriate to the execution of individual processing steps. According to choice of alternately available machines, the number of necessary operations will vary, so that in the example presented, without yet considering variations in the order of the process, there are already four different variants.

task / variant	**cutting to length, centering**	**turning, 1. side**	**turning, 2. side**	**drilling, flange side**	**milling, peripheral**
var. 1	cutting machine	universal-lathe	revolver-lathe	multi spindle drill	universal milling machine
var. 2	cutting machine	lathe with synchronized 2nd spindle		manual drill	universal milling machine
var. 3	cutting machine	universal-lathe	lathe with powered tools		universal milling machine
var. 4	cutting machine	lathe with synchronized 2nd spindle		machining center with lathe	

Fig. 1 Alternative choice of machines (following REINHARD 1995).

Here, shop floor management faces, among others, the following decision situations:

- For scheduling purposes, should the criteria be the machine which achieves the shortest manufacturing time, in order to produce the product as fast as possible?
- Or should the guiding criteria be the machine in which the most capital has been invested, so that it can be amortized as quickly as possible?
- To what degree should rate of capacity utilization of individual machines determine choice?
- Are short processing times counteracted by long waiting periods, as shown by past experience?
- Should routings and machines be chosen according to the number of workpieces to be produced in the same way in order to reduce set-up times?
- Should progression from machine to machine be dictated by the realization of the shortest transportation distances?
- What assurances and experiences exist to the effect that the necessary raw material and tools can be readied at the desired time and in the required quality and amount?

Depending upon which objective is chosen, different schedules result, which themselves then affect non-prioritized goals. Additional areas of decision arise with regard to the choice and distribution of responsibility for tasks among various persons:

- Which machine choice, or order of sequence, will require qualifications which are held by only a few persons?
- Are such specialists already charged with carrying out other tasks during the time period foreseen for the manufacturing process?
- Should scheduling aim for an even workload among workers in order to avoid overtime and underwork?
- Should the job be given to the worker who has the most routinized command of the tasks involved, or to a person who can gain further qualifications by taking over the tasks?

Considerations such as those, illustrated by the above examples of the advantages and disadvantages of individual decisions, show us clearly that it is not a simple matter to make the 'optimum' choice. The individual considerations themselves contain contradictory aspects as long as the focus is upon the technical or the social field, and become all the more so once one attempts to integrate them all. The art of shop floor management consists to a considerable degree in the ability to find a viable compromise solution.

2.2. Economic goals and evaluation

The various areas and alternatives in decision-making from the perspective of individual processes are embedded within economic goals which transcend the processes. The goals of high capacity machine utilization, low inventory in production and warehouse, short lead times and high delivery reliability contain mutually contradictory aspects. Production enterprises strive to achieve full utilization of their resources because the capital sum expended for production is only amortized while work is done on the formal resources, e.g., machines. The individual components of the sum may be distributed among the manufactured products as piece costs. Measures which increase capacity utilization have a negative effect upon the other economic objectives: in order to achieve high capacity utilization, it is useful to operate with a considerable stock of materials, so that one has the flexibility to even out fluctuations in capacity by means of additional contracts and shifting of scheduled due dates. But this measure causes an increase in inventory and, thus, also storage costs. Reduction in set-up times can be achieved through a clever combination of various manufacturing tasks (batching). This , however, then requires a sufficient stock of materials and semi-finished products. And at the same time, batching increases stock and the costs involved, whether incurred in storing stock, in production or in storing finished products, as some work processes will be carried out too early in order to minimize set-up times. Set-up optimization has, in addition, a negative effect on the lead times of other, competing contracts, which will tend towards overstepping their planned finishing schedules.

Through low inventory, the least possible capital becomes bound where material is in the form of raw material or semi-finished products and not being processed or where no finished products await monetary realization. To reduce stock, various methods of calculating stock can be implemented and purchasing can be delayed, according to 'just in time' principles, until the materials are to be used. But low inventory limits flexibility for capacity regulation, because, for example, work-process sequences with optimum set-up times fail due to a lack of available raw materials. Low stock also lowers the possibility for prompt filling of express orders. The same holds for repeat orders which may be required due to insufficient production quality, if amounts of material are needed which were not initially foreseen. Low inventory of finished products places limits on delivery capability and raw material stock calculated too closely will lengthen lead time for finished products, in so far as the need for replenishing raw material falls near delivery dates. Stock for production calculated too minimally can even lead to periods of machine standstill and to underwork on the part of production personnel.

From the perspective of short lead times, friction can also arise. As already shown, measures to reduce lead times can come into conflict with the goal of full capacity utilization, if schedules are to be changed in order to optimize set-up times. A possible alternative – to ready more material and hire more personnel –

would of course lead to a decrease in the degree of utilization and active working time. Short lead times may also be reached by means of stocking greater quantities of raw material and semi-finished products, directly contrary to the goal of decreasing stock outlined above. Figure 2 points up this contradiction among the economic objectives characterized.

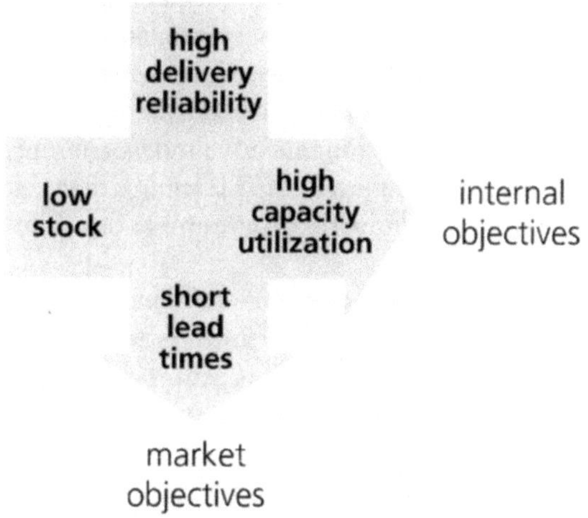

Fig. 2 Goal conflicts in shop floor management.

These conflicts, known as dilemma or 'polylemma' in production management, show clearly that given a certain organization of production and work, one goal is achieved only at the cost of not achieving another (see KURBEL 1993; STUBER 1997). Indeed, active pursuance of one goal not only leaves another out of consideration, but actually produces friction with regard to the current status of progress towards other goals.

The balance, then, regarding the objectives and challenges of shop floor management can be summed as follows: in order to achieve the optimum in any individual enterprise, case-specific opinion-forming, evaluation of objectives and conflict resolution and management on the basis of technological, social and economic assessment criteria will be a prerequisite. The character of the decisions is anticipatory and beset with many insecurities, so that evaluations in terms of an optimum, as well as the variables which themselves influence the optimum, change within the course of the decision-making process. The change is dynamic and also a matter of chance. In addition to the question of realization, the economic success of the chosen alternatives in particular will ultimately only be proven in the long-term – through success of the enterprise in an international, ever more turbulent arena.

3. Efforts to automate shop floor management

3.1. Approaches within operations research

Traditional models of operations research

The attempt to grasp the problems of shop floor management in formal models of automated generation of optimum plans has a long tradition particularly in the context of operations research. Partial and total, or simultaneous models may be distinguished. Partial models aim to automate part problems such as determination of lot size, optimum set-up planning or sequencing. Simultaneous models attempt to integrate within one planning approach numerous relevant planning areas, in order to take the mutual dependencies among individual determinations into account.

In the context of mass production, for example, a number of models were developed to ascertain the 'economic lot size'. The problem definition results from the fact that building up order amounts and size of manufacturing contracts involves costs which increase with amounts on the one hand (for example, storage costs), but also involves costs which decrease with increasing amounts (for example, set-up costs) on the other hand. The task is to find a cost-favorable balance between rising stock costs with increasing lot size and rising set-up costs with decreasing lot size. The models developed meet certain assumptions with regard to the planning field, upon which the mathematical optimization model is constructed[1]. However, whether or not these premises and model elements correspond to any real planning situation is purely coincidental. Relevant decision parameters of the real enterprise are cut out in order to make calculation according to the model possible, and/or the actual content of real connections are submitted to reduction until they finally 'fit into' the models' automatically-feasible formalism. For these reasons, the history of the development of optimum lot size models consists mainly of a series of refinements of premises and modifications of the calculation methods reflecting them. After half a century of research and model development, a textbook on production planning and control states today that:

> „The deluge, branching out in all directions, of lot size models – which arose from the ambitions of the researchers to overcome the theoretical difficulties and, possibly, from the necessity of inventing ever-new topics for theses and dissertations – stands in no commensurate relation to the practical usefulness of lot size planning."
>
> (KURBEL 1993, p. 41, see also WIGHT 1974, p. 24).

---

[1] The classical calculation, first published by HARRIS (1913), and known as „economic order quantity" (EOQ) (see PLOSSL 1985, 35f, WIGHT 1974, 24), optimizes order amounts in view of costs. The lowest limit for order amount and replenishment time are set at zero, endless stock access rates and constant stock movement rates are assumed.

That, indeed, the practical relevance of lot size calculation models tends towards zero, is demonstrated in a recent study:

> „Of 167 studied production planning and control software systems, only 17 systems contain special multiple-step[2] lot size methods according to the descriptions of the software producers. But as these products name no method, or could not name a method known from the literature, we can assume that even many less production planning and control systems, or perhaps none, actually have a special multiple-step lot size method at their disposal, and that rather, only one-step, one-product methods are applied successively across manufacturing stages." (FANDEL ET AL. 1994, p. 168f).

Scientific research on automation approaches favors, by far, specific aspects of job sequencing and machine scheduling. Within the framework of so-called 'scheduling theory', the crux always deals with one variation or other of wrestling with the following defined problem: Given m manufacturing orders, which are filled at n number of machines or workplaces, and according to whether all orders will be filled in the same sequence of work (assembly line manufacturing) or in a variety of sequences (idealized job shop manufacturing), the following are calculated: the most favorable manufacturing due dates and stages with regard to one chosen objective (for example, minimal lead time, maximum utilization, highest delivery reliability) (CONWAY ET AL. 1967). Up to now, no satisfactory calculation method has been found to solve even this reductionistic formulation of the problem. In the literature, this is usually explained by a reference to combinatory explosion. The following is representative of many:

> „In one shift, 99 orders are to be worked on a machine that should be set up alternately as little as possible. In order to find the best plan, all manufacturing possibilities would have to be investigated. That would make 99! sequences to determine and evaluate. It is assumed that sufficient computing capacity is at our disposal to solve the problem within one day. But if two new orders are added, this means that 101! sequences would have to be studied, which would require additional computing time of 100 x 101 days, or approximately 27.7 years."
>
> (SCHMIDT & MEYER 1995, p. 12).

If one now attempted to extend the model through formalization of conditions relevant to practice, such as shift planning, varying manufacturing duration or manufacturing alternatives, the sheer complexity of the problem so formulated would become clearly apparent. Even if it is assumed that the necessary 'computer

[2] Multiple-step lot size methods take account of level of decomposition or manufacturing stages in their calculations.

power' were available for the volume of the task when practice-relevant – the speed of hardware development continues to surprise – there are a number of problems which hinder the undertaking:

- Projecting automatic optimization of machine reservation is based upon the assumption of a deterministic data situation: in order to have relevance for practice, (1) all model parameters would have to be not only known, but they would have to (2) remain unchanged throughout the course of time between making the plan and realization. The optimum calculated result is after all based on them. Thirdly, any disturbances in carrying out the plan would render the optimum, calculated at great cost, invalid.
- A further difficulty, which has received little attention in the literature, also enters into the picture: automated optimization with regard to a objective function assumes unrealistically that this exists in pure form, so to speak. It is assumed that an unambiguous decision has been made by the enterprise favoring one of the competing economic objectives (minimal lead time, minimal stock, maximum capacity or high delivery reliability) rather than any other. Additional relevant goals, such as product quality, development of personnel qualifications or the innovative contributions of a decision are not even considered. Only under these assumptions can automated optimization be carried out according to the objective striven for in the model. In actuality, however, these contradictory goals are followed *simultaneously* in the real world of shop floor management, and the optimization of one of these goals has a negative effect on all the others.

The difficulties incumbent in quantifying conflicting planning objectives, entering them into a formal model and performing computerized calculations so that they have meaning in real practice has led to the failure of attempts to apply the methods of linear programming, decision-tree methods in dynamic programming and approaches known as 'branch and bound' methods. The latter seek, by means of dexterous branching, to eliminate from further calculations early on all partial solutions whose non-optimality is recognized (see KIM 1993; VASKO ET AL. 1993). The principle remains ever the same: the undertaking isolates and reduces one of the diverse and contradictory chains of effect to a calculable model, which is optimized according to a certain objective function. The difficulties sketched briefly here are also particularly valid for those attempts which use *simultaneous models* to connect the automated optimization of lot size and scheduling (for example, JOHN 1994) as well as concepts of *hierarchical production planning* based upon simultaneous approaches (for example, HAX & MEAL 1975; KISTNER & SEVEN 1991).

Scheduling rules

For the transition from a completed production plan to order initiation, there are various concepts of so-called priority, or scheduling rules. With their use, automated choices are generated as to which orders or operations should be executed next (see HAUPT 1989). Scheduling rules have a differing effect on competing economic objectives. The 'shortest processing time' rule, for example, supports high capacity utilization and short lead times for orders having relatively short operations. At the same time, however, it has a negative effect on delivery reliability, as long-range orders are left undone. Delivery reliability is advantaged through the 'shortest remaining processing time' rule which favors orders with the smallest time buffers to delivery date and tends to lengthen general lead times. To counteract and balance out overly strong effects of individual scheduling rules, diverse simulations with various combinations of such scheduling rules are carried out. In the literature, nearly two dozen (combined) scheduling rules have become known (PINEDO 1995).

Their practical usefulness is a matter of considerable controversy. The reason is that the general difficulty of applying scheduling rules is that, on the one hand, the isolated application of one scheduling rule often too strongly favors one of the competing planning objectives at the cost of the others, and the connection to the negative effects on other goals which arise gets lost. On the other hand, combining several scheduling rules does not help: their implementation does not allow economic effects to show up as cause and effect. Through the combination of more than one rule, their effects tend to cancel each other out, and transparency of manufacturing control is lost. A more recent approach to production management notes the following with regard to the usefulness of scheduling rules:

> „A complicated control system with difficult scheduling rules... can only prefer some urgent production orders against [!] other 'less important' orders, the waiting times of which will be increased and so the whole production system works in an endless vicious circle."
>
> (KOSTURIAK & GREGOR 1995, p. 495).

From a scientific perspective, there is nothing more to add to this 'endless vicious circle'. Even if attempts are currently being made to complement the concept of scheduling rules with so-called *interaction rules*, for the reasons discussed no lasting improvement in shop floor management will be achieved by means of the application and refinement of scheduling rules.

3.2. 'Artificial intelligence' approaches

Approaches towards automated machine reservation in particular have produced diverse experiments within the context of research on artificial intelligence (AI)[3]. Here we often find that no longer is the goal followed to produce *the* optimum plan. Rather, an attempt is made, by means of various approximation techniques, to generate a nearly optimum plan and to combine scheduling and rescheduling. Rule-based and constraint-driven approaches are meeting with the increasing competition of approaches which are based on 'fuzzy logic', the technique of 'case-based reasoning' or 'nature-analogous' methods (for an overview, see SZELKE & KERR 1994, FOX & ZWEBEN 1994).

The popularity of these latter techniques with regard to the question of automated machine reservation is the result of the difficulty – in the face of the complexity, unsteadiness and contradictory nature of the matter – of finding an adequate *explicit representation* and *algorithm*. While the rule-oriented expert system approach held that planning can be reduced to a number of 'if ... then' relationships, which need only be duly processed by an inference machine for the problem solution to be reached, these new techniques place different accents. They involve experimentation with more recent search, select and generating methods and their refinement.

Fuzzy logic

Fuzzy logic, which is based on the theory of fuzzy sets, transposes percentile values into so-called 'linguistic variables', whose values are not numbers, but rather words. These are defined, by means of a membership function in the interval (0,1), as a fuzzy set (see ZIMMERMANN 1987). The reason that fuzzy logic is applied to shop and floor management:

> „Motivation for this choice was the ease of formulating knowledge that reflects the complex non-linear behavior expressed by the human experts. Additionally, fuzzy sets are well suited to model knowledge containing vague human-like formulations. Such formulations can often be heard from human experts explaining their domain.“ (DORN & SLANY 1994, p. 637).

[3] This paper does not deal with the philosophical assumptions of research on artificial intelligence. I will say that one would have to begin with the 'artificial intelligence community's' popular definition of the computer as a 'physical symbol system', as formulated by SIMON (1988), for example: „The computer is a member of an important family of artifacts called symbol systems, or more explicitly, physical symbol systems. Another important member of the family (some of us think, anthropomorphically, it is the most important) is the human mind and brain“ (SIMON 1988, 26f.). Even his philosophical starting point of a 'physical symbol system' asserts a *contradictio in adjecto*: symbols are, as we know, things which *represent something other* than they themselves *are*. In so far as symbols suffer or produce *physical effects,* this would be according to their physical constitution or nature, and certainly not according to their *meaning* as symbols in a particular context.

There are certainly fields where fuzzy logic may be applied meaningfully, as shown by its success within control engineering. However, implementation in the not only technically-determined decision arena of shop floor management poses questions to its promoters. To the extent that a 'fuzzy' formulation of attributes strives to modify the concept of scheduling rules, the same critique is valid that was made of traditional scheduling rules. For what happens with computer-internal handling of fuzzy rules is nothing other than a shift in choice points determining when one or another rule should go into effect. In this way, application of the fuzzy technique to priority rule-controlled decisions does not represent progress. What the advantage to shop floor management – or 'model knowledge containing vague human-like formulations' – is supposed to be is described as follows:

> „Naturally all possible due dates should be met, lead times should remain short and available resources should be used as well as possible. However, only through linguistic variables can such 'fuzzy' goals be described and automatically processed.“ (GLADIS ET AL. 1993, p. 5).

Just how it is that reformulation and representation of the conflicting goals in the form of 'fuzzy' and 'linguistic' variables might cancel out or do away with the *real conflicts* remains the authors' secret. Their message clearly shows the *will* to master the difficulties involved in automating conflicting planning goals through means of modeling „the vague information in natural human language into a form accessible to electronic data processing“ (GLADIS ET AL. 1993, p. 4). A mistake in a thought content such as an assessment of 'medium or low utilization' and its computer *notation* is, however, indispensable: to the computer, this evaluation will gain no *meaning* when it is expressed differently for the computer's internal operations.

Case based reasoning

The case based reasoning approach start outs from the fact that in shop floor management there are often problems to be handled which were already existent and were dealt with in the past (KOLODNER 1993). This program of research aims to structure available data in such a way that, as known 'cases', they can be utilized in automated derivation of a new plan. In order to achieve this, formalisms are developed with regard to the 'similarity' among data attributes and their values, which are computer-evaluated for comparisons between the real task and stored case data. The fact that evaluation of two sets of facts is lost through such representation as a mechanical operation in the computer and that instead, one – however cleverly constructed – operator conducts the comparison of two variables in reference to distance measures already defined at the start is noted by researchers of case based reasoning, but in a very curious manner:

„Finally, a serious problem is the issue of *how* memory can be realized at all and, in particular, how such a system can be brought to remember the *correct* cases. Mainly there is the danger that solutions to old cases be applied blindly, even if they are completely inappropriate to the problem at hand.“ (LENZ 1994, p. 19, original italics).

Here, the difficulties for automated planning sketched out do not lead to the consideration that appropriate selection criteria could be developed for potential users who *themselves* could assess which would be the 'correct' cases. Rather, future case based systems are even attributed with learning ability – at this point, the origin of such programs in theoretical AI is all too apparent:

„One of the significant advantages of case based reasoning lies in the coupling of problem-solving and learning. While classical expert systems have as a prerequisite knowledge acquisition which is completed and merely [!] deal with problem solving, methods of computer learning aim mainly to represent explicitly the knowledge contained implicitly in case examples... The system remembers the old solution and can put it to immediate use.“ (LENZ 1994, p. 17).

If we for the moment leave aside the enthusiasm of the above citation that the system has 'knowledge' at its disposal or even can gain knowledge itself[4], 'machine learning' consists in giving the knowledge 'hidden' in the data a helping hand, so that an adequate representational form can be developed. Special comparative operators efficiently sort and filter data on the basis of quantitative distance measures and generate new data. What is left in the end is nothing more than a programming technique.

A general limitation of the method is the fact that such operators operate by definition upon *past values* and thus ultimately cannot replace evaluation of adequate procedures in real 'cases' (see also MCKAY ET AL. 1989).

Experiments with 'nature analogous' heuristics

The so-called nature-analogous methods set remarkable new accents as they increasingly distance themselves from computer-aided representation of *known* connections and conditions. The methods are interesting with regard to shop floor management, because they attempt, to varying degrees, to dispose with explicit algorithms and favor experiments. In the context of machine reservation and utilization, diverse experiments have been carried out in recent years using methods of 'simulated annealing', 'genetic algorithms' and 'artificial neural nets'.

---

[4] NAKE (1992) makes this comment: „Representations of knowledge are very different than knowledge itself. They refer to knowledge, evoke it perhaps, but are not *themselves* knowledge. For knowledge is not delivered in buckets which then stand around, as much as some like to think of books and similar storage medium in this manner.“ (NAKE 1992, p. 199, original italics).

Simulated annealing refers to a method which was developed in answer to the challenges presented by physical cooling processes. Application to machine utilization bases upon a *random* shift in the work steps of a valid initial plan. The new schedule is then evaluated according to a objective function chosen in advance (minimal lead time, for example) and either rejected or retained as the starting point for the next shifts. Further, the method operates with the parameters 'dynamic acceptance probability' and 'temperature', which are decreased at each step of a certain cooling plan (compare KURBEL & ROHMANN 1995). The price of doing without an explicit choice of the next work process is that influencing the optimization process through the set parameters may only be tested experimentally.

> „There are no generally valid rules for the design of plans, so that 'good' decisions usually may be made only on the basis of tedious testing by comparing alternative parameter sets – and without the possibility of determining the general validity of the quality of the decision reached."
>
> (SIEDENTOPF 1994, p. 25).

The same difficulties confront approaches which use *genetic algorithms*. These methods imitating genetics consist in a multiple-step, randomly determined ('mutations') process of choosing among the produced problem-solution candidates ('chromosomes'). The principle of genetic algorithms seeks to produce new solutions through selection and reproduction of existing solutions. In selection, intermediate results are chosen according to relative 'fitness'. For this the interaction of selection with reproduction must be controlled by an external 'acceptance' rule.

Models using genetic algorithms also suffer from the fact that doing without an explicit solving method yields only apparent help. Namely, in order to suppress the inadmissible solutions that are also necessarily generated by the method, a „repair algorithm" (KURBEL & ROHMANN 1995) must also be introduced into the process, which ensures that 'genetic development' proceeds along the desired pathway.

The final approach to be discussed disposes completely with *knowledge* of how optimal machine utilization is achieved. *Artificial neural nets* arose from a 'sub symbolic' variant of AI ideology which asserts that human and machine are functionally equivalent and mistakes the material bases of thought with cognition itself. Artificial neural nets conceive of the developer (and thus of course the user) as a 'black box', thereby making a *principle* out of that which others lament: the lack of transparency of software systems. This modeling technique is no longer consistent with a traditional understanding of systems development. Here the developer becomes an *experimenter*, who either accepts or rejects the results calculated by the system.

> „How the system changes within itself in order to produce a hopefully acceptable result cannot be influenced by the experimenter, never mind understood." (HOEPPNER 1991, p. 155).

In order to free the method from its initial *total arbitrariness*, diverse formalisms and so-called 'learning functions' were developed to evaluate the stochastic output data (backpropagation). HOEPPNER (1991) pointed out the fact that the activation functions of the 'neurons' are indeed algorithmic elements – computing rules, in order words. Some success in applying the connectionistic paradigm, particularly in the fields of technical pattern recognition and the production of spoken language, inspired various attempts to model scheduling problems with the aid of artificial neural nets and to compare them to other 'nature-analogous' methods as well as traditional optimization methods (for example, see KIM ET AL. 1995, SINGH ET AL. 1995).

Results to date have been somewhat sobering, raising the question of whether *replacing* knowledge of certain cause and effect relationships with random-based methods will indeed present the solutions needed. A recent study has come to the following conclusion:

> „The best results were reached by a human scheduler. Trained in machine utilization planning with electronic leitstand[5] systems, the human, who attempted an optimization by hand, achieved the shortest cycle time by far!" (KURBEL & ROHMANN 1995, p. 592).

In addition to the difficulties mentioned, the reservations outlined with regard to OR methods apply equally to the new heuristics:

- the abstraction of practice-relevant situational conditions, in order to make the model 'calculable',
- the necessity of an underlying deterministic data situation,
- the foreseen failure of the optimum when there are disturbances in the execution of the schedule,
- the unrealistic assumption of isolated optimization of one objective against the others and without taking into account the resulting negative effects on these others.

There is also the fact that 'nature-analogous' methods dispense with computer-internal representation of known connections as algorithms in favor of the

[5] The 'leitstand' is a computer-aided decision support system for interactive shop floor control (ADELSBERGER & KANET 1989, KANET & SRIDHARAN 1990). The concept of the leitstand was originally developed in Germany in 1980 and is available commercially as stand-alone software system or as an integral part of a larger production planning and control software system. The term has become commonplace in English language literature.

delegation of 'input data' to a randomly determined selection – in the case of artificial neural nets programmatically, even, to a *non-comprehensible mechanism.* This leads to a strong emphasis on data as the results of earlier planning. Even if success is reached in forming recommendations for actual planning by means of extraction and generating methods, these always reference back to the past. Shop floor scheduling, however, consists in *anticipating* processes yet to be realized. As to the radicality of the procedure, in particular, in the case of artificial neural nets, we must ask if not the knowledge acquired in many decades of research on interrelationships in shop floor management is here being thoughtlessly thrown 'overboard'. While the connectionistic paradigm is of doubtless interest with regard to research and development on simulation of natural phenomena as well as technical processes, it fails to be of service when automating goal-oriented procedures which are neither natural in origin nor of purely technical content nor free of contradictions.

A paradox situation arises: the very striving to automate shop floor management has led to the current tendency to give up on algorithms in favor of experiments. With this, the research approaches have worked their way forwards to a contradiction: they conduct optimization of shop floor management with the aid of non-transparent and arbitrary methods and recommend them to production practice.

4. A plea for reorientation in computer-aided production management

Although the 'nature-analogous' methods also did not bring the breakthrough hoped for, the research scene discussed in the paper seems to remain abstinent towards the real tasks and problems of production management. In a recent textbook on scheduling management, we find, alongside pure theory, astonishing things. For example, there is a list of 12 (!) reasons why the techniques do not function in reality, but from this no conclusion is drawn for the future of research (PINEDO 1995, p. 254ff). In another instance, in a chapter on 'Case Studies', the author describes a production control system in which not one (!) of the methods described in the theory section was implemented. And that is not all: at the time of publishing the book, the system had already been taken from the market five years previously.

The failure to deal with the reality of shop floor management results from holding fast to the goal of full automation to be achieved at some point in the future. An acknowledgment that shop floor management cannot be mastered mathematically in a comprehensive way is long overdue. What has not been heeded, or perhaps what has been ignored for ideological reasons, is the fact that the transposition of planning and control tasks as formalized by work organization to a mechanism which can be realized in computer software results in something of a very different character. Therefore, in conclusion, five suggestions for future research and development are as follows:

1 In order to computerize mental operations such as shop floor management, two different abstractions are required: first, a formal model must be designed. Here quantified units of real or imagined actions are determined and directions for the combining of their quantitative values are established. In computerizing, a second step follows, in which the formal model is transposed into an artificial mechanism. What previously belonged to the spectrum of tasks of the work, now becomes a part of means of work. On the basis of the principles of construction of the modern computer, there is a general universality – in so far as a subject of study may be transformed into data – which allows combination rules to be given in order to structure data. An algorithm may also be given for processing the data according to the type of calculation desired. This universality is however subject to the limitation of calculability: a problem can only be delegated to the computer if it can be formalized. The model chosen thus determines the type and the degree possible of the computerization of mental contents. NAKE (1992) aptly termed this process „the mechanization of thinking“[6].

2 What tasks of shop floor management may be expressed as data and whether algorithms can be found for particular processes depends to a considerable degree upon the particular *design of an organization.* Production management organized in departments far from the production process will present the designer with a very different view than a production organization in which the shop floor is given relevant powers of decision. For this reason, the extent of the division of work, whether at present or as planned, must serve as a first orientation when considering whether mechanization should be conceived as automation or interactive usage. The development of a model which can fit the relevant aspects and their representation as data and algorithms into a description that works, is thus confronted with various work organizational contexts. The modeling process must meet the demands of the particular context. One can tell upon looking at a systems concept the type of tasks it was conceived for and the type of organizational form with which it corresponds.

3 As was illustrated in part 2, shop floor management does not just consist of industrial engineering processes. Rather, the planning and control of work processes operates with technological, economic and social objectives and variables. There are numerous values to be quantified, such as planned processing time in dependency upon the number or geometric design of the processing object, money values, delivery times, capacity, and so on. But there are also elements which play significant roles and which are difficult to

---

6 „Mechanization of thinking refers to the process through which the thinking parts are separated out of the work process (given or planned) and transferred to the machine.“ (NAKE 1992, p. 193).

formalize completely. If, for example, the appropriateness of a chucking device must be judged, diverse calculation methods might be implemented to assess the necessary strength, overall size or behavior under pressure. Whether or not the optimal values calculated by the methods will actually be realized, however, depends on further decision parameters: resulting costs, real availability of materials, time needed as compared to other tasks, whether an existing, similar device could be slightly modified in this instance of planning, etc. Similar difficulties would arise in connection with evaluating the prognosis of the advantage of a further training measure or in connection with choosing the correct action in the face of unplanned disturbance.

4 Although referred to generally as 'intelligent' programs, what computer-aided problem solving actually achieves lies not in the area of mental judgment, but rather in the fact that *judgements made in advance* on characteristic features and connections of shop floor management are put into a form that can be processed *by the computer*. Skillful computer-aided modeling and calculation may, for example, produce a blinking diagram indicator, telling us the 'The job is critical!', when it has determined a delivery delay or that too little material has been calculated. Computer control of material supply or the meeting of due dates *eliminates* the need for constant manual time-checking and stock-taking of inventory. In this sense, judgements on the necessary connection among individual facets are computerized. But the blinking indicator is the result of calculating processes, not of understanding, in which the judgements manifest themselves. It is different if we speak of judgement processes themselves: in the context of judgements made in shop floor management ('the job is critical!'), the human abilities of mental identification do not, formally, express the very same content.[7] The indicator gains *meaning* only for the user who *knows* and *understands* the situation and context causes[8].

5 It is particularly important in regard to computer-aided shop floor management that the necessary modeling not only be expended in relation to expected gain. As shop floor management moves within a field of conflicting goals and

7 In the process of mental judgment formation, the special ('job' in the example) is determined through and by thinking, and the general ('critical') made into something that is valid for the special. In the identity asserted or shown here, the grasping of a fact is expressed as judgment. The mental practice of determining identity and differentiation at one and the same time is not only a case for 'expert judgments', but for every decision and evaluation: and it represents a significant contrast between human thought processes and computer-based combinations of data and algorithms. This principle difference between human and computer is noteworthy, because in the context of information technology research and development, the functionalistic view presumes equivalency between live thought and computer-aided problem solving. A recapitulation of general determinants of human thought processes on the basis of a critical analysis of Hegelian judgement theory is presented in FISCHER ET AL. 1995.

8 Along with a certain entertainment value we can find examples of this circumstance in regularly published marginalia of the 'computer age', such as when the pornography identification program in the World Wide Web blocked the White House Children's Page, because the President and his wife were introduced as a married 'couple'.

demands, computerization comes up against absolute limits: elements of the decision process, when transposed for computer processing, are stripped of their conflicting character, in so far as they have taken on the function of the means of work. As long as quantitative measures and value proportions can be found for conflicting goals, comparisons can be carried out and desired optimum among the values can be calculated. The qualitative problem of optimization of one goal producing friction with the achievement of other objectives is reduced here to quantitative relations. But which of the conflicting decision criteria should take priority is at its core a *qualitative* question, and not a question of quantity. The dialectic quality of human judgment formation, which makes decisions on the basis of known contradictions and does not follow previously determined behavioral strategies, stands in contrast to computer processing. Computer processing irons out the real goals and tasks with their inherent demands and conflicts, in order to make them accessible to formal representation and calculation.

In modeling, these peculiarities of computer-aided shop floor management must be taken into account and consciously designed. Research efforts should focus on how to adapt the disparate decision field to dialogue-oriented systems and on showing users the consequences of their decisions within the field of conflicting objectives and interests. In this way, more of a contribution towards mastering real tasks in shop floor management and towards defusion of operations planning and control problems will be achieved. And perhaps this will also serve to defuse the tug of war that MARK FOX described:

> „... the future direction of scheduling will be a tug of war between seeking scheduling solutions and altering production processes. He believed that the latter will occur to simplify the former." (cited in: HSU ET AL. 1993).

5. References

ADELSBERGER, H. H, KANET, J. J. (1989): The leitstand – A new tool in computer-aided manufacturing scheduling. In: STECKE, K. E., SURI, R. (Eds.): Proceedings of the 3rd ORSA/TIMS Conference on Flexible Manufacturing System. Elsevier, Amsterdam. p. 231-236.

CONWAY, R., MAXWELL, W. L., MILLER, L. W. (1967): Theory of scheduling. Addison-Wesley, Reading.

CORBETT, C. J., WASSENHOVE, L. N. VAN (1996): Crisis? What crisis? Four decades of debate on Operational Research. In: FORTUIN, L., BEEK, P. VAN, WASSENHOVE, L. N. VAN (Eds.): OR at wORk. Practical experiences of Operational Research. Taylor & Francis, Bristol. p. 289-304.

DORN, J., SLANY, W. (1994): A flow shop with compatibility constraints in a steelmaking plant. In: FOX, M. S., ZWEBEN, M. (Eds.): Intelligent scheduling. Morgan Kaufmann, San Mateo. p. 629-654.

FANDEL, G., FRANCOIS, P., GUBITZ, K.-M. (1994): PPS-Systeme. Grundlagen, Methoden, Software, Marktanalyse. Springer, Berlin.

FISCHER, M., JUNGEBLUT, R., RÖMMERMANN, E. (1995): Jede Maschine hat ihre eigenen Marotten! Instandhaltungsarbeit in der rechnergestützten Produktion und Möglichkeiten technischer Unterstützung. Donat, Bremen.

FOX, M. S., ZWEBEN, M. (Eds.)(1994): Intelligent scheduling. Morgan Kaufmann, San Mateo.

GLADIS, A., WEGENER, R., WEILER, A. (1993): Unscharfe Mengen. Termintreue verbessern und Durchlaufzeit verkürzen mit Hilfe der Fuzzy-Logik. MM – Maschinenmarkt, 33.

HARRIS, F. W. (1913): How many parts to make at one factory. The Magazine of Management, 10:2, p. 135-136.

HAUPT, R. (1989): A survey of priority-rule based scheduling. OR Spektrum, 11:3, p. 3-16.

HAX, A., Meal, H. (1975): Hierarchical interpretation of production planning and scheduling. In: GEISLER, M. A. (Ed.): Studies in Management Sciences. Vol. I, Logistics. North Holland, Amsterdam.

HEEG, F. J., HIRSCH, B. E. (Eds.)(1995): Effektiver EDV-Einsatz in der werkstattnahen Produktion. Bremer Schriften zu Betriebstechnik und Arbeitswissenschaft. Verlag Mainz, Aachen.

HOEPPNER, W. (1991): Konnektionismus, Künstliche Intelligenz und Informatik – Informationsverarbeitung und die Semantik von 'ist'. Informatik Spektrum, 14:3.

HSU, W.-L., MCKAY, K.N., OW, P.S., PINEDO, M. (1993): Integration of disciplines for applied scheduling: a workshop report. Production Planning and Control, 4:4, p. 416-421.

JOHN, U. (1994): VeRPlan II – Verteilte rückgekoppelt-zyklische Produktionsplanung. Arbeitspapiere der GMD, 861. Sankt Augustin.

KANET, J.J., SRIDHARAN, V. (1990): The electronic leitstand – A new tool shop scheduling. Manufacturing Review, 3:3, p. 161-170.

KIM, Y.-D. (1993): A new branch and bound algorithm for minimizing mean tardiness in two-machine flowshops. Computers and Operations Research, 20:4. p. 391-204.

KIM, S.-Y., LEE, Y.-H., AGNIHOTRA, D. (1995): A hybrid approach to sequencing jobs using heuristic rules and neural networks. Production Planning and Control, 6:5. p. 445-454.

KISTNER, K.-P., STEVEN, M. (1991): Operations Research in der hierarchischen Produktionsplanung. OR-Spektrum, 13:3.

KOLODNER, J. L. (1993): Case-Based Reasoning. Morgan Kaufmann, San Mateo.

KOSTURIAK, J., GREGOR, M. (1995): Total production control. Production Planning and Control, 6:6. p. 490-499.

KURBEL, K. (1993): Produktionsplanung und -steuerung. Methodische Grundlagen von PPS-Systemen und Erweiterungen. Oldenbourg, München.

KURBEL, K., ROHMANN, T. (1995): Ein Vergleich von Verfahren zur Maschinenbelegungsplanung: Simulated Annealing, Genetische Algorithmen und mathematische Optimierung. WI – Wirtschaftsinformatik, 37:6. p. 581-593.

LENZ, M. (1994): Fallbasiertes Schließen für die Produktionsplanung. Arbeitspapiere der GMD, 863. Sankt Augustin.

MCKAY, K. N., SAFAYENI, F. R., BUZACOTT, J. A. (1988): Job-shop scheduling: What is relevant? Interfaces, 18:4, p. 84-90.

McKAY, K. N., BUZACOTT, J. A., SAFAYENI, F. R. (1989): The scheduler's knowledge of uncertainty: the missing link. In: BROWNE, J. (Ed.): Knowledge based production management systems. Elsevier, Amsterdam. p. 171-189.
NAKE, F. (1992): Informatik und die Maschinisierung von Kopfarbeit. In: COY, W., NAKE, F., PFLÜGER, J.-M., ROLF, A., SEETZEN, J., SIEFKES, D., STRANSFELD, R. (Eds.): Sichtweisen der Informatik. Vieweg, Braunschweig.
PLOSSL, G. W. (1985): Production and inventory control. Principles and techniques. 2nd Ed. Prentice-Hall, Englewood Cliffs.
PINEDO, M. (1995): Scheduling. Theory, algorithms, and systems. Prentice-Hall, Englewood Cliffs.
REINHARD, T. (1995): Wirtschaftlicher produzieren aufgrund modifizierter Arbeitspläne. wt – Produktion und Management, 7:8, p. 347-350.
SCHERER, E., SCHÖNSLEBEN, P., ULICH, E. (Eds.) (1996): Werkstattmanagement – Organisation und Informatik. vdf, Zürich.
SCHMIDT G., MEYER, J. (1995): Theorie der Werkstattsteuerung. CIM Management, 11:1. p. 11-14.
SIEDENTOPF, J. (1994): Anwendung und Beurteilung heuristischer Verbesserungsverfahren für die Maschinenbelegungsplanung – Ein exemplarischer Vergleich zwischen Neuronalen Netzen, Simulated Annealing und genetischen Algorithmen. Leipzig.
SIMON, H. (1988): The Sciences of the Artificial. Cambridge.
SINGH, K., SCHNEIDER, B., KURBEL, K. (1995): Solving sequencing problems using recurrent neural networks and simulated annealing – A structural and computational comparison. In: DERIGS, U., BACHEM, A., DREXL, A. (Eds.): Operations Research Proceedings 1994. Springer, Berlin. p. 400-405.
STUBER, F. (1997): Rechnerunterstützung für arbeitsprozeßnahes Planen. Software-Innovation im Kontext von Ökonomie, Organisation und beruflicher Bildung. Donat, Bremen.
SZELKE, E., KERR, R. M. (1994): Knowledge-based reactive scheduling. Production Planning & Control, 5:2, p. 124-145.
VASKO, F. J., WOLF, F. E., STOTT, K. L., WOODYATT, J. R. (1993): Adapting branch-and-bound for real-world scheduling problems. Journal of Operational Research Society 44:5. p. 483-490.
WIGHT, O. (1974): Production and inventory management in the computer age. CBI Publishing, Boston.
ZIMMERMANN, H.-J. (1987): Fuzzy sets, decision making, and expert systems. Kluwer-Nijhoff Publishing, Boston.

The Link to the 'Real World': Information, Knowledge and Experience for Decision Making

Jürgen Fleig and Robert Schneider

In many companies the planning and control of production is a very complex process involving several different operational functions and people. In the 1970s and 1980s, in order to fulfill planning and control tasks effectively and efficiently, complex planning and control algorithms and computer based systems were designed and used in industry. As part of a comprehensive CIM philosophy, it was thought that these systems would help to create the fully automated, 'man-less' factory. This was a global paradigm (MILLER, DE MAYER & NAKANE 1992). Frequently, however, the hoped-for successes did not materialize.

Implementers of CIM had failed to recognize the great importance of planning and control to workers on the shop floor, who 'experience' the production process at first hand (material flow, warehousing and processing) and influence it by their actions. This is one important reason why interest has been focused on organizational innovations tending towards 'lean' structures, in which the factory worker once more plays a central role. The worker again takes on much of the responsibility for processing orders fast, efficiently and in adequate quality – thus also assuming responsibility for the planning and control of material flow in production. It can therefore be said that a change of paradigm has taken place over the last few years (RUIGROK & VAN TULDER 1993).

If they are to take on the tasks of planning and control, workers on the shop floor need adequate information, knowledge of the relevant interrelations and above all – their own experience. In psychology, research has been done for many years on determining the importance of implicit and explicit „knowledge from experience" for people and their actions (POLANYI 1967). This also applies to actions in the factory (FLEIG & SCHNEIDER 1995). In the following contribution, we would like to demonstrate the importance of knowledge from experience on the factory floor, show what additional information is needed for the decentral planning and control of production, and indicate the important factors involved in shaping organization.

1. The Importance of experience in decision making

1.1. Production needs knowledge from experience

The development of the concept of experience-guided work, and thus the transfer of research findings on knowledge gained through experience to industrial production began in the second half of the 1980s. BÖHLE & MILKAU (1988) and BÖHLE (1989) described the importance of complex sensory perception and the importance of an 'affective' relationship to the work object in the mastering of work processes. Using the example of work with CNC tool machines they demonstrated the efficacy of experience-guided work processes. In a first study on „Computer-aided experience-guided work", completed in 1989, it became clear that the observable developments in computer use in production might inhibit this important form of action (MARTIN & ROSE 1992): For instance, the progressive encapsulation of tool machines was not accompanied by any technical developments which could compensate for the lost opportunities for cognitive sensory perception of the work process.

The trend towards computer-integrated production, dominant at that time, also failed to take account of experience as a source of information. It targeted the types of information and information sources that are accessible to computer processing. The re-organization of production was oriented exclusively towards technical requirements and neglected the important 'human resources' altogether.

However, experience-guided adjustments and interventions are necessary for the regulation of production, particularly if external influences such as fluctuating demand or client-specific product requirements create a need for changes in production situations that occur at short notice and cannot be planned for in advance. Moreover, due to the complexity of industrial production systems, it is not really possible to foresee exactly the impact generated by even small variations and interferences. The production process is thus dependent on continual regulatory interventions by humans.

People 'function' here as stabilizing 'regulating elements'. However, by their intervention they may also alter the predefined goals of the production system, thus changing and renewing the whole system.

1.2. 'Theory of Action' and knowledge from experience

Since the beginning of the 1970s, the theory of action in work psychology has addressed the issues of the initiation and control of activity and the transfer function of activity between the subject and his environment. Work activities basically pursue the ideal of rational action, planned and executed along a hierarchy of goals. The theory of action regulation, as formulated by HACKER (1973) and VOLPERT (1974, 1980) and modified and extended over the years, identifies a hierarchical-sequential activity structure. It shows that action regulation, in order to be complete, is dependent both on elements of planning and on elements of execution and control. This theory is one important basis of the concept of experience-guided work.

Another basis of the concept is LEONTJEW'S (1977) description of work activity as an element of transfer between the person and his environment[1]. Activity always changes the acting subject as well as the object. By effacing the clear delimitation between the person and the environment, the interrelation between work design and organizational form is rendered visible. Moreover, the process of exchange described by LEONTJEW makes it possible to explain such phenomena as motivation and the assignment of meaning and to describe the impact of work activity on the person. It becomes clear that notions such as intuition, 'hunches' and feelings not only have to be a part of our understanding of the realities of production, but also have to be taken into account when planning production; these notions with their subjective, corporeal connotations are poorly regarded in the rational world of technology, but are essential for describing human work in general and experience-guided work in particular[2]. The model concepts based on these theories and findings are briefly described in the following section[3].

1.3. The basis of experience-guided work

Knowledge through experience is an autonomous mode of knowledge (BÖHLE & MILKAU 1988). It is a type of knowledge based on practical, concrete experience. It is accumulated through action and is largely linked with the situations in which the action took place. It includes the conscious or unconscious use of memories of the physical sensations associated with these situations. One of the important features of experience-led work is that this knowledge, together with corresponding sensations, is activated in similar situations and is transferred to

---

1 A brief overview of the various transitions is given, for example, by FRIELING & SONNTAG (1987, p. 39ff.).

2 On the importance of physical activity as a source of experience, VOLPERT (1990) refers, among other sources, to the philosophical notion of HEIDEGGER. He appeals to employers to create work tasks which involve „manifold bodily activities" (VOLPERT 1990, p. 31). More 'concrete' indications are to be found in empirical memory research. Thus ENGELKAMP (1990) demonstrates the existence of a motoric „memory effect".

3 The following remarks on the basis and models of experience-led work are largely based on the descriptions in CARUS, NOGALA & SCHULZE (1992), SCHULZE & CARUS (1994) and WITT (1994).

them. Thus the similarity of situations is directly 'established' rationally and analytically in the work activity and, at the same time, it is also intuitively 'lived' and 'experienced' via complex associative links of cognition and perception.

This *associative cognition* is a basic element of experience-guided work. The similarity of a current situation to a previous one, or with one expected to occur, is established through associative, sensory cognition than potentially makes use of all sources and channels of information. The sources of information are both the concrete and conceived environment, knowledge and also feelings. The channels of information are the senses relating to the environment (sight, hearing etc.) as well as the cognition of 'inner' feelings, states of mind and moods[4]. An intuitive assessment of the current situation, guided largely by feelings, takes place on the basis of a comparison with earlier experienced situations. A role is also played in this comparison by the situational development *expected* (anticipated) at that point in time, possibly in conjunction with a desired divergent development. The associative linking of 'images', individual sensory perceptions, feelings, items of knowledge and patterns of action can give rise to amazing combinations, making a high degree of flexibility possible in the 'identification' of similar situations and in the utilization of previous experience.

In situations where time is a critical factor, the intuitive processing of manifold individual perceptions enables a degree of process mastery which an analytical-rational approach can not achieve. The unconscious transposing of perceptions of the situation into feelings allows the production process to be efficiently monitored and influenced (a sequence of noises sounding 'good' or 'bad', a situation that looks 'good' or 'bad').

The result of an investigation by KONRAD (1992) on strategies for dealing with interferences substantiates the effectiveness of associative perception: according to this study, experts were found to use mainly symptomatic patterns, direct sensory impressions and historical information whereas, by contrast, beginners tended to operate according to patterns of minimal effort and the use of documentation.

A very flexible and rapid transfer takes place of knowledge from experience to new situations, but this is not without risks[5]. This dual effect makes it clear that experience-guided work is dependent upon continuous *evaluation of efficiency and risk* and an *inner willingness to assume responsibility.* These aspects of work performed by employees constitute further basic elements of experience-guided work. One prerequisite for the utilization of these abilities is worker participation

---

4 This perception of the formation of experience is to be found in similar form in the neurobiological cognition model of MATURANA & VARELA, when they write: „When we speak here of the relationship between action and experience, it would be a mistake to confine this only to our environment, that is, to the 'physical' world. For this characteristic of human action applies to all the dimensions of our lives." (MATURANA & VARELA. 1987, p. 32).

5 MATURANA & VARELA (1987, p. 33) possibly mean something similar when they write: „Usually this conjuring up of a world through cognition is treated as a difficulty, a mistake or an unexplainable residue, as something which should be removed."

in the process of updating the goals and aims of production. From this, the worker can derive the information needed for risk assessment.

The knowledge gained through experience necessary to accomplish a predefined task represents, together with theoretical knowledge, a substantial, but not sufficient, information base for employees in production, where an intuitively correct reaction to interferences has to be put into action under time pressure. Other information bases are knowledge of the purpose of the work tasks relative to production as a whole, and knowledge of the purpose of the production system as a whole.

Direct and indirect contact with people employed in the same process or production system can generally be regarded as sources for the definition and updating of goals[6]. However, information about how the production system is perceived 'from outside' is also relevant.

1.4. The acquisition of knowledge from experience

Perception of the result of an action ('active observation' is also regarded as an action) leads to the new acquisition, modification or consolidation of knowledge from experience. The accumulation and utilization of experience are thus interdependent. A prerequisite for building up experience is a degree of 'spontaneous' variability and transformation of the situation into a state which is new, but which must possess sufficient similarity to that which has been experienced before. This 'new' state, if it is attributable to outside influences, requires *reactive* adaptation of the action pattern and thus constitutes a 'barrier' to goal attainment. Experience-guided work as described so far consists in overcoming these barriers efficiently, while adapting one's actions to accommodate risks. Experience-guided work stabilizes the processes. However, modification of experience can also be accomplished by changes in the subject's own behavior in a situation which, in terms of objective characteristics, remains constant in other respects. The significance of this 'exploratory behavior' for learning processes has long been recognized in educational psychology (BRUNER 1966). If changes in behavior are made for a specific purpose, they represent the *active* exploitation of experience and are thus an important source of innovation.

This function of experience-guided work also gains significance if it is placed within the broader context of the production process and thus includes processes of exchange with other people (c.f. KIM 1993). An *exploratory dialogue-based* mode of action and interrelated perceptions are the central elements of experience-guided work.

---

6 Relationships through work represent a substantial part of the 'social and societal embedding' which is important for individuals' perception of meaning; on this principle and on the implications for the shaping of work and organisation, cf. VOLPERT (1990, p. 34).

In view of these considerations, it becomes understandable that many of the actions of experts at work cannot be explained or described in terms of algorithmic or heuristic rules[7]. To describe and explain action, notions such as *intuition, experience* and *feelings* are needed; terms which designate the unconscious use of perceptions, knowledge, attitudes, motives and behavior patterns. It also becomes clear that a rather static term such as 'knowledge' does not adequately explain the processes of the acquisition and utilization of experience that are linked with a person in action. The use of unconscious sources of knowledge, the 'process' character of the updating of experience, and the physical, bodily element of perception are conveyed by 'experiencing oneself' in the work situation. This makes it difficult to convey and transfer experience fully to other people, particularly by the use of teaching methods which are exclusively oral and written. Due to the strong situational context and the importance of physical sensations, even the parts of knowledge from experience that are accessible to the conscious mind can only be conveyed incompletely to other people. In the final analysis, therefore, people to have the experience themselves.

> Experience-guided work is based on knowledge and behavior patterns that are only accessible in part to the performer of the work intuitively and in the context of the concrete situation, i.e. when doing the work. Large parts of such experience cannot be expressed and are linked to one individual. The following elements are fundamental to experience-guided work: interrelated perception as well as the intrinsic motivation to take responsibility in exploratory dialogue-based work actions.

At the level of work organization, these elements are influenced by measures aimed at integration of work activities. These include the components of:

- planning (differentiation of goals, designing procedures, deciding on them and anticipating them),
- operating/performing (cognitive and sensory-motor operations and actions),
- evaluating (perceiving and interpreting characteristics, perceiving/recognizing interrelations, feedback to the anticipated state) and
- participation in developing the goal tasks and production system tasks.
- the aim must be to provide the individual worker on the shop floor with numerous and varied opportunities to learn and thus acquire knowledge through experience.

7 This fact also sets limits to the possibilities of using so-called 'expert systems'.

1.5. Implications: Knowledge through experience and the organization of production

The representation shown in Figure 1 of the feedback processes important for experience-guided work follows the systems model outlined by MARKS (1991). As well as feedback that regulate action, it also includes feedback for determining goals and for assigning associative meaning. This 'systems model of the working person' is intended to clarify the 'adjustment points' that can be used to shape experience-guided work. The *feedback processes* should not, however, be confused with the feedback principles used in control engineering, which are valid for precisely defined processes in closed systems. An open socio-technical systems model should be used for production as a whole, as well as for the working individual. The system boundaries here are basically variable. The transitions between the resources belonging to the individual and those belonging to the individual's environment are fluid.

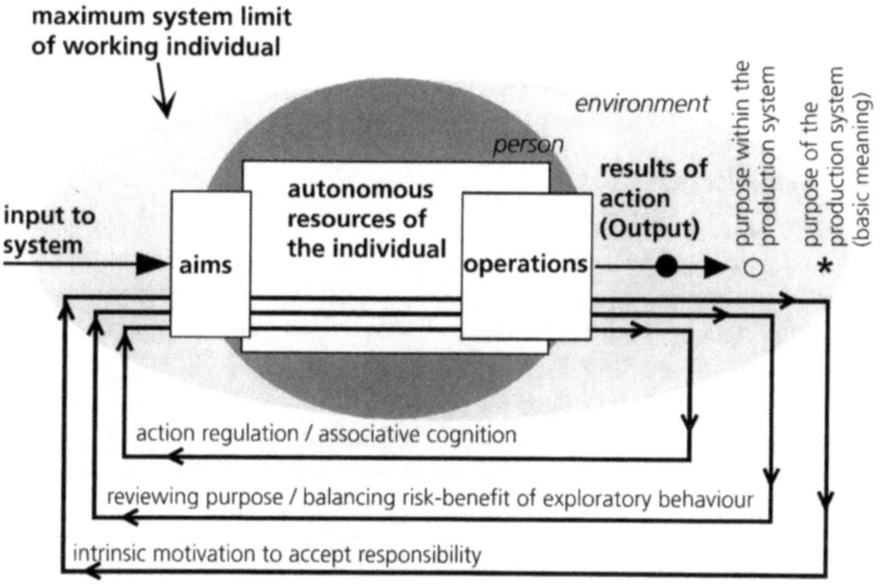

Fig. 1 Systems model of the feedback processes important in experience-guided work.

At the level of operations, the situation currently perceived is compared with one recalled or anticipated, and the comparison is used to correct behavior. In the main, this process takes place unconsciously, according to the pattern of the perception of interrelations described above. The determining of purpose, however, tends to take place in dialogue-based negotiation processes. The purpose (or reason for existence of the system) is defined through two-sided contracts (MARKS 1991). These may, for instance, be agreements about the quality and prospective use of the results of work between the working individual and his environment (involving colleagues and superiors as 'customers' for the results of the work, and as suppliers of the means of work, right on through to external customers and suppliers). The further the relationship of purpose extends beyond individual work tasks, the more a societally-relevant associative meaning can be assigned. This fosters an (intrinsic) willingness to take responsibility which arises of its own accord due to interest in the job.

In summary, it can be stated with regard to the use of experience in a production system that:

> The systems elements important for experience-guided work are those individual areas of freedom of action and feedback processes which are necessary in each case for exploratory dialogue-based action. This action yields associative perceptions and information on the goal and meaning of the task.

2. The information environment of the shop floor

2.1. Information on the shop floor and its importance

The central task of a skilled worker in production is still to effect one or several changes to a predetermined object (which may be a material, a part or a group of components), to process it in order to make it available for a subsequent process. However, the skilled worker increasingly has to take into account – often simultaneously – various different criteria which may be partially contradictory. There are basically two ways of doing this (Figure 2):

- Other people in the planning department plan all the necessary activities of the skilled worker (and his colleagues) in advance and issue the results of this planning as a clear set of instructions to the skilled worker.
- The skilled worker receives the necessary information, with which he can himself make the necessary decisions based on his knowledge, his skills and his experience and on the basis of which he can derive and perform the necessary activities.

A comprehensive analysis in twelve industrial enterprises (FLEIG & SCHNEIDER 1995) showed that in the area of shop floor control real 'pre-planning' is only possible in a few exceptional cases. In most cases, the planning departments themselves have to rely on the knowledge and experience of skilled workers. Moreover, on the shop floor numerous interferences crop up which cannot be anticipated in preliminary planning – WILDEMANN 1994 therefore suggests a change of paradigm in production planning and control, from „management oriented" towards the „elimination of interferences'". In particular, the skilled

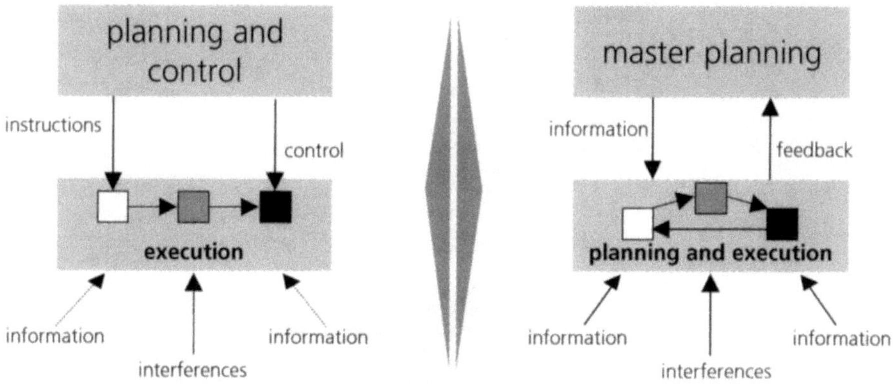

Fig. 2 Central and decentral forms of planning and control.

worker receives not only information on due dates, work resources and quantities – but also technological information (often specific to the situation) on the product, on set-up processes and on processing procedures (Figure 3). These numerous items of information have to be combined and processed correctly for the specific situation. Thus the skilled worker, with his knowledge, experience, and the decisions and actions derived from them, will (also) always play an important role in production control.

It is thus of particular importance that a concept for the technical support of shop floor control should also take skilled workers into account. Skilled workers must receive sufficient information to be able to make optimal decisions and perform the correct actions; unnecessary preliminary planning may narrow down the freedom to make decisions to such an extent that the optimum can no longer be achieved (FLEIG & SCHNEIDER 1995, p. 42). Analysis of the twelve enterprises in the survey showed that the following categories of information are relevant for skilled workers.

2.2. Information on the order

When the skilled worker receives an order, i.e. the materials, parts or components, on the shop floor, they are usually accompanied by a job card. This is a print-out of various data, including data on the article on which he must perform one or more processes, data on when his work should begin or be completed, on the customer receiving the product, a priority indicator, the number of articles he is required to produce and further technical information about the work sequence, instructions for quality assurance (QA), a drawing number (information on the product, which may, however, be specific to the order), etc. Many of these data are in coded form.

The skilled worker interprets these data in the light of his 'interpretation schemas', which have arisen situationally from his knowledge, his experience or from other additional information gathered from quite different sources. He knows, for instance, that on the due date for his order, he must be finished by

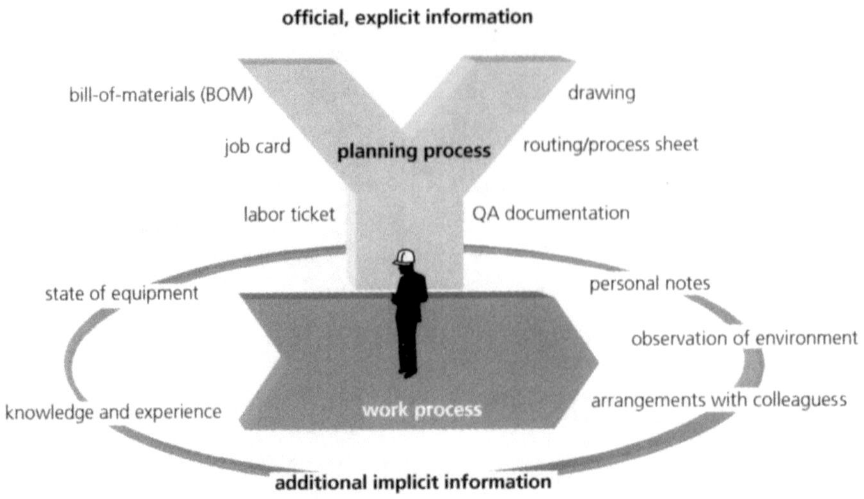

Fig. 3 The flow of information to the skilled worker.

eleven o'clock in the morning, because the carrier will be taking the parts to another company for surface processing. He may also have heard from a colleague that the order due date is no longer so urgent, due to missing parts that have to be bought in, and that priority is to be given instead to another order; he knows that the customer has particularly high requirements for the front face of the turned part, and that another customer does not have these requirements for the same part (he therefore has to select a different cutting speed).

2.3. Information on the product

At the same time order documents are issued, the skilled worker receives data on the product (which may or may not relate specifically to the order). This is generally technical information in the form of drawings, parts lists, instructions for assembly and for quality assurance, etc. Depending on the complexity of the product, the mass of data may be very substantial. In this case, too, numerous data can only be given in encoded form.

Again, the skilled worker has to interpret these data. It may happen that the data prove to be incomplete or ambiguous, so that he cannot interpret them clearly. A measurement has been left out on the drawing, or changes in design have not been followed up in the parts list, the instructions for assembly do not give information about the exact positioning of a cable. The expert worker has to complement the data with information from other sources. Again, his knowledge and experience come to his assistance. Often, however, he keeps his own notes, jotting down 'tips and tricks' or drawing up his own product files. Notwithstanding, he has to make inquiries of other people about the missing data; he asks colleagues who have already produced the part in question, he asks the engineering department for complementary information, he checks with the company on the future use of his product (where does the part fit in), and he consults technical handbooks and tables.

2.4. Information on the technological process

Lastly, the skilled worker receives data concerning the work process. These may be technical set-up data, information about setting fixtures, CNC programs or maintenance instructions. These data may also be in coded form.

Again, the skilled worker interprets these data in the light of his own expert knowledge and his experience, taking into account the present availability and performance potential of the resources he is working with. Thus it may turn out that CNC programs (written by another person) need to be altered to fit the specific situation. He procures additional information from machine handbooks on their use, on special settings and the frequency and details of the maintenance required. He also makes his own notes on setting attachments, possibly takes photographs or keeps a note of 'tips and tricks'. Discussions with colleagues can also be helpful here. While performing the work processes he keeps and uses data on the state of the process and the machine, to avoid and remedy interferences.

2.5. Information as a basis for planning and control on the shop floor

In day-to-day business, all these data are very closely linked and are interdependent to a large extent. The skilled worker can only interpret their informational content through their interaction, with situational and planning information complementing one another – as described above. Information also includes 'diffuse', informal information and sensory experiences ranging all the way to feelings and 'intuition'. The skilled worker can interpret them, for instance, by looking at the order shelf, where he can gauge the amount of work pending at his own workplace and in his department at the time and can assess what is urgent; at work, he can hear the change when tools become blunt or when there are irregularities in the properties of the material; when talking to colleagues he may happen to hear that a particular part is always causing problems in assembly, etc. (FLEIG & SCHNEIDER 1995, p. 52).

In fact, these factors also play a central role when the skilled worker makes decisions and performs actions. Very concrete issues in the planning and control process always lie behind these. Here are some examples:

- What orders are pending?
- What do the due dates look like?
- What are the order requirements?
- Is everything that I need available?
- What state are the available tools in?

- What points in the drawing are particularly important?
- What setting was used here?
- Where can I get the tools?
- Where is the relevant CNC program stored?
- Who has done this already?
- Isn't there a better way to do it?

These questions in turn have a special influence on production control, because they continually create a new reality in production, which only skilled and experienced workers can control in such a way that goals can be optimally fulfilled in production.

3. Creating an experience-enhancing environment

The following design requirements were developed on the basis of comprehensive empirical analysis of production enterprises (FLEIG & SCHNEIDER 1995).

Since knowledge from experience, and even more the utilization of experience, are inseparably linked to individuals, the role of computer support has to be defined quite differently from its role in the familiar CIM concept. It does not suffice to use - as the integrating components of a production process - only those items of information that are accessible to computer technology, nor can it suffice to optimize the production system only with these in mind. On the contrary: the design of computer technology support must be primarily oriented towards a broad exploitation of the production resource of experience, since only this contains the source of a continual, self-controlling renewal of the production system.

It is not computers that create the links between tasks in production companies, but the employees. An engineer is responsible for the functioning of a component in a design, a person in manufacturing is responsible for turning out the component in sufficient quality, on time and at reasonable cost. There is a clash of contradictory aims and different levels of consideration which have to be responsibly balanced in each individual case. Moreover, modern production is coming close to single-piece production, since this is required by the market. A *responsible* weighting of contradictory aims is the type of production integration that can only be effected by people, not by computers. It certainly makes sense for computers to support the *production function* of engineering, e.g., constructing the description of a component; many successful applications bear witness to this. The part that has not yet succeeded so well is computer support of the *act* of construction, which might, for instance, also include discussion with production experts about cost targets. Tasks which can be solved by dialogue do not require any algorithms produced by computers. The computer might possibly be brought into play here to act as a supplier and processor of information, as an aid to memory, as a supporter of communication and possibly also as a provider of tips. However, it cannot be used as a decision-maker. Moreover, the resources used to feed the computer with data should not exceed the benefits.

1 Thus three things are necessary to create experience-guided work:

The computer is not particularly well-suited to supporting activities at the shop-floor level that promote the acquisition of experience. There are several organizational recommendations for the division of labor and communication / cooperation that are better suited to this purpose.

2 So-called 'conventional' (non-computerized) means of accessing information are much more important to the shop floor than the computer, especially when we consider that the direct physical environment is a provider of information.

3 Notwithstanding, suitable computer support should be provided as quickly as possible, *particularly for skilled workers* on the shop floor, because they are not sufficiently involved in the flow of information, which in many companies is now almost completely computerized. And it is precisely this involvement which determines the possibility of realization of some organizational alternatives – those which allow enough freedom of action for workers to gain experience.

Thus the organizational options relevant to a person's work activity have to be identified first, and only then the technical means of realization. This sequence also emphasizes the higher priority of the organization, which, in its broader sense, includes the use of technology.

Division of labor

The creation of freedom of action in industrial production necessarily depends on the *division of labor.* It would seem advantageous to make direct use of human associative cognition, which is rapid and efficient, in the value-added process. This would mean assigning responsibility in a decentral manner for whatever processes possible. On the other hand, there are limitations here due to the increasing burden placed on individual employees if the supply of information and the area of responsibility becomes excessive.

The minimum degrees of freedom of action necessary to experience-guided work have clearly not been achieved if boredom and monotony signal that the activation level is too low for learning to take place. Indifference signals a lack of motivation ('inner resignation'). Thus there are limits in both directions to the extent of the situational variation to be mastered; these limits need to be modified again and again.

Division of labor should ensure that the feedback information necessary for mastering frequently recurring operative tasks can be directly derived from performing the work action itself. This supports explorative action and associative cognition. Division of labor is not favorable if it results in the segmentation of planning, execution and control components of tasks whose content is interrelated

(Figure 4). If this happens, regulation of goal attainment becomes more difficult. The information necessary for mastering a process cannot be gained through intuitive cognition embedded in action if task components that belong together are assigned to various different people. If, at the same time, there is no opportunity for dialogue-based processes of exchange through direct communication and cooperation, the only regulation that can take place will be based on and limited to indirect information.

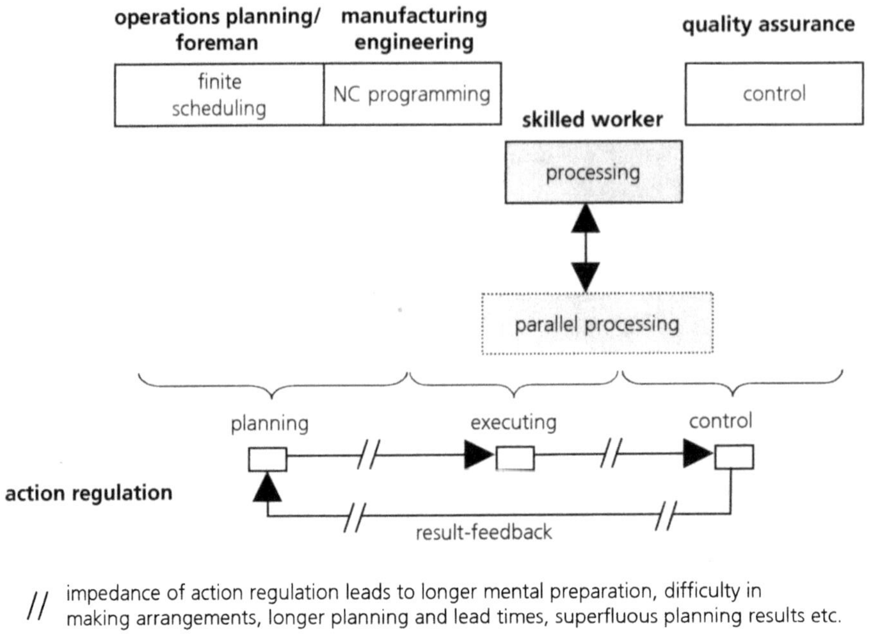

Fig. 4 Action regulation inhibited by erroneous task integration.

Communication and cooperation

In industrial production involving the division of labor, the goal of a task and any changes in goals are largely determined by situating the work within a larger general context – and thus, above all, by *communication and cooperative relations.* Therefore, as well as the possibility to engage in relations that help to ensure fulfillment of the operative task, there should also be an opportunity for relations concerned with innovative consideration of goals. For this purpose, contacts should be possible which are not confined to a narrow area of work and which, although they are linked with the person's own work, are not absolutely necessary for operative mastery of the task. Experience-guided work can only be permanently sustained in flexible production if the individual can place his own activity within a broader context. The innovative potential of a company is decisively strengthened by the application of experience gained through communication processes.

Participation in production innovations, in changing the way the production system performs, is indispensable to work guided by experience. This implicitly involves the important process of negotiating the purpose of the individual tasks. This purpose can serve as a desirable information base for ascertaining the appropriateness of explorative behavior. For most people working in industrial production, a renewal of purpose of this kind is only a secondary task. For this very reason, it is recommended that activities of this kind be specifically encouraged, for instance by an appropriate 'praise and reward culture'. This will provide recognition of the fact that use of one's experience in a way that goes beyond the individual workplace is the very personal achievement of every individual employee.

Each person's area of work is thus more comprehensive than the definition of the workplace would suggest, and it constitutes the smallest complete unit in the socio-technical system of production. Potential flexibility of delimitations is an important component of flexibility, and it enables very rapid adaptation to variable situations - when supported by appropriate conditions. As a change in the delimitations of one person's activity also always affects the areas of activity of colleagues, processes of negotiation are necessary. Since changing the delimitations requires not only factual information, but also understanding and assent, *direct contacts* are extremely important, with their possibilities for conveying additional messages by means of body language and facial expression quite apart from the exchange of spoken information. In *direct*, person-to-person contact, a negotiation may be successfully concluded between people who know each other well by as little as brief eye contact, for example.

One obstacle to these direct, informal arrangements and to cooperative task fulfillment is spatial remoteness. Another hindrance is posed by rigid regulations that only allow arrangements to be made via direct superiors (Figure 5). This discourages skilled workers from obtaining necessary information; colleagues from other departments come to production less often to avoid or clarify problems. Instructions are simply accepted, skilled workers try to help themselves in their own way and turn only to their foreman. Then, either setting-up is repeated in a time-consuming process, or NC programs are quietly corrected (and with repeated parts, corrected time and time again).

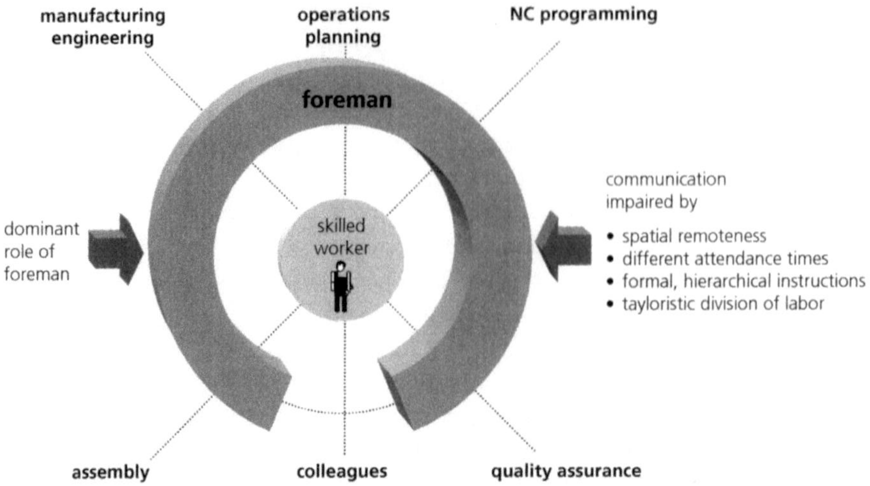

Fig. 5 Factors that may inhibit communication and cooperation.

Authorization and responsibility

Due to their exploratory and intuitive character, the acquisition and utilization of experience always represent a risk for the enterprise and for the acting individual. Only the consequences of an action in a new situation will ultimately show whether the goal, the decision and the action were appropriate. The inevitability of risk in the acquisition of experience and in its application means that *freedom of action* has to be granted by superiors and cooperating colleagues, and that associated *authorization* must be viewed by the person concerned as *responsibility*. Inappropriate degrees of freedom, or freedom which is not felt to be binding, represents an uncertainty factor in the production process. For the working individual, this implies the necessity of *willingness to assume responsibility for intuitive action,* to bear a risk (Figure 6).

This willingness should be rewarded as far as possible by the backing of colleagues and superiors and should spring from intrinsic motives. Thus to support experience-guided work, it is necessary to create conditions which, on the one hand, tolerate 'well-meant mistakes' (backing), and on the other, strengthen emotional participation in the object of the work, by enabling individuals to situate their own work within a broader, meaningful context. A suitable way to achieve this would seem to be to create opportunities for freedom of action which make participation in renewing the goals of production. This requires activities which encourage employees to place the company in a societal context and make them aware of this aspect – that is, activities that are externally oriented[8].

---

8 This imparts particular significance to contacts with customers and suppliers, but also to business events involving external guests and to 'open days'.

Fig. 6 Personal strategies in situations where there is a difference between authorization and responsibility.

Spatial proximity

In view of the importance of associatively interrelated cognition for work guided by experience, and in order to ensure availability of the necessary resources for exploratory dialogue-based action, it is particularly important to have potential sources of information within spatial proximity. The spatial distribution of resources (tools, measuring instruments, materials) must support direct sensory cognition. Objects that have to be arranged in configurations appropriate to the work activity should at the same time be arranged so that they are „accessible" to the sensory organs (Figure 7). Spatial proximity of cooperation partners opens up the possibility for spontaneous direct contacts and dialogue-based negotiation processes, and thus probably constitutes the most important 'resource' for both regulations and innovations.

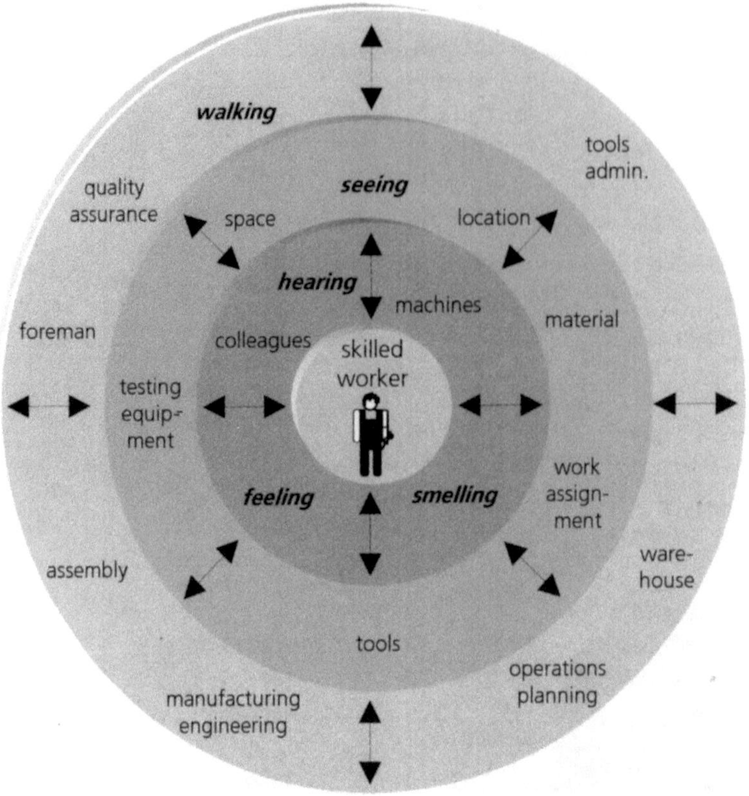

Fig. 7 Spatial environment of the skilled worker (workplace and area of activity).

Assigning work resources to individuals

A prerequisite for the mastery of complex working processes is to be able to intuitively separate process-related variations in the situation from other influences. Thus transparency of the state of resources is important, particularly when working with complex resources. *Assigning all work resources to one single person* fosters this transparency. By applying suitable measures, something similar to this advantage of assignment to one individual can also be achieved by assigning resources to several people (by arrangements with colleagues, arrangements made at shift changeovers, intensive communication and cooperation through teamwork concepts, etc.).

On the other hand, experience acquired with one set of work resources can be transferred to similar resources. Sacrifices with regard to speed and safety, and an increase in the amount of attention fixed on the work resources – and therefore not concentrated on the object of the work – are inevitable when the same resources are used by several people, however, and have to be weighed in the balance against gains in flexibility.

The division of labor between man and machine (automation)

The use of the expression *division of labor* applied to *man and machine* is intended to create awareness of the fact that production technology used for the automation of production processes has progressively lost its tool character. Automation represents a higher degree of technization, rather than mechanization. Using information technology, relatively complex procedures, and thus larger segments of production processes, can now also be automated at reasonable cost. Thus complete action sequences can also be transferred to machines (automatons), a development which represents a potential limitation to the freedom of action which is important for experience-guided work.

The monitoring of automated processes, which also allows for sporadic interventions into these automated segments, should thus build on experience gained with the process concerned at a non-automated level. Automation replaces, by a few steps, the previous subdivision of the production process into the many successive work activity components of planning, operation and control. The number of points at which regulatory action can intervene in the production segment is reduced. However, this creates the opportunity to control bigger process segments. One example of this is the case where the control program for a CNC machine tool for a complex component is completely installed with the help of a NC programming system. Here, the total planning process is summarized and decided upon prior to commencing work, which then takes place mainly automatically. However, with the NC programming logics being used at present, it seems doubtful whether newly-structured process regulations of this kind sufficiently correspond to the hierarchical-sequential structure of action regulation for optimal utilization of experience to take place (BOLTE 1993).

Thus automation always potentially impairs the acquisition and utilization of experience in the automated process segment. The mastery of regulatory action in the longer process segments arising through automation is impeded, if the type of automation merely corresponds to technical rationality and does not take into account the action structure of the employees.

Computer-technological functioning

Direct sensory perception is often better accommodated for by conventional support. Consideration should be given to the way a computer function will influence the shaping of the workplace. It should be clarified whether the re-structuring will also eliminate sources of information directly accessible to the human senses, and whether the option to communicate with colleagues will be reduced or impeded (Figure 8). The bits of information that are really important with regard to process mastery should not be obscured by too many additional items which are of limited or questionable value. The absolute menu of available functions offered should also be a factor when evaluating computer technology.

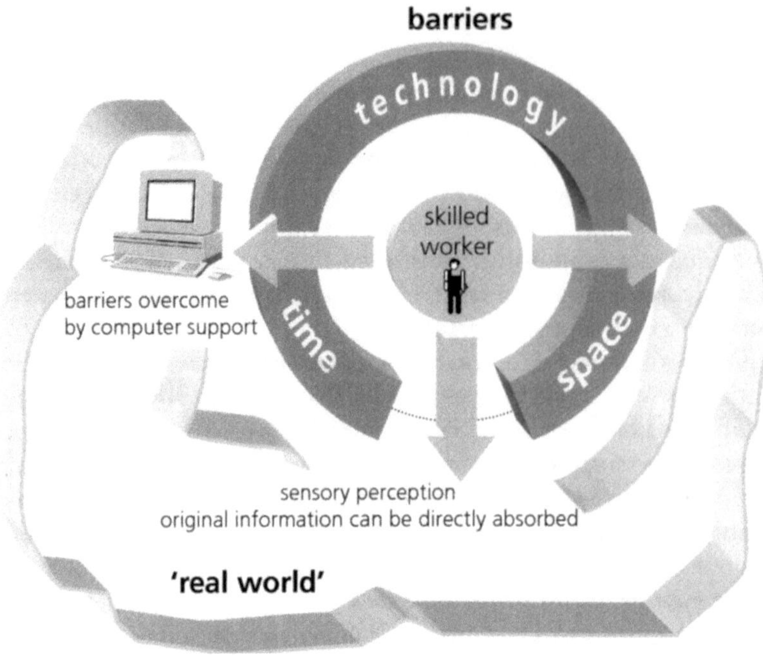

Fig. 8 Opening up the 'real world' by computer support.

Computer functionalities can and should also be designed deliberately to enlarge the degrees of freedom to act and to complement the supply of information used for the control and regulation of a process.

The user interface of computer systems

In principle, the transparency of production processes in computer-integrated production is impeded whenever the computer-internal processes embedded in a work process cannot (any longer) be perceived by the senses. Thus the concept of experience-guided work requires software ergonomics to take appropriate measures to compensate for the loss of perception in these 'blind' processes. One possibility for this could be multimedia feedback by the system on its current state, linked with corresponding intervention options. However, *transparency* has to be maintained. Adequate process stability, for instance, by deliberate limitation of the flexibility of automation, can reduce the volume of necessary system feedback and facilitate *clear* interpretations based on experience. Clarity can also be supported by constancy in the type of output and the device used.

Since work guided by experience is bound to individuals, but still allows, and is even fostered by, cooperative activities, activities of this kind should also be promoted intentionally by technical support. For tasks that have to be tackled cooperatively, the user interface of the technical system must also be specifically tailored to the cooperating groups involved.

Data storage and data access

One important characteristic of computer-integrated production is that individual data access can be expanded to include selected information from other departments of the company. The creation of larger feedback loops enabling this can be used not only for more efficient coordination with the other subsystems, but also to examine and modify the goal of one's own tasks. Particularly in flexible production, which often involves new products and processes, it must be assumed that there will be changes in predefined goals of one's own work and also in the results of that work. Access to information which is not, strictly speaking, necessary in order to master the operative tasks, but which contributes to establishing the context of use, yields the option of changing one's own actions, or even one's own task, exploratively, while taking due account of risks.

However, the emphasis must be on judicious selection of such indirectly used information. The selection must take place with the agreement and understanding of all those who generate, modify and use this information. Appropriate organizational 'rules of the game' must make it possible to review the needs for and rights of access from time to time and to adapt them to possible changes in the situation (Figure 9). Information should in any case be such that it does not replace, but rather stimulates direct personal contacts.

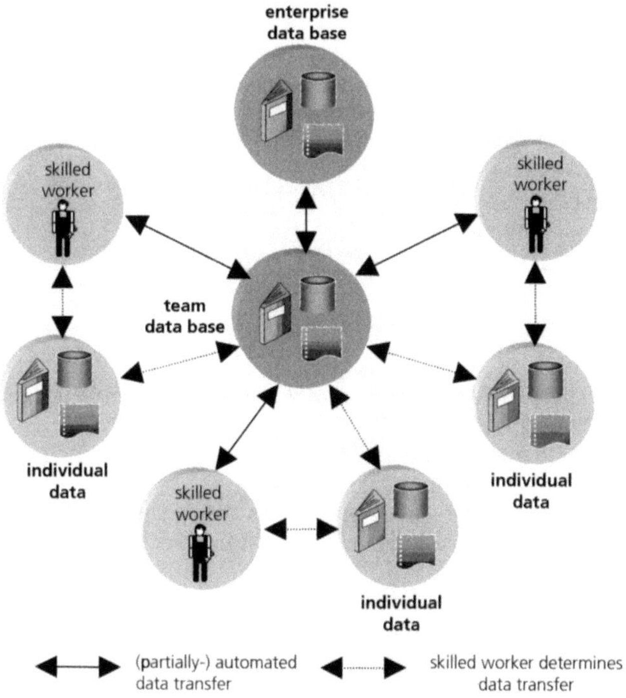

Fig. 9 Data storage and data access for the skilled worker.

Links exist between the 'personal elements' of experience-guided work and the general elements of the production system. The design of experience-guided work means that these elements be shaped deliberately. These statements indicate the importance of various organizational and technical dimensions of production for the design of experience-led work. If work on the shop floor takes place in an environment that fosters experience, the skilled worker will become a competent partner of the company (Figure 10).

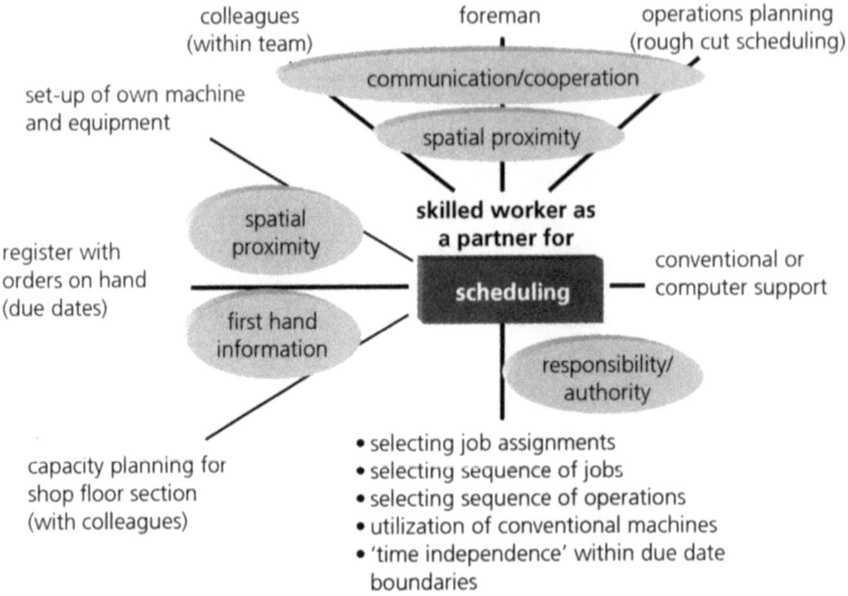

Fig. 10 Workplace fostering the planning experience of the skilled worker.

4. References

BÖHLE, F. (1989): Körper und Wissen: Veränderungen in der soziokulturellen Bedeutung körperlicher Arbeit. Soziale Welt, 40:2, p. 497-512.

BÖHLE, F., MILKAU, B. (1988): Vom Handrad zum Bildschirm: Eine Untersuchung zur sinnlichen Erfahrung im Arbeitsprozeß. Campus, New York.

BOLTE, A. (1993): Planen durch Erfahrung: Arbeitsplanung und Programmerstellung als erfahrungsgleitete Tätigkeiten von Facharbeitern mit CNC-Werkzeugmaschinen. Institut für Arbeitswissenschaften, Kassel.

BRUNER, J.S. (1966): Towards a theory of instruction. Belknap Press, Cambridge.

CARUS, U., NOGALA, D., SCHULZE, H. (1992): Erfahrungswissen im CIM-Umfeld. In: FHG ISI – FRAUNHOFER-INSTITUT SYSTEMTECHNIK UND INNOVATIONSFORSCHUNG (Ed.): Zwischenbericht des Forschungsverbundes CeA. 2 Vols. unpublished research report. FhG ISI, Karlsruhe.

ENGELKAMP, J. (1990): Das menschliche Gedächtnis. Hogrefe, Göttingen.

FLEIG, J., SCHNEIDER, R. (1995): Erfahrung und Technik in der Produktion. Springer, Berlin.

FRIELING, E. SONNTAG, K.-H. (1987): Arbeitspsychologie. Hans Huber, Berlin.

HACKER, W. (1973): Allgemeine Arbeits- und Ingenieurpsychologie. Deutscher Verlag der Wissenschaften, Berlin.

KIM, D. H. (1993): The Link between Individual and Organizational Learning. Sloan Management Review, V, p- 37-50.

KONRAD, U. (1992): Ingenieurpsychologische Untersuchungen zum diagnostischen Problemlösen in technischen Systemen. Working Paper A2-3/92, Sonderforschungsbereich 187, Ruhr-Universität Bochum.

LEONTJEW, A.N. (1977): Tätigkeit, Bewußtsein, Persönlichkeit. Klett, Stuttgart.

MARKS, S. (1991): Gemeinsame Gestaltung von Organisation und Technik in soziotechnischen, kybernetischen Systemen. VDI, Düsseldorf.

MARTIN, H., ROSE, H. (Eds.) (1992): CNC-Entwicklung und -Anwendung auf der Basis erfahrungsgeleiteter Arbeit. Wirtschaftsverlag NW, Bremerhaven.

MATURANA, H.R., VARELA, F.J. (1987): Der Baum der Erkenntnis. Scherz, Bern. (quotes translated from the German, English translation: MATURANA, H.R., VARELA, F.J. (1992): The Tree of Knowledge : The biological roots of human understanding. Shambhala Publications, Boston.)

MILLER, J. G., DE MAYER, A., NAKANE, J. (1992): Benchmarking global manufacturing. understanding international suppliers, customers, and competitors. Business One Irwin, Homewood.

POLANYI, M. (1967): The tacit dimension. Routledge & Kegan Paul, London.

RUIGROK, W., VAN TULDER, R. (1993): The ideology of interdependence: the link between restructuring, internalisation and international trade. Dissertation, Amsterdam.

SCHULZE, H., CARUS, U. (1994): Spezifische Merkmale erfahrungsgeleiteter Arbeit und deren technische Unterstützung anhand der Arbeit von Facharbeitern mit Werkzeugmaschinen. Universität Hamburg.

VOLPERT, W. (1974): Handlungsstrukturanalyse als Beitrag zur Qualifikationsforschung. Pahl Rugenstein, Köln.

VOLPERT, W. (1980): Beiträge zur psychologischen Handlungstheorie. Hans Huber, Bern.

VOLPERT, W. (1990): Welche Arbeit ist gut für den Menschen? In: FREI, F., UDRIS, I. (Eds.): Das Bild der Arbeit. Hans Huber, Bern.

WILDEMANN, H. (1994): Paradigmenwechsel in der Produktionsplanung und -steuerung. VDI-Zeitschrift, 136:10, p. 45-48.

WITT, H. (1994): Erfahrungsgeleitete Arbeit – Ein empirisch begründetes Handlungskonzept. Arbeits-, Betriebs- und Umweltpsychologie, Universität Hamburg.

Making Use of Information Technology: Human-Machine Interaction

Part Five

The Application of Information Technology at Shop Floor Level: Use and Acceptance

Eric Scherer

Order processing and production management has virtually become unthinkable without computer-aided information technology (IT). Even though industry has gained long experience with the application of IT, the undesired effects of IT in an organization are often neglected. Studies show, that in many cases the desired effects of the introduction of new IT systems were overshadowed by unwelcome dysfunctions. To achieve a sustainable improvement, the question of use and acceptance of IT has to be closely examined.

1. A case example

Within the framework of company-wide reorganization, the management of a medium-sized enterprise decided to purchase an integrated software package to support order processing. Following a successful initial phase – the implementation of the software in the purchasing department – modules for all other enterprise functions were introduced in a second phase. In the production area, a very efficient finite scheduling module was installed, while at the same time, the old central planning department was reduced, and the floor supervisor now took over more short-term planning tasks.

Shortly thereafter, things became uncomfortable on the shop floor. In addition to the problem of a growing inventory, the floor supervisor began to show signs of general discontent. They were beginning to suffer under the high demands of computer processing and the simultaneous increase of 'fires to be put out'.

What had happened? The computer system, designed to improve the situation, had pushed the shop floor to the limits of its ability to function. The problems were numerous, but they could all be traced back to a single cause: no one had given any thought to the way in which the new computer system could be used appropriately within their own company. Instead, management's maxim was along the lines of 'We paid for this system, so every function will be used' and 'If a function is missing, then it will simply be programmed'. In consequence, structures which had grown through the course of practice, were replaced by standard processes according to the systems developer's 'reference model'. With this, the enterprise lost its role as its own motor of organizational innovation. The 'key competency of innovation' was 'out-sourced' to the software developers and to the consultants they recommended.

2. Information technology – The motor of organizational innovation?

As a support to order processing and production management, it has become virtually unthinkable to do without the implementation of the computer. A study found that in 1994 in Germany, 87.2% of all firms utilized computer-based production planning and control systems (PPC systems[1]), and it reported euphorically, in view of numerous further projects to introduce such systems, that „The spread of PPC systems is approaching 100%" (STADTLER, WILHELM & BECKER 1995). Currently, more than 200 suppliers of PPC packages have stormed the German-speaking market. In addition, for the area of shop floor control, the 'electronic leitstand', developed for finite scheduling and shop floor control and sold commercially since the early 1980's, is of particular importance. In the meantime, there are probably 15-20 systems on the market which are really productively installed.

The introduction of computer-based systems has, to a far greater extent than all primarily organizational approaches such as KANBAN, segmentation or team work, characterized the innovation of the organization in production in the last 15 years. Companies and software suppliers alike postulate a positive relationship between computer and the success of an enterprise and, in this way, generate the idea of a 'universal remedy', which must only be swallowed to take effect. However, a close look at business reality is sobering: the goals which one hopes to achieve in production management through the aid of information technology (IT) have not been reached in the majority of cases (Figures 1, 2).

1 PPC systems are often also called MRP II (Manufacturing Resource Planning) or ERP (Enterprise Resource Planning) systems. In industry and even software development, the terms are used as synonyms even though from a theoretic perspective 'MRP II' describes a planning model while 'ERP' and 'PPC' refer to a core function in an enterprise. In general, commercial software packages include a range of function many of them beyond the production domain.

Company goal	Goal considered 'very important'	Goal achieved
Increase in delivery reliability	68%	29%
Decrease in lead times	65%	27%
Reduction of material inventory	48%	14%
Improved capacity utilization	41%	31%
Improved calculation bases	40%	41%
Increase in internal flexibility	39%	28%

Fig. 1 'Very important goals' and their achievement through the use of IT systems for production planning and control (KIRSCH, STROHM & ULICH 1994).

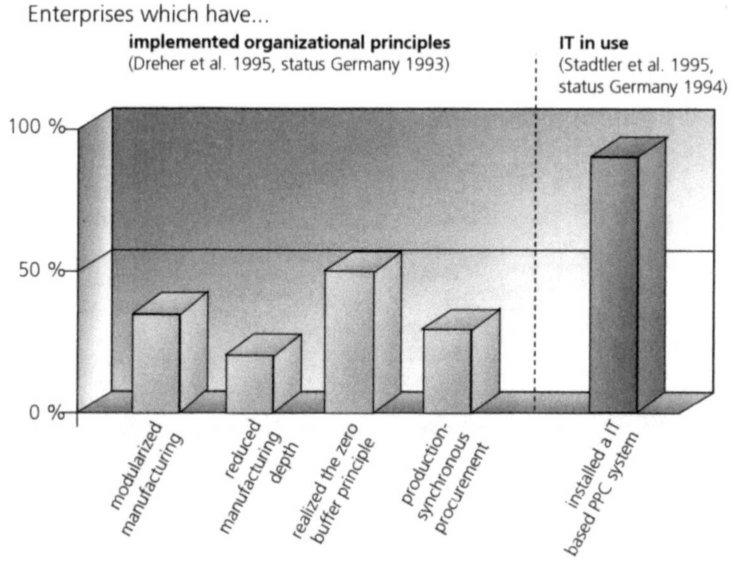

Fig. 2 The spread of new organizational forms in production (DREHER ET AL. 1995, STADLER ET. AL. 1995).

3. Diagnosis: Concepts and lack of conceptualization

Close examination of current concepts of the utilization of and of developmental tendencies of computer support of order processing yields unsatisfactory findings. On the one hand, general organizational concepts are sweepingly extolled – as catchwords – but they leave a multitude of detailed issues unanswered. On the other hand, the software market presents ever more extensive, greater capacity software packages, which are hardly transparent to the business practitioner. Still, industry maintains the 'automation paradigm'. As OLIVER WIGHT already noted in 1974, „computerizing a system that works well manually will seldom generate significant improvements." (p. 4). This leads to two alternatives: (1) reorganization has to be driven by IT capabilities, here IT is seen as an enabler of new forms of organization (this approach is key to the 'reengineering philosophy', see HAMMER & CHAMPY 1994, p. 83ff.), or (2) IT systems design and implementation has take into account the complex and closely coupled relations of reality to ensure an 'organization oriented' application of IT. This means, that IT can not be imposed on an organization but has to take into account the human users involved to achieve acceptance as prerequisite for an efficient use (see KURBEL 1995, p. 267). For industry, both approaches are important and have to form an equilibrium. Still, while the first is widely respected and present in marketing, consulting, and press, the second is often neglected but crucial in an organization where physical activity and manual labor – operating machines, handling material and maintaining the system – is still a mayor duty. The question as to how market-trendy software can be linked with a flexible organization to achieve effective and efficient order processing remains unexplained (Figure 3).

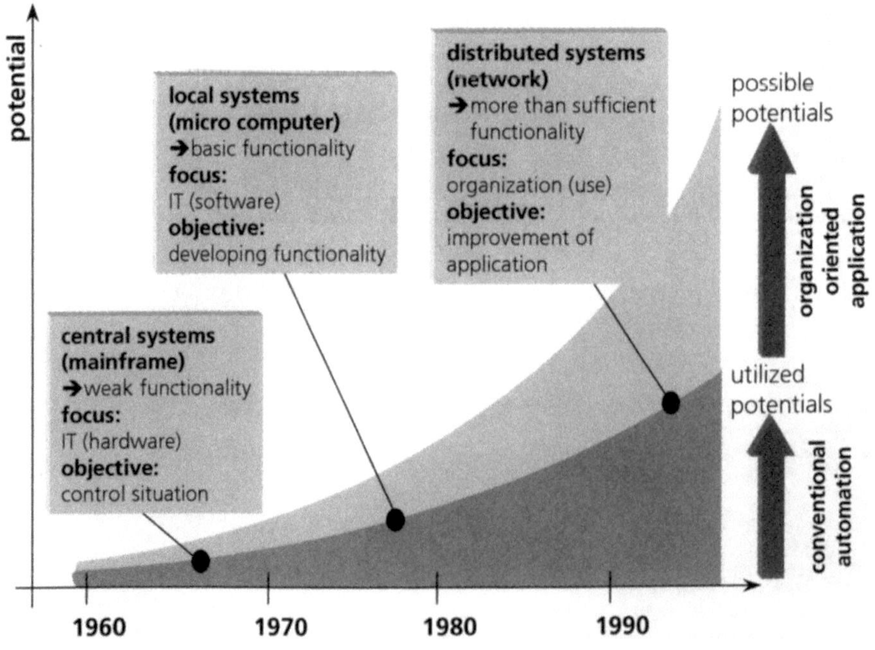

Fig. 3 Potential utilization of computers in production management.

The problems that grow out of this lack of conceptualization are clearly revealed by the rather small success of projects implementing information technology. One of the difficulties involved is the fact that in many companies, there is a lack of knowledge of the multitude of effects of IT. Some of the effects diverge greatly from the users' expectations and the information given by the software provider, and the positive potential uses of computer processing turn into negatives (Box 1). In the case of production management this is exacerbated by the fact that the computer on the shop floor is only a support tool which, in comparison to the 'physical reality of the shop floor' (machines and materials), plays a minor role. It is just this 'physical reality' which is often ignored by project teams 'bent on' computer-aided systems.

Box 1 Use of IT – Where do the problems lie?

Problems of traditional IT application in social-technical production systems

- The field of tension between operative independence and IT based integration leads to an unresolved dilemma for systems design. In the world of practice, serious conflicts between decentralized shop floor domains and upstream, central planning areas result. The implementation of IT – which actually aims towards integration – can result in additional segregation.
- The constraints of implementing 'standard' software block the way towards individual organizational potentials within each enterprise.

Problems of IT systems design methodology

- Within most design approaches aimed towards IT implementation and utilization, organizational aspects are usually given insufficient consideration.
- IT-oriented design approaches in particular often ignore the fundaments of organizational development and change management in social systems.

Problems of project management

- The dimensions and consequences of increased self-organization within the work and project organization as means towards flexibility and conquering complexity are frequently overlooked or underestimated in systems design.
- The 'big bang' philosophy for project management for introducing complex and distributed IT stands in fundamental contrast to the need of continuos and evolutionary social systems development.
- The company-political field is given insufficient consideration.
- True participation of affected workers in the reorganization of production areas is seldom initiated.

4. Socio-technical systems: Human-machine interaction

While IT makes certain forms of organization possible in the first place, its implementation leads to the danger of unnecessary and inefficient mechanization. For decision making, such as sequence planning or processing release, we must take into account that individual intelligence is still required, which ultimately can only be brought to the system by humans. The person is, in operational organization, the only independent and active resource, because people may choose their actions independently. It is the human being who can design and guide informal processes, such as those often needed in the processing of prototypes and rush orders. Shop floor control, normally standardized and computer-aided, can adapt to special situations and interferences through human action. This requires that functions be meaningfully divided between human and computer system. This is division of labor that takes into account both the requirements of the design of the local human-machine system and the particular qualities of decentralized, distributed organization.

> „Shop floor control systems should always be computer systems designed to be used by people. The success of this system depends extensively on the human. It is the human who makes the system work."
>
> (MELNIK & CARTER 1987, p. 65f)

In order to achieve a meaningful symbiosis of human and machine, every work system, such as production, a planning department or everyone and every functional domain involved in order processing, can be viewed as a socio-technical system (SUSMAN 1976). In a socio-technical system, people perform various activities and actions. In order to fulfill the tasks assigned to them, people utilize the machines and procedural techniques which form the technical subsystem. The technical subsystem is not restricted exclusively to physical-material technologies, but instead, comprises all mechanistic functions and elements of an organization, such as the type of order processing, independently of the fact that these are computers or other kinds of aids. At the same time, the person is embedded within a social network and stands in contact to other people. Thus, in addition to technical aspects for the design and introduction of new IT applications the social aspects have to be considered.

The technical subsystem determines the essential demands upon the social subsystem, while at the same time, the social subsystem is the source of organizational flexibility. In reality, the technical and social subsystems are bound together closely. If design attempts aim to optimize both subsystems, then the two must be viewed and designed in their common interaction. The starting point for such systems design is the work task, which represents the interface between the social and technical subsystems.

5. Using IT – Potential use of IT

IT is a resource that can be designed relatively flexibly, that allows diverse design possibilities and that can be adapted well to differing enterprise situations. IT allows the separation of space and time in closely cooperating organizational units and also puts previously difficult and only locally accessible knowledge into a simple form and at the disposal of all. IT plays the role of an 'enabler' to organizational structures: the degrees of freedom for design become ever greater in comparison to previous forms of organizations (DÖRR 1991) (Box 2).

Box 2 Potential uses of IT

- Make processing information accessible.
- Support processing through targeted functionality.
- Control processing through an automated processing system (work flow).

In a certain sense, then, IT forms the backbone of the process organization. Once implemented, however, it becomes rigid and can be adapted to new structures only with considerable difficulty. We should be reminded that IT, in the form of an installed PPC system, has a relatively long life span in comparison to the organization of an enterprise. We can assume that in many companies, a PPC system will be used for a period of 10 to 15 years without much change. In the same period, however, the organization of the firm will change several times.

IT maintains – not to use the more negative term 'dominates' – structures and has a stabilizing effect on the system. With regard to the demand for flexibility, IT should fulfill the task, among others, of maintaining a certain stability and robustness of the total production management system.

6. From function support to integrating information supplier

In future it will be necessary for management to evaluate more positively the grown, 'organic' structures which grow out of self-organization. Typical signs of such self-organization are easily found within the informal system of an organization. An example of this are individual Excel lists which are produced redundantly on the shop floor without being linked to the official IT within the firm. These can be found in most companies. While at first glance this does seem to disregard the aim to integrate into the comprehensive IT system, it does represent a certain integration into the logical system.

In order to secure more extensive integration of computer-aided information systems and individual information systems, much more weight will need to be placed upon ergonomics and the 'task-related practicability' of functions ultimately used. Ergonomics does not mean here that the typical 'manager surface' should be further improved, but that the computer system, along with other aids which are utilized in task performance, should provide meaningful and effective support. The optimal design of the human-machine division of function must stand in the foreground. In some cases this will mean that comprehensible – meaning mainly not too much – information is provided quickly and easily. In other cases, what is needed may be that a large body of data become more precisely presented in the form of an extensive simulation (Box 3).

Box 3 Requirements for the implementation of IT systems (HAFEN, WÄFLER & FISCHER 1996)

- IT (in implementation) must not cement structures, but rather support them. The total socio-technical system must be flexible and adaptive.
- IT must accord with the particular character of the tasks it supports.
- The design of task execution must not be 'coincidental'.
- IT must accord with the particular character of the tasks it supports.

7. Application in the world of practice

To guarantee a positive and demanding field of tasks for employees, it is necessary to design tasks holistically and in a humane manner. Tasks designed according to a work-orientation lead to employees' development of intrinsic motivation and thus to their willingness to assume responsibility, gain experience knowledge and self-esteem. Such tasks allow a work-oriented perspective in which the use and fostering of human resources stands at the fore. At the same time, it must be ensured that employees receive complete information support. This means that all information needed for task fulfillment should be accessible, simply and ergonomically, to employees. The holistic design of tasks makes it possible for employees to apply their potentials to processing and to the further development of their abilities. Through this, the behavior of the whole system improves, and its flexibility, quality and its economics are promoted. In the end, we must remember that IT systems are a part of a socio-technical system, within which people, IT, and organization, implementation and utilization, and use and acceptance cannot be separated (SCHERER 1996).

Practitioners are, as a rule, technical experts. Their organizational knowledge is something they have gained through the course of their work lives. And so, within reorganization projects, they tend to focus upon the problems of technical resources and to work on these in great detail. Design plans with regard to human resources do not usually follow a concept or an objective. As in our case example above, the failure lies in project management and not in the heads of managers. The basis must no longer be that which is technically possible with regard to the computer, but rather that which can be achieved through people. And this can be – where there is motivation and a positive attitude – a great deal.

8. References

DÖRR, G. (1991): Die Lücken der Arbeitsorganisation – Neue Kontroll- und Kooperationsformen durch computergestützte Reorganisation im Maschinenbau. Edition Sigma, Berlin.

DREHER, C. ET AL. (1995): Neue Produktionskonzepte in der deutschen Industrie: Bestandsaufnahme, Analyse und wirtschaftspolitische Implikationen, Physica-Verlag, Heidelberg.

HAFEN, U., WÄFLER, T., FISCHER, D. (1996): Arbeits-, Prozess- oder Informatikorientierung? In: SCHERER, E., SCHÖNSLEBEN, P., ULICH, E. (Eds.): Werkstattmanagement – Organisation und Informatik. vdf, Zürich. p. 31-58.

HAMMER, M., CHAMPY, J. (1994): Reengineering the corporation. paperback edition. Harper Business, New York.

KIRSCH, C., STROHM, O., ULICH, E. (1994): Design of computer aided manufacturing systems: work psychological concepts and empirical findings. In: KIDD, P.T., KARWOWSKI, W. (Eds.): Advances in Agile Manufacturing. IOS Press, Amsterdam. p. 214-217.

KURBEL, K. (1995): Produktionsplanung und –steuerung. Oldenbourg Verlag, München.

MELNIK, S.A., CARTER, P.L. (1987): Shop floor control – Principles and practices and case studies. APICS, Falls Church.

STADTLER, H., WILHELM, S., BECKER, M. (1995): Entwicklung des Einsatzes von Fertigungsleitständen in der Industrie. Management & Computer, 3:4, p. 253-266.

SUSMAN, G (1976): Autonomy at work: A socio-technical analysis of participative management. Praeger, New York.

WIGHT, O. (1974): Production and inventory management in the computer age. CBI Publishing, Boston.

Supervisory Control Principles for Dynamic Operations Management of Manufacturing Shop Floors

Elizabeth Szelke

In intelligent real-time scheduling and control of advanced manufacturing systems, two trends can be observed: one is the full automation of these processes with the elimination of human agents, and the other is the semi-automation of these processes with explicit integration of humans through implementation of the supervisory control paradigm. Discussion in the paper is devoted to the second alternative with the main concepts, principles, models and design considerations that enable a supervisory control system to contribute to dynamic operation management of manufacturing shop floors with their many uncertainties. Discussion is also extended to an example system (called SUPREACT) that focuses on the reactive – and proactive – scheduling functions involved in the system as the major intelligent supervisory control functions besides learning.

1. Introduction: Impetus and background

While advanced manufacturing and information technologies have gained a competitive edge, there seem to be at least *two major factors* in operation management at manufacturing companies today that often cause the successful implementation of advanced automation to fail. *One of them* lies in the dynamic, stochastic nature of the live manufacturing environment, which is due to the rapid change of market demands and to the many external/internal uncertainties that emerge during the execution of scheduled activities. *The other* is human resource integration – from an organizational perspective – into shop floor decision-making and control required when new technology is introduced in companies (EDISON

1991, WALL, CLEGG & KEMP 1987, KIDD 1990, SZELKE & MARKUS 1993b). As the increase in manufacturing competitiveness forces organizations to use more sophisticated and complex software, system performance depends on clever system design and efficient real-time scheduling and control of the related processes on the shop floor. With these advanced manufacturing systems, the dependence on human competence is greater. However, previous studies indicate that the human aspects of successful implementation of such systems have been neglected.

Both of the above problems are the focus of attention at today's companies while attempting to achieve management objectives during the execution of advance scheduled manufacturing operations on the shop floor. With respect to shop floor control over the execution of scheduled activities, two tendencies seem to dominate:

1. Systematic removal of humans from the control loop, as a side-effect of increased automation, or
2. human integration through supervisory control of advanced manufacturing systems, with semi-automated control of processes.

Discussion in the paper attempts to answer the question of why the use – instead of the abuse – of supervisory control is the more viable alternative to managing manufacturing operations on the turbulent shop floors of today's companies. For the further course of discussion, a definition of *supervisory control systems*, adopted from SHERIDAN & JOHANNSEN is given here:

> „Systems that explicitly include a human operator to oversee the predominantly automatic system operation are called supervisory control systems." (SHERIDAN & JOHANNSEN 1976).

More recently, human-aided control systems that include explicit decision support for the human operator have been called *human-machine cognitive systems* (WOODS 1986). An important attribute of a supervisory control system is that somewhere in the control loop there is a human operator responsible for supervising automation, monitoring system flows and outputs and intervening to diagnose and either correct/prevent or compensate for system failures and other unanticipated events (AMMONS, GOVINDARAJ & MITCHELL 1986).

1.1. Impact of the manufacturing environment and operational complexity

The dynamically changing and uncertain manufacturing environment has a great impact on the increasing preference to implement the supervisory control perspective in real-time scheduling and control on live shop floors. Companies with advanced manufacturing systems operate in complex environments rife with uncertainties. Some external/environmental uncertainties arise from changes in customer request dates/qualities/quantities or some late deliveries by suppliers.

Many internal/executing uncertainties occur due to the breakdowns of machines/equipment, operator absence and erratic production yield (SZELKE & MARKUS 1994b).

The growing complexity manifests itself not only in the manufacturing systems, but also in processes and company structures (WIENDAHL & SCHOLTISSEK 1994, GUIMARAES, OWEN & MARTENSSON 1997). Success in meeting the actual targets of production will depend on the reliable performance of processing equipment. In today's manufacturing systems, these technologies will work not as stand-alone units but depend on each other's behavior in a complex manner. Manufacturing systems of such complexity have unfortunately in many cases shown unsatisfactory performance. It is an observed fact that many companies have to re-schedule a high percentage of their scheduled and prepared manufacturing activities (NA 1990) due to the rapidly changing production goals and constraints on the factory floors. Good performance depends on clever system design, on efficient scheduling and on competent real-time running of the system. However, activities during scheduling and real-time running rely mainly upon human competence (ROUSE 1988a, NAKAMURA & SALVENDY 1994). The complexity of real-time managing of shop operations arises from the novelty of events and symptoms and the multitude of interactions that arise when attempting to control the execution of various scheduled shop floor activities. The coordination of these activities by real-time scheduling and control represents a complex task for supervisory control. Some control specialists, among them DAVIS, JONES &SALEH (1992), KIM (1995) and PASSINO (1995), suggest that the situation can be solely remedied by an *intelligent control* methodology *with* severe *knowledge integration* in the shop floor and by continuous *learning to adapt* shop floor scheduling and control strategies *to changes*.

A wide variety of real-time scheduling and control systems is used in manufacturing practice today (VILLA 1991, HARMONOSKY & ROBOHM 1991, SHUKLA & CHEN 1996). However, conventional scheduling and control approaches have severe limitations when implemented in live shop floors. This is because they lack abilities to address the multifaceted goal structure and the transient/dynamic nature of live shop floor processes (SZELKE & KERR 1994), i.e., to achieve the required level of fidelity in modeling the controlled manufacturing environment. They often lack responsiveness, are not capable of real-time response to some disturbance(s) in the simplest sense (i.e., that an executable solution should be available by the time it has to be enacted in shop floor processes). In practice, a solution should be imminently available in order to maintain the continuity of executed (previously prepared) manufacturing processes.

Among researchers in the field of shop floor control, there is general agreement that control effectiveness should be enhanced by making the automated systems more intelligent and responsive. In a recent NIST proposal, DAVIS, JONES & SALEH (1992) discuss the *functions* an *intelligent controller should perform* while controlling resources/operations in the shop floor. These are: planning vs.

scheduling, monitoring of execution, error recovery, interfacing to communicate/exchange information and learning. As real-time intelligent controllers have been developed, however, knowledge-based aspects (and real-time issues, to a minor extent) have been the focus of interest, as pointed out by SHUKLA & CHEN (1996). Lack of attention to the limitations that may impact the system's flexibility (e.g., in the presence of multiple, often conflicting objectives in the shop floor, the need for trade-offs) and lack of consideration of limitations on solutions feasibility, (i.e., an insufficient degree of fidelity in modeling)result in 'intelligent' scheduling and control systems that are not of real utility in a live manufacturing environment (KEMPF ET AL. 1991, ATABAKHSH 1991).

Management of operational complexity, changes and disturbances is of great importance if a manufacturing organization is to be competitive. Companies have found different ways to cope with these issues (RUGE ET AL. 1995). The most important ways according to (WIENDAHL & SCHOLTISSEK 1994) are:

1. decentralization of company functions;
2. exploitation of the creativity, experience and competence of employees;
3. concentration on the core skills of the company.

In manufacturing operation management, there are obvious, although overlapping, ways to overcome operational control complexity, such as:

a. Developing different distributed manufacturing architectures, such as fractal manufacturing (WARNECKE 1993) and holonic manufacturing (VALKENAERS ET AL. 1994, KADAR, MONOSTORI & SZELKE 1997), which account highly for the inclusion of human resources/operators as counteracting decision-making agents in shop floor decision-making and control processes.

b. Enhancing control responsiveness by new reactive (SZELKE & KERR 1994) and proactive (SZELKE & MONOSTORI 1995, MONOSTORI, SZELKE & KADAR 1997, SZELKE & MONOSTORI 1997) scheduling approaches, and also coupling them with learning to ensure adaptivity. For reactive schedule repairs, MYASHITA & SYCARA (1994) and SZELKE & MARKUS (1994a) have coupled reactive scheduling with learning from experience (Case Based Learning, CBL) during the execution of scheduled manufacturing activities, while re-optimizing a current schedule by also capturing humans' best preferred choice from alternative solutions.

c. Implementing the supervisory control approach via a realistic and operational integration of human operator(s) can best be achieved by the design and engineering of a real-time scheduling and control system with human orientation from the outset. This approach, occasionally overlapping with the two above, can best exploit the creativity, experience and competence of human operators and bring the needed flexibility to disturbance management on live shop floors while increasing the whole system's efficiency towards the actual target of production (MITCHELL & MILLER 1986, SZELKE & MESZAROS 1990).

The ways to manage complexity in manufacturing operations management outlined above may concern (a) and (b), but (c) explicitly implies human orientation, as shown by the scheme of Figure 1.

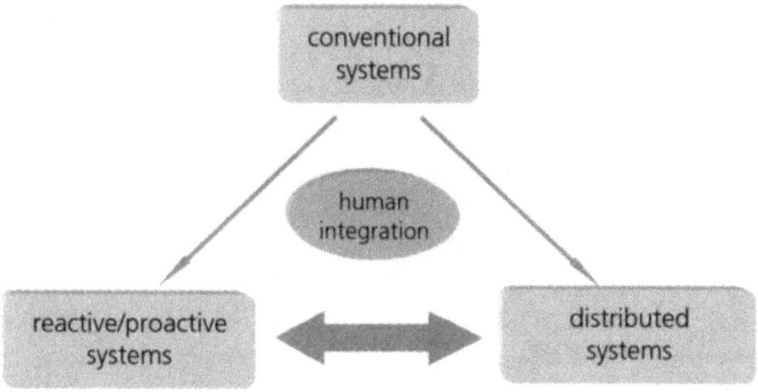

Fig. 1 Ways of managing complexity.

1.2. Human factor impact

Advances in applied information technologies, support of the cooperation of autonomous distributed systems/agents and some degree of responsiveness have greatly increased the potential for automating decisions while people interact with the system or with other agents involved in manufacturing operations management on the modern shop floor. The mere availability of more information via networks, however, will not necessarily lead to better quality of decisions.

Successful implementation of advanced automation in manufacturing practice may also fail due to a lack of systematic integration of human factors into shop floor control when companies introduce new technologies. As RASMUSSEN describes:

> „Modern technology and the rapid succession of designs, materials and processes require flexible systems by inclusion of human; tasks and functions of people are not stable, and introduction of advanced tools will influence work procedures, roles of people and organizational structures."
>
> (RASMUSSEN 1990)

Departure from classic centralized systems and moving towards decentralized shop floor control will lead to the decentralization of responsibilities among working teams, the autonomy of small working groups/teams, and also a re-integration of shop floor activities with a process orientation (BULLINGER ET AL. 1994). This requires that some scheduling-related decision support be offered at the needed location and that there be distributed control with ergonomic design of human/system interaction and user interfaces (UI) oriented toward the work procedures in order to achieve effective use of human capital/know-how on shop operations management.

Manufacturing operation management should neither be based on normative work procedures nor on past role allocations. The design of effective control systems to manage shop operation should rather aim at creating a resource envelope within which people can adapt their work strategies to current requirements and personal preferences without losing support from the automated control system (RASMUSSEN 1990).

Even advanced software, developed to support the scheduling and control of shop operations, is not likely to be very useful without effective operator and supervisor training and/or without an effective human/machine interface (GUIMARAES, OWEN & MARTENSSON 1997) because of often missing user acceptance. As discussed by MARTENSSON & STAHRE (1992), operations execution in most manufacturing systems relies mainly on the performance of the technology and makes very little use of human competence, in spite of the fact that the systems are likely to deviate considerably from the ideal behavior anticipated during its design. The overall result is that even a moderately complex system will show frequent stops and, unless very competently operated by the very same experts who designed it, will yield poor availability.

The importance of human issues behind the implementation of manufacturing technology should not be underestimated. Studies have clearly indicated that plants with the highest levels of productivity are the ones that effectively combine technological improvements with a human orientation (HORTE & LINDBERG 1994) and decision aid. It has been revealed in an overview series on general shop floor problems by OWEN & SPROW (1994) that „shop operators often get a trapped feeling as manufacturers push responsibility down to their level: to the shop floor", and due to their overload and the many uncertainties in processes, „operators lack ability to do performance evaluation and draw conclusions". It is likely that human operators are often not given effective decision aids by the automated scheduling and control systems in use. To reduce losses in aspects such as manufacturing quality and customer satisfaction, the human supervisors' or operators' abilities should be augmented by computerized expertise as a decision aid to assist them in turbulent and often ambiguous scheduling and control decision-making situations.

The above and similar problems have directed researchers' interest during the past years towards the use of a cognitive engineering approach in order to

- identify the problems and problem-solving requirements of the challenging changes on modern shop floors;
- improve problem-solving performance via developing real-time scheduling and control systems in the form of joint man-machine cognitive systems (WOODS, ROTH & BENETT 1990), based on cognitive modeling techniques.

2. Supervisory control: Paradigm, concepts, models

2.1. What is supervisory control?

The term supervisory control is derived from the close analogy between the characteristics of a supervisor's interaction with subordinate human staff members and a person's interaction with intelligent automated subsystems (SHERIDAN 1987). A supervisor of people gives general directives that are understood and translated into detailed actions by staff members. In turn, staff members aggregate and transform detailed information about process-execution results into summary form for the supervisor. The degree of the intelligence of staff members determines the level of involvement of the supervisor in the process. Automated subsystems permit the same sort of interaction to occur between a human supervisor of a machine (i.e., the operator) and the process (SHERIDAN & HENNESSY 1984).

2.2. Why use supervisory control?

The goal of automated scheduling and control systems, using either analytic procedures or knowledge-based approaches (or both), is the automation of the entire coordination process (KREBS, PLATZMANN & MITCHELL 1989). The realities of implementation, however, require the presence of shop floor managers/operators to monitor operations' execution. Regardless of the level of automation, at some points there are human operators who are responsible and accountable for the safe and efficient operation of the manufacturing system. Given the complexity of real-time scheduling problems, human decision makers are needed to fine-tune the scheduling system's decisions in order to compensate for the effects of unexpected changes in system dynamics, execution failures, and other occurrences for which automation does not fully account. Automatic scheduling systems, thus, require the person responsible for effective operation of the system to intervene in the automatic process (as discussed in detail, e.g., by AMMONS, GOVINDARAJ & MITCHELL 1986 and SZELKE & MARKUS 1993b).

A recognized viable alternative to the goal of complete automation, i.e. the human removal that underpins most analytic and knowledge-based/qualitative scheduling and control approaches, is the design of scheduling and control systems that integrates the human responsible for the system operation into the overall architecture of the control system (KREBS, PLATZMANN & MITCHELL 1989).

2.3. Supervisory control paradigm

The above perspective of scheduling and control is called the *supervisory control paradigm*. It explicitly addresses the role of automated scheduling and control tools, both analytic and knowledge-based, and the human, and it integrates the three factors by design in order to enhance the whole system's effectiveness towards the actual target of production. The above definition has been adopted here with its possible extension of integrating all the human operators who contribute to decision-making and control in a distributed manufacturing environment.

The implementation of the paradigm requires the integration of the automated scheduling and control tools into a workstation through which the human supervisor(s) can monitor the processes executed, tune parameters, select from available alternatives and compensate effectively for the deficiencies of the automated scheduling and control system (AMMONS, GOVINDARAJ & MITCHELL 1988, KREBS, PLATZMAN & MITCHELL 1989). In shop operation management such activities may include: selecting the proper scheduling rule in view of the actual goals of shop operation management, expediting unforeseen high priority jobs, compensating for machine failures, etc.

In essence, the *supervisory control paradigm* is *a model for engineering design* that addresses all the relevant pieces of the control system: the hardware, the software and the human supervisor. The paradigm demands *two objectives* for the design process (AMMONS, GOVINDARAJ & MITCHELL 1988):

1. The first is that designers of automation, algorithms and knowledge-based controls do so with a clear understanding of how each piece relates to the human operator who will manage the whole system's operation.
2. Secondly, before a scheduling and control system is built, all the pieces of the system must be integrated into an efficiently functioning entity which makes the best possible use of both human and other resources.

The supervisory control paradigm itself has been developed during the past decades and specialized for the control of complex systems such as dynamic manufacturing systems, which can best be grasped by comparing the above adopted definition to the initial concept given by SHERIDAN (1976): „The supervisory control paradigm applies to situations where a person allocates his attention among various graphical or alphanumerical displays and intermittently communicates new programs to a computer which itself is in continuous direct control of a physical process".

2.4. Modeling aspects

By investigating the role of the computerized system and the human operator in person-supervised, hybrid automated systems, SHERIDAN & FERRELL (1974) found that many aspects of the human operator can be described by well-defined experimental paradigms or tasks. Some of these, such as the signal detection task, the absolute judgment task, or compensatory tracking task (SHERIDAN & FERRELL 1974) have been found to be relatively sophisticated, thus needing well *validated decision, information, and control models* to accompany them.

In the supervisory control literature, there is general agreement on the fact that the „supervisory control task", considered either in continuous control systems (SHERIDAN 1976) or discrete systems' control, e.g., flexible manufacturing systems

(FMSs') (AMMONS, GOVINDARAJ & MITCHELL 1988) is hardly a single task. It is doubtful that any single experimental paradigm or mathematical model can encompass the whole of it. It is more likely that *a variety of models* will be *useful* in order to describe and predict different aspects of supervisory control.

If any *model or modeling framework* were viable, it would have to meet several tests. First, it would have to *accommodate the different functions* performed by either the person or machine or both and, to help keep account of the exchange relation, it would have to be a model of a *complete man-machine system,* not solely that of the human operator. Secondly, it would be convenient to have *the model organized in terms of functions* which are have been classically differentiated in the literature (TZAFESTAS 1989):

a. sensory mechanisms for acquiring information about the world
b. internal cognitive mechanisms for evaluating alternative action which might be taken in response to what is sensed/observed, and
c. mechanisms for implementing action.

Different models of supervisory control, for example of FMS, are available in the special field literature (DUNKLER ET AL. 1988). Their common purpose is to describe the appropriate role and functions of the human operator given certain kinds of underlying system automation and configuration.

SHERIDAN'S initial model (SHERIDAN 1976) conceptualized the supervisory control paradigm by involving five actors: (1) the human being, (2) the computerized system, (3) controls for commands, (4) the controlled physical system and finally, (5) the display. In this first concept a computerized supervisory controller can be said to operate within the four modes of: planning, teaching, monitoring and intervening.

According to a newer conception manifested in SHERIDAN'S *human supervisor model* (SHERIDAN 1987), the human operator interacts with an *„intelligent" computer* (human-interactive computer) that *supports* five basic *cognitive operations or roles*:

1. Goal-setting and planning, means gaining enough understanding of the controlled process to define a resource allocation strategy. Constraints affecting available resources as well as production requirements must be assessed by the supervisor. The resulting plan might later be revised if problems occur or the understanding of the process changes. Included in planning is also decisions on and testing of proper control actions.
2. Teaching/implementing a plan to reach a goal, or programming in manufacturing systems' context, means communicating the necessary commands to the supervised system, making it act according to the plan. The methods of programming might vary from pushing a button to abstract high-level command languages.

3 Monitoring the automatic execution of pre-planned and pre-programmed system actions forms a closed control loop. This is the normal state of an automatic system running without problems. Monitoring includes the acquisition, estimation and evaluation of process information. It also includes drawing conclusions in relation to previous knowledge and desired system goals in order to diagnose possible failure states. In the case of any deviation from the plan, the supervisor should react. Minor adjustments to the process are part of the monitoring role.

4 Intervening when needed for error recovery is conditional upon a failure or a system halt. If the stop is due to a known fault, the operator should intervene and correct the problem (i.e., do minor re-programming). If the stop is caused by the normal termination of a pre-programmed task (i.e., the end of operation on a batch of products), the supervisor should take action by simply starting the next process step. Both actions generally mean looping back to the teaching/programming role.

5 Learning from the results of previous actions is invoked when an unknown error occurs, or during the introduction of new processes. Supervisors should record the immediate events surrounding the unknown error, thus accumulating experience. The events should be properly documented, and the documentation should be accessible to the operators. The analysis of cumulative experiences within the controlled process closes a peripheral control loop back to re-planning of system behavior because of the supervisor's change in process understanding.

The five cognitive roles, which Sheridan attributes to an 'intelligent' computer, are roughly conform to the functions an 'intelligent controller' should perform while managing the manufacturing operations and resources that were included by the NIST proposition of DAVIS, JONES & SALEH (1992). As manufacturing systems become increasingly automated, intelligent and complex, there is an obvious need for intelligent supervisory control systems to operate and control manufacturing systems in an adaptive manner. These control systems should inevitably be provided with *expert supervisory controllers* that incorporate all the functions involved in the NIST proposition for intelligent controllers in order to be of benefit *as an effective decision aid to human operators* in controlling turbulent shop floors. Of the functions mentioned above, the importance of two activities, scheduling and error recovery, has been stressed by many (e.g., O'GRADY 1987, AMMONS, GOVINDARAJ & MITCHELL 1986; DUNKLER ET AL. 1988) for FMS control. SZELKE & MARKUS (1994b) emphasize the reactive/proactive scheduling ability of an expert supervisor for error recovery and for preventing the occurrence of error states during schedule execution in manufacturing. With these reactive scheduling functions and an interfacing function for communicating reactively with the environment, SZELKE & MARKUS have complemented the „roles" given in SHERIDAN'S later model above. These together, as intelligent supervisory functions, are incorporated in the Expert Supervisor Unit (ESU) of a Supervisory Reactive Scheduler called SUPREACT (SZELKE & MARKUS 1995), developed to accommodate supervisory control situations and actions common to discrete part manufacturing, as discussed in detail in Section 4.

The performance and utility in decision support of an expert supervisory controller are influenced by the degree to which the chosen knowledge-representation scheme, control strategy, and explanation capabilities reflect the user's knowledge and problem solving style, i.e., these are compatible from the cognitive aspect. By incorporating good mental models that reflect the way that skilled human operators understand and use dynamic complex manufacturing systems, the expert supervisory controller does not simply imitate a human's reasoning process. Rather, its knowledge-representation schemes and control procedures must be able to capture the fullest possible range and power of the human expert's ability in the supervisory control domain (ROOK & DONNEL 1993).

2.5. The human supervisor's role in supervisory control of discrete part manufacturing

One can find a strong and long tradition of work (since SHERIDAN & JOHANSSEN 1976) on the role of the human supervisor in continuous process control that is being translated by many authors to apply to automated manufacturing systems.

However, a discrete part manufacturing system, in contrast to continuous process manufacturing where the product is indisjunctive, produces parts in different lots to orders. The supervisory control for discrete part manufacturing processes is difficult for a variety of reasons: many different parts may be processed at the same time; these parts may require different processes, tools, fixtures, technological routes etc.; even identical parts may take different routes through the system due to dynamic scheduling problems; perspectives of monitoring may change in time (e.g. between orders and resources). The scheduling objectives and constraints can be characterized as dynamic and time-varying in nature. For discrete part manufacturing it is hard to visualize a steady-state equilibrium which is naturally assumed in the case of controlling a continuous process plant (BARFIELD, HWANG & CHANG 1986).

Let us mention only some of the kinds of problems and obstacles a human scheduler/operator must face in discrete part manufacturing systems: (1) individual events might be quite rapid, but their outcomes may not be clear for hours or days; (2) operators must watch several things simultaneously and mentally integrate their locally independent states in order to estimate overall system state; (3) there are many multiply-determined distinct events; (4) trends cannot be easily determined from a set of discrete states set widely apart in time, (5) the system components are tightly coupled and interact with each other. A means-ends approach to goal attainment, therefore, is difficult in discrete control. System states do not stand for the whole, so subgoals have no privileged perceptual relation to the overall goal. These factors make it hard to determine the effect of a single action on the system as a whole and also make the discovery of cause and effect relations difficult.

Despite the overwhelming complexity often encountered in shop operation, skilled and experienced human schedulers often achieve very acceptable levels of performance. Such experts can reduce the combinatorial complexity of the situation to a set of satisfying options and find solutions that satisfy multiple criteria. The field studies reviewed, for example in HWANG & SALVENDY (1988), attest to this ability.

The *human supervisory role in* computerized *discrete part manufacturing*, according to HWANG, SHARIT & SALVENDY (1984) can be characterized by the following features:

- almost solely relying on indirect control, and the carrying out of situation assessment for the system state is rather qualitative, based on pattern recognition; managing disturbances is rather a question of plan recognition for action;
- anticipating the performance of executed activities and the effects of control actions is difficult, and it may be supported by using stochastic behavioral models (HOLLOWAY ET AL. 1991; SZELKE & MONOSTORI 1997) in computerized intelligent assistance;
- control strategies being heavily altered by the effect of plant size and complexity;
- severe disturbances to the processes may not necessarily jeopardize the overall system operation; there may be recovery from failed states by, e.g. reactive scheduling of the affected parts of a current schedule;
- there is no direct impact of system output on the formulation and selection of ongoing control strategies (apart from considering system throughput as a performance measure with milestones to reformulate the scheduling/control problem's model).

With increased automation, the human supervisor's/operator's role in shop operation management has shifted from skills in manual control to cognitive skills (manifested in, e.g., monitoring, planning, re-scheduling, disturbance management), which are all demanded by supervisory control activity. The *emphasis* has shifted towards *efficient decision support* for highly skilled workers in performing the above supervisory control activities on the shop floor. This support is provided by information technology (IT) systems that follow the ideas of work enrichment and work enlargement and use a so-called shop floor supervisor metaphor and graphical user interfaces with standard tool kits. In the opinion of BULLINGER ET AL. (1994), the shop-floor oriented production support concept may represent a future concept that takes into particular consideration human factors and human-computer interaction. According to their concept, lean production refers to a current leading production philosophy in which it is possible to better coordinate and integrate man, organization and technology so that performance and acceptance come into being – something that is lacking in today's conventional shop floor scheduling and control systems (BULLINGER 1992).

3. Human-centered design issues of supervisory control systems

3.1. Design philosophy

Manufacturing operation management in advanced manufacturing systems needs a design philosophy based on objectives and centered on the human user, rather than being technology-driven (ROUSE 1988b). Design objectives should be oriented towards providing means to help users achieve the operational objectives for which they are responsible. This design philosophy results in viewing the operators or shop supervisors on the shop floor and the strategic planners of a company as all part of the same system for company operation management.

In order to make this design philosophy operational with respect to supervisory control of a manufacturing shop, a user-centered design approach imposes demand for:

- understanding and defining the role of human operator(s) and the automated control system in terms of functional responsibilities, to ensure their integrated good performance in shop floor control;
- systematic mapping of the defined functional requirements into decision support concepts concerning human-system communication via intelligent interfaces/information displays.

3.2. Cooperation concerned design principles

Controlling and problem solving in the broadest sense are the main activities in a complex man-machine system for the supervisory control of manufacturing processes. Both activities have to be accomplished by human operators and automated systems (machine) in an integrated way, as outlined in Figure 2. However, both the human operator and the machine have their own capabilities:

- Computers are able to perform precise actions and measures, to store and rapidly process large amounts of data and to perform repetitive tasks; these capacities can easily be fixed and maintained during design to enhance human operators' decision abilities, to decrease the human load, to take into account human performance limitations in memory, and be decisional with regard to the problems' combinatorial complexity, human response capabilities and attention abilities.

- Human operators are able to set out goals, to work out strategies, to reason upon imprecise and uncertain data, to adapt themselves to difficulties, i.e., they have exceptional cognitive capabilities and experience.

The above complementary capabilities are necessary to the supervisory control of complex manufacturing processes, performed through cooperation between man and machine. Some *principles* have to be followed *in cooperation design*, depending on human capacities:

1. The automated system (machine) has to assume the tactical actions in order to allow the human operator to set out goals, to work out strategies, i.e. to exploit the human exceptional cognitive pattern-matching capabilities.
2. The system representation of the human operator has to be respected by the machine, i.e., be compatible in terms and procedures; then the machine has to provide relevant information which can accomplish or improve this representation.

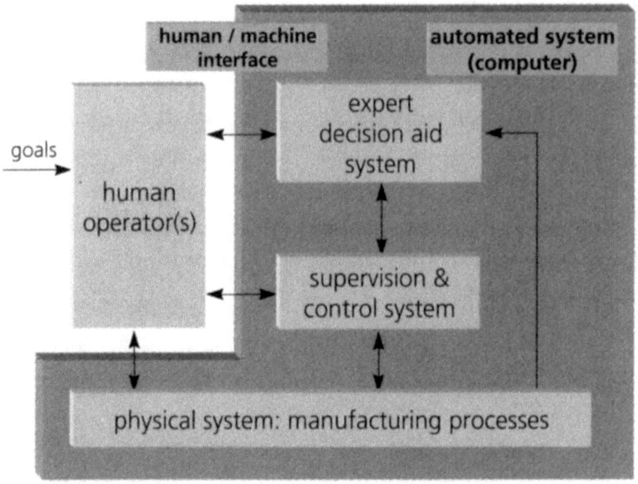

Fig. 2. Man-machine system in supervisory control.

A human-centered approach to engineering design of appropriate integrated automation should take into account different paradigms and technologies, such as the supervisory control paradigm, intelligent control, decision support and expert systems, human-computer interaction, and cognitive engineering.

Cognitive engineering issues have recently become accepted as central to the design process. With the introduction of automation in manufacturing systems, important human concerns arise at the levels of the organizational effects of new technology, the displacement and retraining of personnel and workers' attitudes to automation. These are all well-documented and widely discussed (WALL, CLEGG & KEMP 1987). Less attention has been paid, however, to the cognitive role of humans in complex ambiguous decision-making situations and to their overload with complex and time-consuming tasks such as performance evaluation (OWEN & SPROW 1994) during the execution of scheduled activities. During the past few years there has also been a growing interest in the cognitive role of human operator in the supervisory control of advanced hybrid or automated manufacturing systems (WOODS, ROTH & BENETT 1990, AMMONS, GOVINDARAJ & MITCHELL 1988). The persons who remain part of such systems have a very demanding role that poses a challenge for human factors' engineering in the design phase.

Concerning the engineering *design of the cooperation* between human operator and machine, it requires a path conveniently composed of top-down and bottom-up design phases (GANDIBLEUX, CREVIS & MILLOT 1993):

Top-down phase is based on the technical features and objectives of the manufacturing process and starts with an analysis of the industrial process allowing definition of all control functions and supervisory tasks. Its first step consists of *cognitive task analysis* (STEINBERG & GITOMER 1993) of the process functioning in all the foreseeable contexts, in order to determine the interactions between the operator and the system and to map the work demands (*what* goals and actions should be targeted). With the focus determined, the next question concerns the way in which these critical goals and actions or plans to reach the goals should be presented (via, e.g. cognitive operator function models) in order to support the supervisory performance of critical actions in a direct and seamless fashion by the interface. The second step aims at determining for each task the information needs and decision aids required by the operator. Then, according to the human capacities, these tasks may be classified as tasks allocated either to the machine or allocated to the operator. This constitutes static task allocation, also called the system automation degree, that also manifests itself in the expert supervisor's capabilities. If tools for assistance are required, the man-machine cooperation could be implemented, and eventually the user-interface has been designed.

The *bottom-up phase* aims at validating the above man-machine system design by analyzing it in three steps: (1) Defining *objectives of evaluation* that may be *performance*, security or *workload* and experimental protocols. They can describe the supervisory situations that the operator will face. (2) Experimenting, i.e., allowing the operators to test the system according to the experimental protocol. (3) Analyzing the data recorded during experiments.

With respect to cooperation between the human operator and the intelligent assistant or supervisor built into the computerized system, the man-machine cooperation concept was brought to life with the appearance of Artificial Intelligence (AI) tools as supports to a human decision maker. It results from the necessity to predict, and moreover to prevent, the decisional conflicts liable to arise between the two types of decision makers: human and AI-based. For this, in a single operator system, two cooperation modes called, respectively, vertical and horizontal, can be defined, as shown schematically in Figures 3a & b.

In *vertical cooperation,* the human operator is responsible for the system while being provided with decision supporting tools. The human operator works with the expert decision-aiding system as a partner in finding solutions. In *horizontal cooperation*, the operator and the decision-aid system have the same responsibilities and abilities, the decision support tools being directly connected to the process control system. Tasks can be distributed dynamically between the two decision-makers – either the operator controls the allocation through a dialogue interface, or the allocation is automatically managed by the AI-based decision support tools.

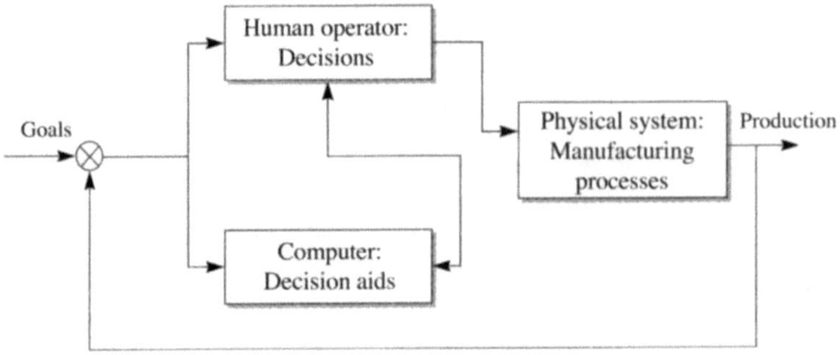

Fig. 3a Vertical cooperation of man/intelligent assistant.

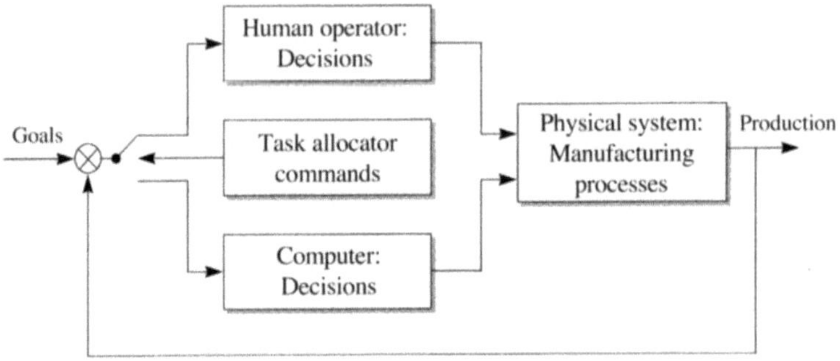

Fig. 3b Horizontal cooperation of person/intelligent assistant.

Intelligent decision support systems (HOLLNAGEL, MANCINI & WOODS 1986), model-based workstations (MITCHELL & SAISI 1987), operator associates (AMMONS, GOVINDARAJ & MITCHELL 1988) and intelligent assistants (SZELKE & MARKUS 1994a) are all design concepts that strive to exploit the strength of both the human and computer component of scheduling and control systems. The notion of the computer-based intelligent assistant or expert supervisor, for instance in SZELKE & MARKUS (1994a), assumes that although some control tasks or functions can be completely automated, many major control functions in complex manufacturing systems require a design that incorporates human override capabilities- amplified, however, by the intelligent and timely decision aid of the expert supervisor.

3.3. Decision support perspective of design

The major steps proposed by ROUSE (1988b) to a user-centered design approach to human-computer task sharing and cooperation in supervisory control are illustrated in Figure 4. These steps involve identifying means for enhancing humans' abilities and overcoming humans' limitations for the tasks of interest. The fifth and final step concerns fostering user acceptance of the means identified for enhancing abilities and overcoming limitations in supervision. General user-system tasks are collected from supervising manufacturing processes in practice and included in groups in Figure 5.

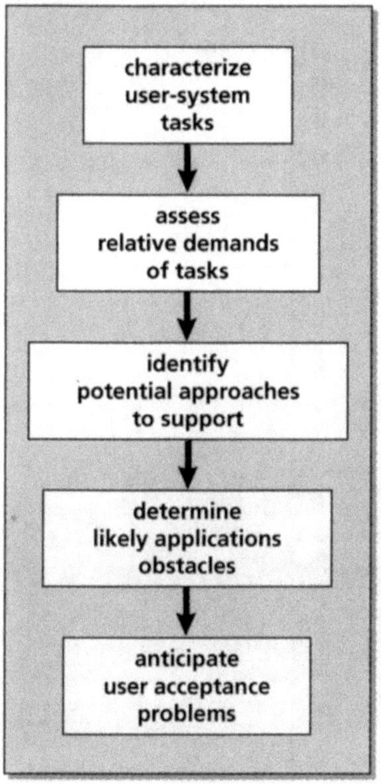

Fig. 4. User-centered design methodology for decision support (following ROUSE 1988b).

While the support of the generation, evaluation and selection of decision alternatives (retrieved as previously relevant and useful, e.g. for a reactive scheduling situation) is central to a human-centered design methodology, these types of support are not sufficient for a more comprehensive user-centered design. It may also be necessary to support implementation and observation. From a system design point of view, supporting observation (by monitoring the execution) involves display design according to the UI community's current propositions (SULLIVAN & TYLER 1991).

3.4. The main issues of cooperation design

The *main design issues* of cooperation contexts for a supervisory control system are

- the allocation of functions, i.e., how the human and the automatic control system should share the tasks (BARFIELD, HWANG & CHANG 1986, ROUSE 1988a, SZELKE & MARKUS 1993) in complementary fashion depending on the workload,
- the human-computer communication in the system (dialogues);
- the information *display/user interface design*.

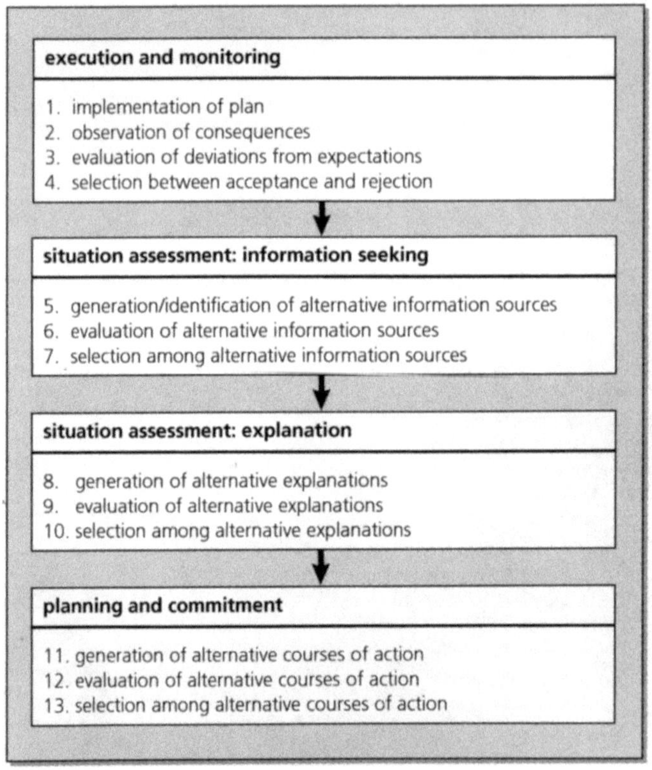

Fig. 5 General user-system tasks in supervisory control (following ROUSE 1988b).

3.4.1. Allocation of tasks/functions

As control systems become more „intelligent" (NIST proposition), the role of the human being in the control structure will also evolve. In manufacturing operation management, man and computer share the responsibilities of supervisory control of shop floor activities for achieving manufacturing management's goals in the short or longer term. It is likely that the optimal system is a symbiosis of the human being and the computer; however, defining an appropriate role for the human as a function of the manufacturing system's objectives may be critical (HETTENBACH, MITCHELL & GOVINDARAJ 1989). Knowledge of the decision-making processes used by humans (more recently as „system/project managers" who expedite one-of-a-kind products/orders in teamwork, rather than „supervisors") in dynamic manufacturing environments can help to define this role and to more adequately describe human control functions.

Research in the field and experimental investigations HETTENBACH, MITCHELL & GOVINDARAJ (1989), for example, on human system manager's performance in guiding the execution of scheduled activities towards achieving different system goals show that humans should be made an 'intricate part' of the manufacturing control process as opposed to 'monitors' of the manufacturing process, since intricate knowledge of the system state and system sensitivity proved to be crucial to human decision making. Experiments in an emulated FMS manufacturing environment, as carried out by HETTENBACH, MITCHELL & GOVINDARAJ (1989.) have indicated that even when in an aggregate role, humans use detailed system status information rather than summary information or system performance history as the basis of their control decisions. Humans and computers, however, do have overlapping capabilities in some respects; thus, they should share decision-making responsibilities (BARFIELD, HWANG, CHANG 1986) made possible by a kind of built-in redundancy in control. This possibility makes the problem of the allocation of functions more difficult to resolve at the stage of control system design (ROUSE 1988a).

If the human is retained in the control loop, one method is to simply let the human control by default those tasks which cannot be accomplished by the computerized system. To do so, the difficulty lies in defining two distinct and coherent subsets of the overall set of supervisory tasks. It is also unlikely that the computerized system can control a manufacturing system full-time without encountering a hardware failure or an unexpected novel event which exceeds the abilities of the computerized system. Humans should have the option of overriding the computer when they feel that this action is necessary.

The level of computer or human control of the manufacturing system can follow a continuum – with the human performing most of the control at the one end or with the computer performing most of the control tasks at the other. CHU & ROUSE (1979) suggest that some level between the two extremes would be most fruitful. Implicit in this continuum is the idea that responsibility at any particular moment will go to the decision-making 'agent' most able at that moment to perform the task at hand well. SZELKE & MARKUS (1993b) suggest that the allocation of tasks be organized dynamically in a complementary way, based on demand-resource matching, rather than separating them in advance between the two decision-making agents (the human and the computerized system). This approach to task allocation suggests an adaptive, dynamic and flexible allocation scheme. Such an integrated human-computer system is adaptive and dynamic in the sense that the allocation of responsibilities depends on the state of the controlled system, as well as on the states of each decision making agent (e.g., on the human operator's overload). The scheme is flexible, as neither the human nor the computer alone would supervise or monitor the execution of manufacturing activities. Exception handling can represent 'exception' (e.g., at computerized system start-up, termination, or breakdown, which all require that control be handed over to the human).

The dynamic task allocation principle assumes the insertion of a task allocator into the control and supervisory loop, as illustrated by Figure 4. According to GANDIBLEUX, CREVIS & MILLOT (1993), dynamic allocation implementation can be used either explicitly, when the human operator acts as a performance and workload estimator (the implementation is rather easy, but the allocation organization and the commands that the operator sends to the task allocator increase his workload) or implicitly, if task sharing is controlled by the computer with the criteria of keeping the operator's workload between two thresholds (a maximum one, WL max, and a minimum one, WL min), as an overload would lead to human error risk and an underload to a decrease in attention.

3.4.2. Human-computer communication design

The human and computerized decision-making agent/assistant interact to realize maximum efficiency and reliability in system operation.

The human agent is provided with a variety of information by the computer, and in turn provides the computer with various kinds of information, including additional data, control instructions and commands and program status alterations. When a computer and a computer-controlled interface are placed between the human and the manufacturing system, the human and the computer share a data-processing task and must necessarily cooperate (RASMUSSEN 1986). The human and the computer will only be able the share the data-processing task in cooperation if they are using compatible processes and strategies for solving the task. The operators should understand and accept the strategy of the computer and not be allocated to a part of the task that exceeds their limited data-processing mechanisms.

ROUSE (1981) indicates that there are several dialogue structures that can be used for human-computer interaction. At one extreme of the interface, the operator can choose from computer-generated menus. At the other extreme, the computer can react to commands initiated by the human. In between these menu-driven and command-driven dialogues, there is an alternative termed by Rouse as mixed-initiative dialogues, whereby both the human agent and the computer initiate dialogues and prompts for each other.

When controlling dynamic systems, it appears that command-driven dialogues are best, with a mixed-initiative approach appropriate for situations where, for example, the human has made a mistake. Dialogues can be organized in form of pop-up menus or can use pull-down windows in implementation.

Whatever the preferred type of communication, implementation should always give consideration to this principle: communication should be designed with the goal of minimizing data collection and data entry by the user/human operator, as both activities are time-consuming and risky due to human errors as attention decreases.

3.4.3. The intelligent information display design

From a system design point of view, supporting observation or monitoring the execution of scheduled activities involves display design which has long been the stock in trade of human factors engineers. Fairly recently, methods based on expert systems technology have been developed for on-line, intelligent information management (ROUSE 1988a). These methods have the potential to filter, transform and format information automatically. This capability will, for example, make it feasible for operators of complex manufacturing processes to cope with the enormous amount of data available via computer and communication technologies (ROUSE, GEDDES & CURRY 1987).

Another performance issue in the design of the user interface is that of usability with the application of the user's mental model (ROOK & DONNEL 1993) and with the available modern interface technologies/tools. The user interface design is not so much a problem of how to interface the machine to the user, but rather of how to use the human as an interface between the machine and the controlled manufacturing environment.

Computer-based displays provide an opportunity to create dynamic rather than static or passive information displays, i.e., displays that adapt to changing task and operator needs. The major design challenge for multipage information displays is the development of schemes which enable the operator to access the required information effectively and efficiently.

Computer-based information display design implies display page organization techniques (menu systems and hierarchies), but also some capability of varying the level of detail of the displayed information, and of displaying data items selectively in order to provide sufficient information in a timely and relevant manner to the given control situation. For operator-oriented adaptive information display design to assist the human with fast and relevant information in the proper form, there must be a *detailed representation of the operator's goals, functions, procedures*. Furthermore, *the representation or model must be dynamic,* modifying the operator's information needs by current system state.

Display systems may be organized around operator decision-making functions. This requires precise specification of what the operator needs, when and in what form. Given a normative model of operator control actions, several critical issues in display system design can be addressed. This level of analyzing the operator's cognitive requirements is a challenging aspect of display design in supervisory control. MITCHELL & MILLER (1986) propose a design methodology based on a discrete control model of the human operator's cognitive and decision-making activities in the supervisory control task.

The discrete control model can be thought of as a possible representation of an operator's internal (mental) model of the system plus a control structure that specifies how the model is used to solve the decision problems that comprise control functions. Specifically, the discrete control model accounts for the way specific control actions are determined given the information about the controlled system, the environment and the context of the situation. The model is constructed in order to guide information display design with the relevant content of display pages.

The use of direct manipulation interface technology may be a valuable design technique for operator interaction and control with real-time, supervisory control systems (BENSON ET AL. 1989).

4. An example system for intelligent supervisory control

4.1. System model and functions

The example system, called SUPREACT, has been designed as a joint man-machine cognitive system for supervisory control of discrete manufacturing processes on turbulent shop floors. The term 'Supervisory Control System', in the case of SUPREACT, denotes an intelligent real-time control system that integrates human operators (by its engineering design), performs the detailed scheduling of shop floor activities and the real-time control over the execution of these scheduled activities.

The primary engineering design of SUPREACT is focused on implementing the supervisory control paradigm and gradually encompassing all the functions an intelligent controller should perform while controlling resources/operations on manufacturing shop floors (i.e., it has adopted the view of the above mentioned NIST proposal). Therefore, SUPREACT is designed to assist the human in performing functions such as: planning/(re-)scheduling, monitoring execution, error recovering, interfacing with the environment to communicate or exchange information and learning from experience. All of these functions have been encompassed in its Expert Supervisor Unit (ESU) based on a Cognitive Operator Function Model (COFM, detailed in 4.3).

With the increased demand for system reactivity in manufacturing operation management on the shop floors of today's companies, SUPREACT is developed (within a national grant project) as a Supervisory Reactive Scheduler, with the dominant intelligent functions of reactive scheduling (RS) and proactive scheduling (PS). Both functions aim at performing disturbance management (SZELKE & MONOSTORI 1995) during an advance schedule execution in manufacturing. RS enables the system to repair a flawed schedule by compensating for the effects of some disturbance(s) which emerge in schedule execution. PS serves to prevent the occurrence of a foreseeable disturbance or performance degradation during schedule execution in the physical system. To adapt the control strategies to changes on live shop floors, the system is being provided with an embedded learning ability by using a CBR/L (Case-Based Reasoning and Learning) approach (SZELKE & MARKUS 1997) to improve RS/PS problem-solving efficiency. Both RS and PS require the simultaneous development of real-time status monitoring of manufacturing processes for situation assessment, supported by stochastic behavioral models (SZELKE & MONOSTORI 1997).

For effective integration and coordination of the decision-making and control functions on the shop floor, the design of the underlying framework of the system has implemented the supervisory control paradigm by explicitly addressing the role of the human, computerized algorithms and knowledge-based system components, for their integrated effectiveness in system performance.

From the aspect of knowledge engineering, the system has a knowledge-base/case-base (KB) for supporting the cooperation of rule-based (RB) and case-based (CB) agents and knowledge sources in RS problem solving via a blackboard (BB). The BB architecture derives from that of the COFM; they both share the same hierarchy of knowledge abstraction, as detailed in SZELKE & MARKUS (1994a).

The system interface is engineered with two options: one for the user/operator and another one for the system specialist to allow for updating the controlled system model and the control system knowledge-base/case-base. The interface option to the system specialist is designed for learning from examples in order to accommodate new cases into the system's case-base. Both interface options have been described in detail in earlier publications (SZELKE & MARKUS 1993a/b).

4.2. System functional structure and RS related operation

Details of SUPREACT, in its functional structure and interrelations of its components may be found in SZELKE & MARKUS (1994a). Its operation follows an adaptive control pattern. This means that it can perform status identification, decision-making and control, i.e., there is adaptive feedback by its main modules, defined by their functions. The system operates in the real-time shop floor control layer, gets input from an MRP II from the plant control layer and results in executable programs in the form of detailed schedules for manufacturing.

With respect to reactive scheduling function, the identification, decision-making and control have been carried out by SUPREACT as follows:

- *Identification* of live status on the shop floor is carried out by its shop floor data/event acquisition subfunction based on on-line monitoring. Subsequent comparison-evaluation and analysis is performed with respect to predicted (scheduled) and executed activities/operations. As the evaluation reveals any divergencies from plans, the results are offered to the human operator to check for the type of disturbance. The human operator can initiate a preview simulation run to reveal the ripple effects of a disturbance on downstream operations. From this analysis for situation assessment, the current decision situation can be characterized by the necessary information (i.e., some global context features reflecting performance trends, by, for instance, the mean/max tardiness of scheduled activities on orders and the desired orientation of schedule repair; some local context features indicate the feasibility of local repair operations).

- *Decision making* on the basis of the analysis might propose modification of the control strategy, e.g., change of performance criteria and the corresponding rules selection from the knowledge base, as attention shifts, e.g., from minimizing work-in-progress (WIP) to expediting a critical order. This can again be checked and overridden by the human operator. With respect to the indicated modifications and the characterized decision situation, the Expert Supervisor Unit yields a set of hopefully appropriate schedule repair strategies from the system's case-base (using previous preference ranking of the human operators) and initiates a reactive scheduling session for each strategy selected. This results in a list of alternative solutions and their outcomes with respect to main performance criteria for the human to compare and choose from. His preferred choice is recorded in a case history file with the preference weight to be applicable to future decisions.
- *Control in the system* makes the invalidated schedules unscheduled and replaced with validated schedules. Then the valid schedule, as a feasible/executable program for manufacturing, is passed to the process control sub-system for execution in the physical process. This is accomplished by operator commands or by the Expert Supervisor Unit, in case of time-out of the operator.

In addition to state identification, decision making and control, the system's 'learning module' allows it to include new schedule-repair cases, the handling of novel situations into the case-base, with the help of the human operator. Clustering of these unclassified new cases is then left to the system specialist – to be performed in a subsequent 'learning session'.

4.3. Design of human-computer cooperation in SUPREACT

During the implementation of the supervisory control paradigm, by design, the human is considered in the system as taking the role of a supervising, monitoring and problem-solving actor, cooperating with the Expert Supervisor Unit (ESU) which continually provides the human with intelligent assistance.

With respect to the allocation of functions in the system, human and automatic system components share supervisory responsibilities in a dynamic way, depending on the system state and human operator workload. The computerized system takes over some complex combinatory decision-making tasks (e.g., in reactive scheduling) and some low-level tasks of execution control (e.g. giving signals for execution, data sensing, data acquisition, acquiring statistics on performance trends). The human is responsible for higher level tasks, such as determining overall system goals, intermittently reformulating the decision strategy, etc.

Man-machine communication is designed to be performed through mixed command and menu-driven dialogues via the intelligent user interface. With normal operating conditions, the communication is organized to proceed via dialogues initiated by the operator. For purposes of user acceptance, the dialogues use terminologies and procedures compatible with human needs and with those used in the computerized system. With abnormal operating conditions, the computerized system is to alert the operator through some alarm signals and then wait for response. The system start-up and termination are considered as exception handling, and here control is passed over to the human.

The interface design incorporates the semantics of the discrete manufacturing control domain and that of the supervisory tasks for stereotype situations (SZELKE & MARKUS 1993a). It is based on the COFM as a user mental model that supports plan recognition to identified operator's goals. COFM also supports a purposeful information display design to provide the human with the right information in the right form and at the right time. The construct of COFM is designed taking into account human cognitive abilities and limitations, as well as some general guidelines of cognitive engineering, such as that for cognitive task analysis (STEINBERG & GITOMER 1993), for plan-recognition capability of intelligent interfaces (GOODMAN & LITMAN 1990), for using mental models and inference explanations (ROOK & DONNELL 1993), and for information display design (MITCHELL & MILLER 1986).

The information display design is organized to capture the relevant data/piece of information and to group the displayed data into sets to match the human operator's decision-making needs, while implementing any expected operator functions. The semantics of the tasks and the domain are incorporated into the content and grouping of the display pages, which has been achieved by using the cognitive operator function model.

We view a central cognitive engineering problem in the whole system design of SUPREACT the achievement of coordinated action between the human and the computerized decision agent in such a way that they together perform better, for example, in a reactive/proactive scheduling situation, with respect to both maintained schedule quality and response time efficiency than either would individually do. For this reason, and to gain better user/operator acceptance, the COFM model with functions and sub-functions has gradually been established and customized from industrial experience gathered in operating FMSs. Thus, COFM can be considered normative to the shop operators' behavior by representing and describing how a skilled operator might organize and coordinate (often concurrent) control actions, while being involved in supervisory activities in the shop floor.

Mathematically, the model is a hierarchic-heterarchic network of nodes in a finite state automata, where the nodes represent operator activities as operator functions/sub-functions/tasks/actions. It is illustrated by the network in Figure 6. Each level in the hierarchy may be a heterarchy, i.e., a collection of activities that are performed concurrently. Network arcs represent system-triggering events or the results of operator actions that initiate or terminate operator activities. In this way, the model accounts for the coordination of multiple activities and can also guide the dynamic focus of attention in information-seeking to these activities. It is an approach to dynamically displaying state-dependent information to the operator. However, COFM is also a prescriptive model of human performance in supervisory control. Given the system-triggering events, it defines the functions, subfunctions, tasks and actions on which the operator should focus. When it is used in a predictively, model-derived way, it generates the expectations of likely operator actions in the context of the current system state. If the model used for the data derived inferencing, it defines likely operator functions as goals, subfunctions as plans, and tasks that can be inferred based on current operator actions and the system state, as is illustrated in Figure 6. Thus, the COFM can describe the entire range of human performance in the control of manufacturing systems. Enriched with some understanding capabilities by the above goal-directed activity performance aspects, COFM can infer the intentions of a skilled operator in supervisory control of manufacturing systems. Such a goal-directed intent inferencing structure can best be implemented by a blackboard architecture, with the same levels of the knowledge representation hierarchy: goals/plans/tasks/actions.

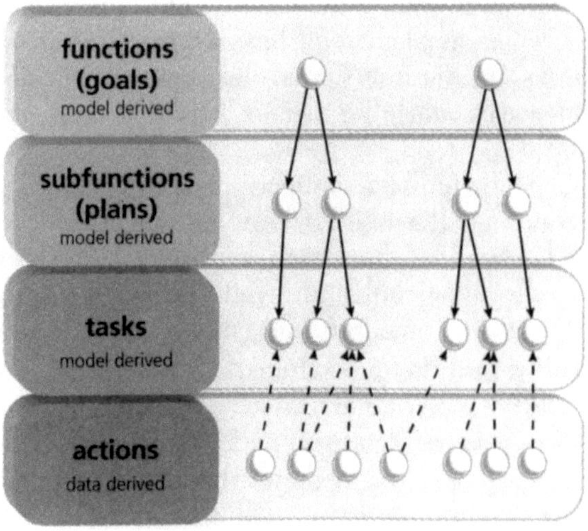

Fig. 6 Cognitive Operator Function Model (COFM).

SUPREACT with the COFM can be conceived as a joint human-machine cognitive system. With respect to the human and the intelligent knowledge sources/computerized decision-making agents in its ESU, it is a multi-agent decision-making system cooperating via a blackboard (BB) architecture, which is derived from the cognitive operator function model and enriched with some interrupt capabilities for the controller of the BB. Its more detailed description can be found in SZELKE & MARKUS (1995).

5. Conclusions

As the increase in manufacturing competitiveness forces organizations to use more sophisticated and complex software, system performance depends on clever system design, efficient scheduling and execution control of the related processes. For advanced manufacturing systems, dependence on human competence is greater. However, previous studies, such as that by MARTENSSON & STAHRE (1992) indicate that the human aspects of successful implementation and operation of such systems on live shop floors have often been neglected.

The integration of intelligence of human and machine on manufacturing shop floors to the success of companies competing now and in the next century (GERHARDT 1997) seems indispensable. Even in Japan, where almost everything is automated, there is an increasing trend towards human factor integration in shop floor decision-making processes. This trend is manifested in the implementation of different philosophies for manufacturing production management (such as KANBAN, holonic systems, bionic systems etc.). Modern manufacturing philosophies, such as lean production, assume the effective integration of human operators' knowledge, skills and competence into the decision procedures of turbulent shop floors.

For the intelligent control with knowledge integration (ALBUS 1992, KIM 1995) on the shop floor, one should take into account the fact that skilled operators carry several forms of knowledge that are required for a satisfyingly good execution of scheduled activities, where it is likely that resources are used in the best possible manner. Their judgments are usually skillful also in situations demanding re-scheduling or reactive scheduling when the available information is mostly incomplete or uncertain. However, for successful knowledge integration via human-machine cooperation, the supervisory control knowledge assumed for the knowledge-based expert supervisor should be expressed by representational forms, should use control strategies, inference explanations and a problem-solving style compatible to that of the human supervisors' head. It can be ensured by an intelligent interface design for cooperation through embedding user/operator mental (cognitive) models that reflect the semantics of the supervisory task and the manufacturing domain. Mental models are based on the notion that the manner in which people understand complex dynamic manufacturing systems will affect interaction with these systems and also the ultimate level of operator/system performance. As studies by ROOK & DONNELL (1993) among others show, the use of graphic-based mental models and inference explanations yielded the highest user/expert system interaction performance. The expert supervisor's knowledge repository should continually be updated to ensure the whole system's adaptivity to changes; i.e., the system should learn from experience to improve its supervisory performance. Updates to the controlled system model and the control system knowledge-base (in learning sessions) also require a good system specialist's interface (SZELKE & MARKUS 1993a).

There is no longer any question that to the effective operation of the advanced automation in today's manufacturing systems, the implementation of the supervisory control paradigm is indispensable during the engineering design of their control systems, in order to then operate the whole system efficiently. With its human factor engineering aspect, this paradigm is a radical departure from current emphases in design and automation. Frequently, the human's role in a highly automated shop floor scheduling and control system is an after-thought. The human agent is there to compensate for the times or events that automation handles inadequately or not at all. From an engineering perspective, the human's role is not designed, but evolves over time as inadequacies of the automated control system arise. In a related manner, user interfaces to such systems also do not reflect intelligent design. Typically, information displays are 'data dumps' where a programmer, unfamiliar with the domain or the human's tasks, displays all data that are collected. User controls are awkward, often concatenations of low level commands, putting operators in a position where they must trick the system into performing necessary functions. Research results in assessing the importance of various complexity factors (such as system complexity, operator tasks complexity etc.) on the effectiveness of the whole system operation show (GUIMARAES, OWEN & MARTENSSON 1997) that training of system operators and a reasonably good quality of the man/machine interface can reduce the negative impact of system complexity on the operational efficiency of the system.

6. References

ALBUS, J. S. (1992): RCS: A reference model architecture for intelligent control. IEEE Computer, p. 56-59.

AMMONS, J.C., GOVINDARAJ, T., MITCHELL, C. M. (1986): Decision models for aiding FMS scheduling and control. IEEE Transactions on Systems, Man, and Cybernetics, 18:5, p. 744-756.

AMMONS, J.C., GOVINDARAJ, T., MITCHELL, C. M. (1988): Human aided scheduling for FMS: A paradigm for human-computer interaction in real-time scheduling and control. In: STECKE, K.E., SURI, R. (Eds.): Proceedings of the 2[d] ORSA/TIMS Conference on Flexible Manufacturing Systems: Operations Research models and applications. Elsevier, Amsterdam. p. 443-454.

ATABAKHSH, H. (1991): A survey of constraint based scheduling systems using an artificial intelligence approach. Artificial Intelligence in Engineering, 6:2, 58-73.

BARFIELD, W., HWANG, S-L., CHANG, T-C. (1986): Human-computer supervisory performance in the operation and control of FMSs, In: KUSIAK, A. (Ed.): Flexible Manufacturing Systems: Methods and Studies. Elsevier, Amsterdam. p. 377-408

BENSON, C.R., GOVINDARAJ, T., MITCHELL, C.M., KROSNER, S.P. (1989): Effectiveness of direct manipulation interaction in the supervisory control of FMS part movement. IEEE, p.947-952.

BULLINGER, H.-J. (1992), Neue Produktionsparadigmen als betriebliche Herausforderung. In: WARNECKE, H.-J., BULLINGER, H.-J. (Eds.): Innovative Unternehmensstrukturen Springer, Berlin.

BULLINGER, H.-J. ET AL. (1994): Human-computer interaction and Lean Production: The shop floor example. Int. Journal of Human-Computer Interaction, 6:2, p. 121-154.

CHU, Y.Y., ROUSE, W.B. (1979): Adaptive allocation of decision making responsibility between human and computer in multi-task situations. IEEE Transactions Systems, Man and Cybernetics, 9, p. 769-778.

DAVIS, W., JONES, A., SALEH, A., (1992): Generic architecture for intelligent control systems. Computer-Integrated Manufacturing Systems, 5, p. 105-113.

DUNKLER, O., ET AL. (1988): The Effectiveness of supervisory control strategies in scheduling flexible manufacturing systems. IEEE Transactions on Systems, Man, and Cybernetics, 18:2, p. 223-237.

EDISON, T. (1991): Intelligent support system for continuous improvement in a dynamic manufacturing environment. Computer-Integrated Manufacturing Systems, 4:4, p. 229-238.

GANDIBLEUX, X., CREVIS, I., MILLOT, P. (1993): Man-machine cooperation in real-time control: Two case studies for future control systems. In: YOSHIKAWA, H., GOSSENAERTS, J. (Eds.): Information Infrastructure Systems for Manufacturing, IFIP Transactions B-14. Elsevier, Amsterdam.

GERHARDT, L. (1997): The integration of intelligence in the 21st century: Human and machine. Keynote speech. Proceedings of the 4th IFAC Workshop on 'Intelligent Manufacturing Systems' IMS'97. Seoul, Korea.

GOODMAN, B.A., LITMAN, D.J. (1990): Plan recognition for intelligent interfaces. IEEE, p. 297-303.

GUIMARAES, T., OWEN, J.E., MARTENSSON, N., (1997): Assessing the importance of manufacturing system complexity and moderating variables. In: MONOSTORI, L. (Ed.): Proceedings of the Second World Congress on Intelligent Manufacturing Processes & Systems. Springer, Berlin. p. 263-271.

HARMONOSKY, C.M., ROBOHN, S.F. (1991): Real-time scheduling in computer integrated manufacturing: A review of recent research. Int. J. Computer Integrated Manufacturing, 4:6, p. 331-340.

HETTENBACH, D.A., MITCHELL, C.M., GOVINDARAJ, T. (1989): Decision making in supervisory control of flexible manufacturing system. IEEE, p. 953-985.

HOLLNAGEL, E., MANCINI, G., WOODS, D.D. (Eds.) (1986): Intelligent decision aids in process environments. Springer, New York.

HOLLOWAY, L.E., ET AL. (1991): Integration of behavioral fault detection models and an intelligent reactive scheduler. Proceedings of IEEE Int. Symposium on Intelligent Control. Arlington, VA. p. 134-139.

HORTE, S.V., LINDBERG, P. (1992): Performance effects human and organizational development and technological development. Int. Journal of Human Factors in Manufacturing, 4:3, p. 243-259.

HWANG, S.-L., SALVENDY, G. (1988): Operator performance and subjective response in control of flexible manufacturing systems. Work and Stress, 2:1, p.27-39.

HWANG, S.-L., SHARIT, J., SALVENDY, G. (1984): Management strategies for the design, control, and operation of flexible manufacturing systems. Proceedings of the Human Factors Society, 27th Annual Meeting, Norfolk, VA. The Human Factors Society, Santa Monica, CA. p. 297-301.

KADAR, B., MONOSTORI, L., SZELKE, E. (1997): An object oriented framework for developing distributed manufacturing architectures. In: MONOSTORI, L. (Ed): Intelligent Manufacturing Systems. Springer, Berlin. p. 548-555.

KARWOWSKI, W., RAHIMI, M. (Eds.) (1990): Ergonomics of Hybrid Automated Systems I-II. Elsevier, Amsterdam.

KARWOWSKI, W., SALVENDY, G. (1994): Integrating people, organization, and technology in advanced manufacturing: A position paper based on the joint view of industrial managers, engineers, consultants and researchers. Int. Journal of Human Factors in Manufacturing, 4:1, p. 1-19.

KEMPF, F., RUSSEL, B., SIDHU, S., BARETT, S. (1991): AI-based schedulers in manufacturing practice. AI Magazine, 11:5, p. 46-56.

KIDD, P.T. (1990): Human Factors, CIM-EUROPE and the ESPRIT Research Programme. Int. Journal of Industrial Ergonomics, 5, p. 105-112.

KIM, S. (1995): Advanced design and manufacturing systems through knowledge integration. In: MONOSTORI, L. (Ed.): Proceedings of the First World Congress on 'Intelligent Manufacturing Processes and Systems', Vol. One. Puerto Rico. CIRP-IEEE Publications. p. 13-22.

KOLONDER, J.L. (1991): Improving human decision making through case-based decision aiding. AI Magazin, p. 52-68.

KREBS, J.E., PLATZMAN, L.K., MITCHELL, C.M. (1989): A real-time AGV scheduling system that combines human decision-making with integer-programming algorithms. IEEE, p. 965-970.

MARTENSSON, L., STAHRE, J. (1992): Operator roles in advanced manufacturing. In: BRÖDNER, P., KARWOWSKI, W. (Eds.): Aspects of Advanced Manufacturing and Hybrid Automation. Elsevier, Amsterdam. p. 155-162.

MITCHELL, C.M., SAISI, D.L., (1987) Use of model-based qualitative icons and adaptive windows in workstations for supervisory control systems. IEEE Transactions on Systems, Man and Cybernetics, 17:4, p.573-593.

MITCHELL, C.M. (1988): Supervisory control: Human information processing in manufacturing systems. In: Concise Encyclopedia of Information Processing in Systems and Organizations. Pergamon Press, New York.

MITCHELL, C.M., MILLER, R.A. (1986): A discrete control model of operator function: A methodology for information display design. IEEE Transactions on Systems, Man and Cybernetics. 16:3, p. 353-369.

MIYASHITA, K., SYCARA, K.P. (1994): Adaptive schedule repair. In: Szelke, E., Kerr, R. (Eds.): Knowledge based reactive scheduling, IFIP Transactions B-15. Elsevier, Amsterdam. p.107-124.

MONOSTORI, L., SZELKE, E., KADAR, B. (1997): Management of changes and disturbances in manufacturing systems. Preprints of the IFAC Workshop on Manufacturing Systems: Modelling, management and control MIM'97. Vienna. p. 27-38.

NA, H.S. (1990): Artificial intelligence in today's factory. Robotics & Computer-Integrated Manufacturing, 7, 315-320.

NAKAMURA, N., SALVENDY, G. (1994): Human planner and scheduler. In: Design of work and development of personnel in advanced manufacturing. Wiley, New York.

NORMAN, D.A. (1986): Cognitive engineering. In: NORMAN, D.A., DRAPER, S. (Eds.): User-centered system design: New perspectives in human-computer interaction. Lawrence Erbaum Associates, Hillsdale. p. 86-124.

O' GRADY, P.J. (1987): Controlling automated manufacturing systems. Kogan Page/Chapman & Hall, London.

OWEN, J. V., SPROW, E.E. (1994): Shop Floor '94 - The challenge of change. Manufacturing Engineering, 112, 33-46.

PASSINO, K.M. (1995): Intelligent control for autonomous systems. IEEE Spectrum, p. 55-62.

RASMUSSEN, J. (1986): Information processing and human-machine interaction. North-Holland, New York.

RASMUSSEN, J. (1990): A model for the design of computer integrated manufacturing systems: Identification of information requirements of decision makers. Int. Journal of Industrial Ergonomics, 5, p. 5-16.

RISSLAND, E. (1982): Ingredients of intelligent user interfaces. Int. Journal of Man-Machine Studies, 2, p. 377-388.

ROOK, F.W., DONNEL, M.L. (1993): Human cognition and the expert system interface: Mental models and inference explanations. IEEE Transactions on System, Man and Cybernetics, 23:6, p. 1649-1661.

ROUSE, W.B. (1981): Human-computer interaction in the control of dynamic systems. Computing Surveys, 13, p. 71-99.

ROUSE, W.B. (1985): Supervisory control and display systems. In: ZEIDNER, J. (Ed.): Human Productivity Enhancement. Prager, New York.

ROUSE, W.B. (1988a): The human role in advanced manufacturing systems. In: DALE, W. (Ed.): Design and analysis of integrated manufacturing systems. National Academy Press, Washington. p. 148-165.

ROUSE, W.B. (1988b): Intelligent decision support for advanced manufacturing systems. Manufacturing Review, 1:4, p.236-243.

ROUSE, W.B., GEDDES, N.D., CURRY, R. E. (1987): An architecture for intelligent interfaces: Outline of an approach to supporting operators of complex systems. Human Computer Interaction, 3, p. 87-122.

RUGE, I., BACHHUBER, W., BAMBERG, S., SCHENK, M. (1995): Coping with complexity. Siemens Review, 62:6, p. 26-29.

SANDERSON, P.M. (1988): Human supervisory control in discrete manufacturing: Translating the paradigm. In: KARWOWSKI, W., PARSAEI, H.R., WILHELM, M.R. (Eds.): Ergonomics of hybrid automated systems. Elsevier, Amsterdam.

SCHNEIDERMAN, B. (1992): Designing the User Interface – Strategies for effective human-computer interaction. 2nd ed. Addison-Wesley, New York.

SHARIT, J. (1985): Supervisory control of a flexible manufacturing system. Human Factors, 27:1, p. 47-59.

SHERIDAN, T.B. (1976): Toward a general model of supervisory control. In: SHERIDAN, T.B., JOHANNSEN, J. (Eds.): Monitoring behavior and supervisory control. Plenum Press, New York. p. 271-281.

SHERIDAN, T.B. (1987): Supervisory control, Chapter 9.6, In: SALVENDY, G. (Ed.): Handbook of Human Factors. Wiley, New York. p. 1243-1267.

SHERIDAN, T.B., FERRELL, W.R. (1974): Man-machine systems. M.I.T. Press, Cambridge.

SHERIDAN, T.B., FISCHOFF, B., POSNER, M., PEW, R.W. (1983): Supervisory control systems. In: Research needs in human factors. National Academy Press, Washington.

SHERIDAN, T.B., HENNESSY, R.T. (Eds.) (1984): Modeling of supervisory control behavior: report of a workshop. National Academy Press, Washington.

SHERIDAN, T.B., JOHANNSEN, G. (Eds.) (1976): Monitoring behavior and supervisory control. Plenum Press, New York.

SHUKLA, C.S., CHEN, F.F. (1996): The state of the art in intelligent real-time FMS control: A comprehensive survey. Journal of Intelligent Manufacturing , 7, p. 441-455.

STAHRE, J. (1994): Evaluating human/machine interaction problems in advanced manufacturing. Computer integrated manufacturing systems. Butterworth-Heineman.

STAHRE, J., JOHANSSON, A. (1994): Decision Support for Flexible Manufacturing. In: KIDD, P., KARWOWSKI, W. (Eds.): Advances in agile manufacturing – Integrating technology, organization and people. IOS Press, Oxford. p.405-408.

STEINBERG, S., GITOMER, D.H. (1993): Cognitive task analysis and interface design in a technical troubleshooting domain. Knowledge-Based Systems, 6:4, p. 249-257.

SULLIVAN, J.W., TYLER. S.W. (1991): Intelligent User Interfaces. ACM Press.

SZELKE, E., MARKUS, G. (1995): A blackboard based perspective of reactive scheduling. In: KERR, R., SZELKE, E. (Eds.): Artificial Intelligence in Reactive Scheduling. Chapman & Hall, London. p.60-77.

SZELKE, E., MARKUS, G. (1997): A learning reactive scheduler using CBR/L. Computers in Industry (In print).

SZELKE, E., KERR, R.M. (1994): Knowledge based reactive scheduling. Production Planning and Control, 5:2, p. 124-145.

SZELKE, E., MARKUS, G. (1993a): Intelligent interface design to supervisory control of dynamic discrete-part manufacturing processes. In: RIX, J., SCHLECHTENDAHL, E.G. (Eds.): Interfaces in Industrial Systems for Production and Engineering, IFIP Transactions B-10. Elsevier, Amsterdam. p. 159-173.

SZELKE, E., MARKUS, G. (1993b): Human integration in intelligent supervisory control for dynamic discrete manufacturing processes. In: TATSIOPOULOS, I.P., PAPPAS, I.A. (Eds.): Advances in production management systems, IFIP Transactions B-14. Elsevier, Amsterdam. p. 81-90

SZELKE, E., MARKUS, G. (1994a): Reactive scheduling – An intelligent supervisor function. In: SZELKE, E., KERR, R.M. (Eds.): Knowledge based reactive scheduling, IFIP Transactions B-15. Elsevier, Amsterdam. p. 125-145.

SZELKE, E., MARKUS, G. (1994b): Contingency and performance driven reactive scheduling in discrete event dynamic environment. In: WALTER, C., KLIEMANN, F.J., DE OLIVEIRA, J.P.M. (Eds.): Evaluation of production management methods, IFIP Transactions B-19. Elsevier, Amsterdam. p. 225-233.

SZELKE, E., MARKUS, G. (1994C): A cognitive engineering approach with AI techniques to reactive scheduling in the supervision of dynamic manufacturing processes. In: LUNZE, J., LEITCH, R., STRUSS, P. (Eds.): Proceedings of the IEE 2nd Int. Conference on Intelligent Systems Engineering, IEE Publication No. 395, London. p.425-433.

SZELKE, E., MESZAROS, I. (1990): CIM related intelligent supervisory control for FMS. In: ELORANTA, E. (Ed.): Advances in Production Management Systems. Elsevier, Amsterdam. p. 271-281

SZELKE, E., MONOSTORI, L. (1995): Reactive and proactive scheduling with learning in reactive operation management. In: STORCH, R.M. (Ed.): Proceedings of the IFIP WG 5.7 Int. Working Conference on Managing Concurrent Manufacturing to Improve Industrial Performance. Washington University, Seattle. p.456-483.

SZELKE, E., MONOSTORI, L. (1997): Reactive scheduling in real time production control. In: VILLA, A., BRANDIMARTE, P. (Eds.): The IFAC Handbook on 'Manufacturing Modelling, Management and Control'. Springer, London.

TZAFESTAS, S.G. (Ed.) (1989): Knowledge-based diagnosis, supervision, and control. Applied Information Technology Series. Plenum Press, New York.

VALKENAERS, P., BONNEVILLE, F., VAN BRUSSEL, H., WYNS, J. (1994): Results of the holonic system benchmark at KU Leuven. Proceedings of the 4th Int. Conference on Computer Integrated Manufacturing and Automation Technology, Troy, NY. p.128-133.

VILLA, A. (1991): Hybrid control systems in manufacturing. Gordon and Breach, London.

WALL, T.T., CLEGG, C.W., KEMP, N.J. (1987): The human side of advanced manufacturing. Wiley, New York.

WARNECKE, H.J. (1993): The Fractal Company: A revolution in corporate culture. Springer-, Berlin.

WIENDAHL, H-P., SCHOLTISSEK, P. (1994): Management and control of complexity in manufacturing. Annals of CIRP, 43:2, p. 533-540.

WIERS, V.C.S. (1997): Human computer interaction in production scheduling. Ph.D. Thesis, Institute for Business Engineering and Technology Application, Eindhoven University.

WOODS, D.D. (1986): Joint cognitive system paradigm for intelligent decision support. In: HOLLNAGEL, E., MANCINI, G., WOODS, D. (Eds.): Intelligent decision aids. Springer, New York.

WOODS, D.D., ROTH, E.M., BENETT, K. (1990): Explorations in joint human-machine cognitive systems. In: ROBERTSON, S., ZACHARY, A., BLACK, J.B. (Eds.): Cognition, computing and co-operation. N.J. Ablex, Norwood.

Acknowledgments

This work was partially supported by the *National Research Foundation, Hungary,* Grant No. T016475, during the period 1995-97.

IT at the Shop Floor – Criteria-based Methodologies for Human-Machine Task Allocation

Eric Scherer and Martina Zölch

Whereas development and implementation of Information Technology (IT) must be carried out by experts, actual organizational design is often left to chance. With regard to 'human resources', organizers as a rule are content to merely propagate the idea of participation and a rather diffuse catalogue of demands and role models as the road to success. This by no means guarantees that the needs of those affected will be accounted for in reorganization projects. Within the project team there are the dangers of power imbalances among different interest groups, of management blindness, and of behavior motivated by social desirability. In addition, there is the problem of the legitimization of needs and proposals, if due to large numbers of affected workers, only a small number of employees, as representatives, are involved in the project (ZÖLCH 1992). It is thus a matter of necessity that a task as comprehensive as the design of shop floor control include methods of project management in addition to participation. Criteria-based methods of designing system content are required that yield exact knowledge and a good description of the reality of production (both social and information reality).

To make such a targeted methodology possible, expert-oriented approaches – such as methods of criteria-based task analysis – were developed by researchers within the field of work psychology. As these approaches supply content criteria for the design of work tasks, they are a valuable complementation to, and aid the evaluation of, the rather unstructured design approaches developed within participatory project management.

1. IT application at the shop floor

Customer focus and strategic target orientation are commonly recognized approaches for structuring an organization. At the operational level of an enterprise, strategic restructuring often results in a complete change of organization: Tayloristic departments are replaced by target-oriented teams. Each team is assigned specific tasks and performs the activities required to fulfill its part of an overall business process. To ensure smooth processing, a well-suited computer-based information and communication system is required. Such a system provides all the information needed to perform a certain activity (e.g., order data, available stock, machine data, etc.), supports processing by supplying specific tools and functions (e.g., calculating sequences, simulation of processing alternatives, etc.) and controls the overall processing of tasks by directing the work flow.

Information Technology (IT) nowadays is seen to enable new forms of organization. Several state-of-the-art integrated software systems provide these basic features and can be adapted to the demands of various industries and enterprises (see BENOY, DEWILDE & VOET 1994). This leads to a large variety of possible system layouts and configurations based on one standard software. In production management, and especially at operational level of the shop floor, activities often become automated which were previously performed manually. In the shop floor domain this leads to a high tension between the automation of technological activities, e.g. numerically controlled machines, and the automation of business activities, e.g. through a manufacturing execution system (MES) or *Leitstand*. The shop floor thereby represents the intersection of the physical value-adding system and the informational management system of the enterprise. Therefore, system specification during the introduction of a new computer-based information system leads to the problems of how to fit generic business processes into an existing shop floor environment.

2. Current approaches to system specification and design

Business processes are information processes. They describe a informational or physical transformation and are performed by human resources or by technological resources themselves. Time-based and target-oriented structuring of these processes leads to the formation of the overall process organization (PICOT & MAIER 1993, 11). Information technology thereby forms the back-bone of organization, whereas human resources are a source of flexibility within the system, since intelligence allows a human being to act independently and creatively. A person can provide flexible reactions and adapt his work methods if the overall system suffers situational variations and disturbances, e.g., in the case of urgent orders or machine breakdown. Human resources therefore are of key value to an enterprise, not only as bearers of an enterprise's know-how and expertise, but as the key to organizational flexibility. This aspect of the relation of human resources to organization is often ignored by many IT projects and by most methodologies for technological and organizational innovation: Humans can not be programmed like computers, and their behavior is not easily determined and controlled.

In the industrial practice of project management the situation is enforced by the emphasis on a primarily technical[1] top-down approach as proposed by American reengineering exponents (e.g., HAMMER & CHAMPY 1994), which leads to a theoretical system engineered on a flip chart or drawing board rather than a organization carefully developed (see VANSINA & TAILLIEU 1995). This type of thinking is similar to approaches common during the CIM euphoria of the 1980s, which – by relying on a primarily technical design logic – mostly led to failure (MANDL 1993). Still, newly designed activities within a business process framework create tasks that are to be performed by real people. Reorganization projects have to consider process organization, information technology and human resources equally as potentials for design and unify them in a holistic system.

2.1. Formal methods for specification of processes and organizations

For the design and specification of business processes, several suitable methods and tools do exist[2] (e.g., SCHEER 1994, CURTIS, KELLNER & OVER 1992, HUCKVALE & OULD 1993), and they have been practically applied. Usually activities are described and linked by discrete events. This forms the basic *control view* of the system model. Each activity is related to the data and information necessary for execution. Additionally each activity is assigned its information resource, e.g., a computer or software system, and the affected organizational unit, e.g., departments or employees. Existing computer-aided tools and methods offer easy support for *formal* process specification even at high complexity. No support is lent to the *conceptual* design of human resources, information technology and organization itself (see KLEIN 1994).

2.2. Human-oriented conceptual design

While information system design is usually performed by IT experts, human resource issues often tend to be a matter of 'trial and error' and 'good faith' and, at best, take participation into account. Participation offers a significant contribution to project management techniques (MORRIS & BRANDON 1994, p. 213ff), leads to a multipersonal approach to problem solving and helps to transform employees

1 The term 'technical' in the context of reorganization refers to the attitude and philosophy in which the respective project is carried out. Already the wording *reengineering* assumes a mechanical way of thinking about organizations.

2 Most methods for the design and specification of business processes are drawn from an IT background. Therefore, they tend to focus on the organizational aspect only as a secondary objective and in the main neglect human aspects all together.

from being affected to being involved (BAITSCH 1985). Still, participation does not ensure that human potentials are truly utilized within a reorganization project (ZÖLCH 1992). Therefore it is necessary to apply methods that offer guidance and expertise to the utilization of human potentials.

Criteria-based methods of system design offer a useful addition to traditional methods of IT design and specification and can provide a distinctive description of the informational and social reality of business processes. Work psychological research has brought up a number of expertise-based methodologies for analysis and the assessment of tasks within administration and production management (see OESTERREICH & VOLPERT 1991, DUNCKEL ET AL. 1993, GROTE ET AL. 1994, WEIK, GROTE & ZÖLCH 1994). Assessment of tasks utilizing conceptual criteria which focus on human strengths can offer advice to business process design at the operational level. This leads to a combined top-down/bottom-up procedure within the framework of restructurization utilizing human resources and IT equally to achieve an ideal process organization, Figure 1.

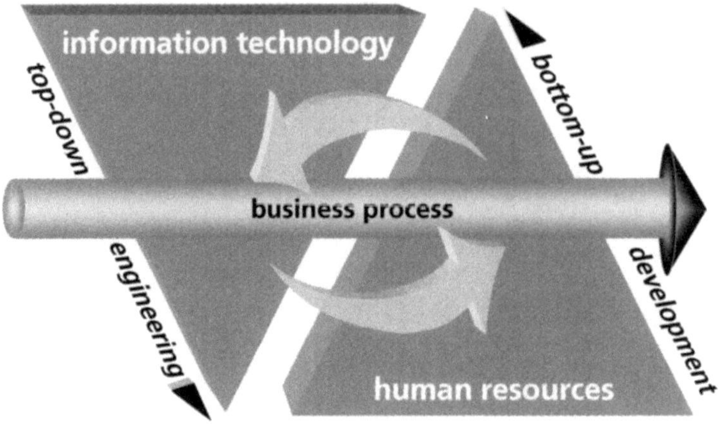

Fig. 1 Top-down engineering and bottom-up development of process organization.

3. Holistic design of control tasks

On the shop floor each employee assumes a number of tasks, each task referring to the processing of a number of consecutive activities, i.e., a section of an overall business process. As part of his duties, an employee deals simultaneouly with activities of several business processes having different characteristics. The perception of a business process by a single employee presents the process as the limited number of activities he has to perform to fulfill the task assigned to him. The summation of all tasks of the various processes forms the sphere of action an employee is assigned to as part of his daily work, Figure 2.

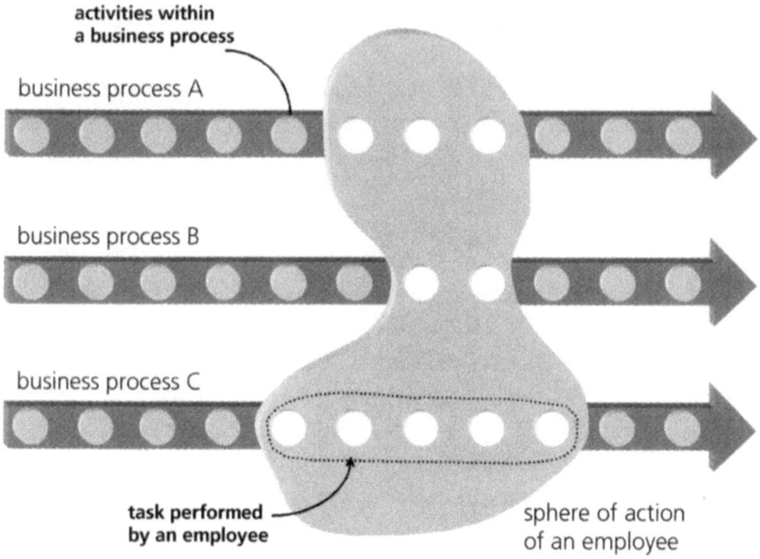

Fig. 2 Process view vs. task view of activities.

In the perception of a single employee, his sphere of action is more a summation of tasks than a subset of several business processes. Hence a task can not be primarily perceived as a function or a duty to be fulfilled as part of a business process or a technical system, but rather as a number of activities to be performed by an individual. According to work psychological research, a task performed by an employee forms the nucleus of the sociotechnical system and links the individual to the organization (ULICH 1994, p. 157ff). The individual becomes an acting agent and adopts certain objectives which guide his own actions. The goal of a task and the respective activities necessary to fulfill a task can be utilized to assess positive requirements regarding the employee affected (OESTERREICH & VOLPERT 1986). This leads to corresponding suggestions for conceptual design to ensure motivation, learning and flexibility.

3.1. The „Ironies of Automation" and human-machine task allocation

To utilize human potentials it is not only necessary to acknowledge their existence and provide an ergonomically-designed work environment but to basically and continuously promote and support human skills and knowledge. It is one of the „Ironies of Automation" (BAINBRIDGE 1983) that automated systems in most cases do not enhance the skills of human operators which are necessary for system regulation in cases of exceptional variations, disturbances or break-down of the automated control. The situation originally observed in automated process control applies to any task of shop floor control distributed between either a human or a computerized scheduler rather than performed jointly. Even basic and repetitive tasks provide expertise for problem-solving in exceptional situations. If not used regularly, skills and knowledge deteriorate. By not performing such simple tasks – such as predictive scheduling – manually, the human controller responsible for variations and disturbances management and reactive scheduling loses the basic knowledge needed. A similar case of 'automation' can be observed in hierarchical organization: 'White collar' human schedulers often lack knowledge of the 'real life' on the shop floor, because they are not involved in 'blue collar' affairs as machine operators.

Therefore, it is necessary to allocate shop floor control tasks in such a way that human resources at the shop floor level gain all the knowledge and skills needed for exceptional situations by performing a sufficient and suitable amount of routine activities. By performing basic tasks they enhance their expertise for reactive decision-making and simultaneously gain information on the actual state of the system. This ensures that the human scheduler has most of the information already on hand in the case of any occurrences of variations and disturbances and results in his ability to solve problems immediately without extensive information acquisition.

In a hybrid shop floor control system, it is the basic task of IT to ensure integrity by modeling relations and constraints and by processing all data with mainly mathematical meaning (not really containing knowledge), e.g., printing of listings or daily compilation of reports. This leaves more time to employees to deal with variations and disturbances.

Figure 3 gives an example of task allocation in shop floor control. While the computerized scheduler is mainly responsible for 'algorithmic' tasks performed

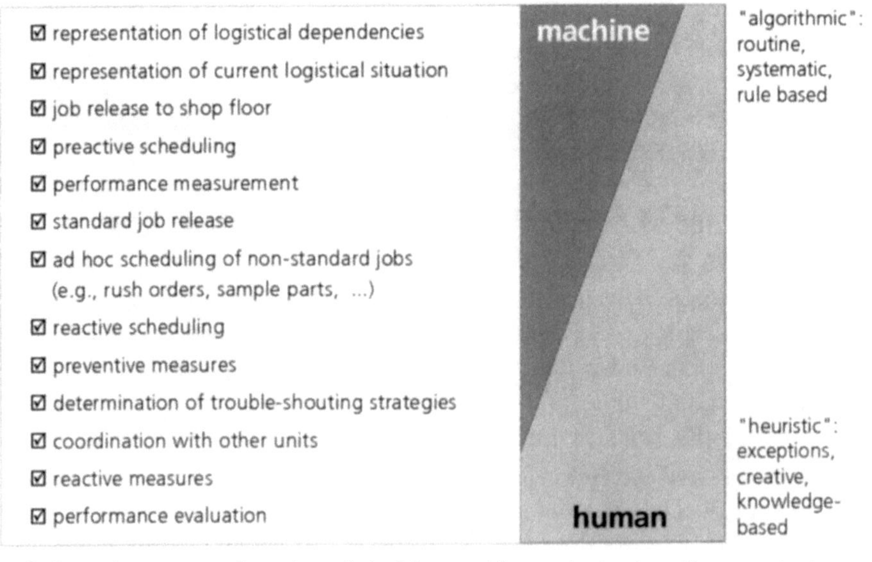

Fig. 3 Complementary allocation of decision-making tasks in shop floor control.

routinely, the human scheduler has an emphasis on 'heuristic' tasks which may present themselves. In considering an existing computerized shop floor control system, the definition of its functionality and parameters has to ensure that a suitable portion of each task is performed by the human scheduler complementary to the computer system, or at least with his knowledge and insight. In the case of more complex tasks, the computer system has to support human activities by providing realistic suggestions rather than obstructing him, e.g., by demanding copious data and information input for knowledge-based decision-making. This will ensure that basic knowledge of system behavior and possible problems and conflicts will not be lost for decision-making purposes.

3.2. Design principles

Design principles include the completeness, variety and meaningfulness of tasks, as well as possibilities for social interaction, autonomy, opportunities for learning and personal advancement, time independence and the opportunity to cope with demands without stress and hindrances during task performance (ULICH 1994, p. 161). Tasks designed according these principles promote employees' intrinsic motivation and their willingness to take responsibility, to gain knowledge through experience, stay mentally alert and develop self-esteem. This results in a work-oriented perspective which attempts to utilize and advance human resources. Still, complete informational support for all tasks has to be ensured and maintained, i.e., all information necessary has to be provided ergonomically and with easy access. The holistic design of tasks thereby gives access to human resource potentials within process-oriented organization and helps to foster these potentials. This enhances the overall performance of the system with respect to flexibility, quality and economy.

A detailed examination of the criteria for task-oriented systems design (Figure 4) makes clear that these may be realized to a great extent through the introduction of teamwork and self-regulation. In order to integrate the aspect of self-regulation into teamwork in a lasting way, complete (holistic) work tasks (ULICH 1994, p. 167-169) are required which allow team members to perform a complete cycle of action, to observe system behavior over long time periods and to, accordingly, gain solid experience. „As a rule, complete tasks [cannot ...] due to their scope be designed as individual assignments, but rather must be designed as team tasks" (ULICH 1994, p. 169). The conception of whole tasks for teams allows the individual employee to 'experience' self-regulation.

Design characteristic	Assumed effect	Means of realization
Holistic nature	■ Employees recognize the meaning and value of their task activities. ■ Employees receive feedback on their own work progress from the activity itself .	■ Tasks having planning, execution and control elements and the possibility to compare work results with task demands and instructions.
Variety of demands within tasks	■ Varying abilities, knowledge and skills may be used. ■ Monotonous demands may be avoided.	■ Tasks involving varying demands upon physical functions and sensory organs.
Possibilities for social interaction	■ Difficulties may be dealt with by the group. ■ Mutual support makes the work load more bearable.	■ Tasks obviously suited to, or demanding, cooperation.
Autonomy	■ Strengthens self-esteem and motivation to take on responsibility. ■ Allows worker to experience self as not without influence or significance.	■ Tasks allowing planning and decision-making.
Opportunities for learning and development	■ General mental flexibility is maintained. ■ Job qualifications are maintained and further developed.	■ Tasks involve problems to be solved, which requires that existing qualifications be applied and expanded.
Temporal elasticity and stress-free regulation	■ Counterbalances inappropriate work intensity. ■ Creates degrees of freedom for stress-free thinking and self-chosen interaction.	■ Time schedules are established which include time buffers.
.Meaningfulness	■ Transmits the feeling the one is involved in producing useful products for society. ■ Assures concurrence between individual and societal interests.	■ Products with unquestionable societal usefulness. ■ Products and technologies whose ecological soundness can be tested and guaranteed.

Fig. 4 Characteristics of task design which promote task orientation and trigger intrinsic motivation (ULICH 1994, p. 161).

4. The KABA method: Criteria-based task analysis and design

The *Contrastive Task Analysis (KABA)* is a procedure to analyze and assess work tasks in administration and offices which was developed by work psychologists (DUNCKEL 1989, DUNCKEL ET AL. 1993, ZÖLCH 1997). It can be utilized to indicate possible consequences of new and existing information technology for employees. Originally designed for qualified work in industrial administration, banking, insurance and government, it proved to be useful to the analysis of administrative tasks in production planning and management and, therefore, was adopted for use at shop floor level.

It can be assumed that human acting is goal oriented, object related, social and flexible. Humans can adopt strategic goals as proposed by strategic management for their own actions, object-related thinking easily reduces complexity, and social behavior leads to the ability to learn and advance. Based on these assumptions, eight key criteria and a number of sub criteria have been derived to assess the human potentials used at a certain system state: Human oriented task design should provide (1) a large *scope of decision,* (2) the *necessity to communicate* (3) sufficient *temporal scope* of work, i.e., *time independence* for decision making and *posobility for time related planning,* (4) *a variation of methods of working* and (5) the *possibility to structure* ones own work, which requires *transparency of the work content and sphere of action* and the *possibility to influence the condition of work*, e.g. through participation in system design. Additionally tasks should be (6) *free of organizational or technical obstruction*, offer (7) *physical activity*, and should be related to (8) *material and social reality*. This allows comparison between the abilities of human resources in contrast to information technology.

KABA is related to the working conditions, i.e., it does not assess the employees themselves and their individual evaluation of the situation, but rather the tasks to be performed and their related sphere of action. KABA can be utilized during *redesign of existing process organization* to identify new human-oriented potentials in system design. Therefore, existing tasks are recorded, and the results of the assessment are used for basic requirement specification, e.g., in case of replacement of an old, computer based MRP II system. During *conceptual design of new organizations*, KABA offers foresighted assessment to various design variations and stages and can be used, valuably, to assess positive and negative effects of proposed software features or computer-based communication systems, e.g., electronic mail or video conferencing. KABA also performs *assessment of existing systems*.

4.1. Application of KABA

Study focuses upon the IT requirements for employees involved in shop floor control. During analysis, the investigator (1) describes the set of tasks related to shop floor management and control to be performed by each employee or each characteristic group of employees. A *work task* is composed of all work units which serve the same goal and can thus be derived from that goal. Here the achievement of the goal must lie within the area for which a certain employee is responsible. *Collateral tasks* are work tasks which make up only a small part of a person's total work time (as a rule, less than 5%). High *task variability* is the case when an employee must process continuously several work tasks which serve differing goals, or where there is the option to move from one task to another. A tasks can by derived based on a real system or based on the prospective of a proposed business process layout. Each task is identified and distinguished from others. Relevant information and control flow is recorded, i.e., how orders are submitted and which work flow information and triggers are used. The assessment of *information flow* is made by determining the need for communication among the workplace under study and other workplaces and organizational units and by pinpointing the tasks at a workplace which are linked to information flow. Then the direction, content, type and frequency of the transmission of information is determined.

Each task identified is (2) split up into the activities necessary to perform the task successfully. All required information and related documents, requirements of information access, processing time and utilized information technology resources are recorded and related to each single activity. This step of KABA principally leads to a formal description of the tasks similar to that achieved though many process modeling approaches, e.g., the event-driven process chains as proposed by SCHEER (1994). Based on the knowledge acquired, (3) the impact of information technology is assessed for each activity. Based on the previously listed key criteria, KABA examines whether the used or proposed information technology has a positive impact on tasks, e.g., easy access to remote information, or a negative impact, e.g., limiting the scope of decision-making. With this knowledge, (4) parameters for human-oriented system design and the utilization of human potentials can be derived.

Evaluation of the work tasks thus examined is performed using the criteria described above. Figures 5 and 6 show the levels of evaluation for the criteria „planning requirements" and „temporal scope". The *planning requirements* determine the degree to which workers may make autonomous decisions on the temporal sequence of work orders in order to fulfill their task. The *temporal scope* of a work task is determined by the number of specific points in time which must be observed when processing an order, as well as the tightness of the due date set for completion. In order to evaluate shop floor planning and control activities, proactive and reactive activities must be distinguished, and at the same time, structural determinants must always be taken into consideration. These cannot be tapped simply according to the criteria described; rather, thorough discussion within the project team and among external consultants and affected employees will be required.

Based on this classification, KABA indicates necessary improvements and measures to access human potentials and permits benchmarking among different teams, organizations and enterprises. Based on job descriptions or similar documents, each level can be interpreted easily in correspondence with the specific activities to be improved. By comparing an employee's qualifications profile with the profile demanded to fulfill prospective tasks, measurements of and measures leading to personal advancement can be derived. This leads employees to view tasks not as a source of stress but rather as challenges; positive incentive is guaranteed.

Level 1: No planning of the sequence of work orders is necessary.

The sequence of work orders is completely and firmly established through instructions, procedure rules, direct orders, etc.

Level 2: The sequence of work orders must be planned.

Due date for completion and processing times for each work order are given. As a rule, only one sequence is possible. The employee must determine this sequence of work orders which will guarantee timely completion.

Level 3: The sequence of work orders must be planned.

Due date for completion and processing times for each work order are given in part. There are several possible sequences. The employee must choose that sequence which will best guarantee timely completion. As a rule, there will be no deviation from this sequence.

Level 4: The sequence of work orders must be planned and regulated in dependency upon success.

Due date for completion is given only in part. The employee himself must calculate each work order time and then sketch out an approximate sequence. This sequence must be regulated during the course of processing, that is, changed in dependency upon actual processing times of work orders, or in dependency upon events such as material deliveries or new information (for example, change in delivery due dates).

Fig. 5 Levels to describe planning requirements (adapted from DUNCKEL ET AL. 1993, p. 36).

Level 1: No temporal autonomy.

Dates for beginning and completion of a work order as well as the majority of completion times for individual operations are established externally and given exact temporal specification.

Level 2: Limited temporal autonomy.

Dates for beginning and completion of a work order are set externally. In addition, time points for certain operations within a work order are given, for example, due dates for semi-finished products or for corrections to machine processing.

Level 3: High temporal autonomy within the framework of external orders.

Dates for beginning and completion of work orders are set externally, but not the finishing times for individual operations

Level 4: Temporal autonomy, times given as orientation only.

Beginning and/or completion of a work order serve only to set general orientating time frame. If times and dates are given, they may as a rule be treated in a flexible manner.

Fig. 6 Levels to describe temporal scope (adapted from DUNCKEL ET AL. 93, p. 48f).

5. Task design for shop floor control: A case study

To provide deeper insight, a short case study is given. For a Swiss manufacturer of optoelectronical instruments, a concept for a new production order processing framework was to be developed and implemented. This included establishing of a number of product-oriented, market-driven segments (cost centers) for assembly of final products and several technology oriented, make-to-order segments for internal supply of parts and pre-assemblies. The processing of all orders on an aggregated level was supported by the introduction of a new MRP II software package. This was to release aggregated production orders, issue routings and control workflow between the segments.

5.1. The electromechanical manufacturing segment

Within the electromechanical manufacturing segment, this led to the decision to introduce a new computer-aided manufacturing execution system (MES), or '*Leitstand*', to replace the previous manual system. The leitstand promised to increase overall production and to meet the requirements as set within the new production order processing framework. Project objectives included among others (1) creation of simple and flexible business processes, (2) creation of more autonomy for operators, (3) increase of reactivity upon work load changes, (4) increase of situational transparency, and (5) humanization of the work. Table 1 gives characteristic numbers for the shop floor segment affected.

Tab. 1 Characteristic numbers.

Staff	44 persons
Number of active product types	110
Number of active assembly types	790
Number of used parts per year	11 million
Number of orders per year (MTO & KANBAN)	6.000
Percentage of KANBAN orders	35%
Average lot size	60
Average lead time (production)	8 days
Average lead time (supply)	50 days
Queuing time/production time	60/40

Analysis included (1) description of order-related business processes at the operational level and within the overall order processing framework, (2) task analysis and definition of human resource requirements, and (3) analysis of information system capabilities (also see BULLINGER & HIRSCH 1994, p. 20ff). For analysis and design, the ARIS modeling approach (SCHEER 1994) was chosen as the formal modeling technique, but KABA methods were utilized for conceptual design of processes and to contrast human potentials and leitstand functionality.

5.2. Results of the analysis

During analysis four classes of employees were investigated at their work place concerning tasks related to shop floor control[3]: floor supervisor, foreperson, skilled and unskilled workers. Figure 7 gives an example of the results of task analysis in the case of skilled and unskilled workers. The reference indicated is based on work

---

3 It should be noted that it is necessary to determine all tasks related to shop floor control for analysis, even if they are neglected by the formal system and management. 'Our workers are not responsible for shop floor control' does not mean that their actions do not influence shop floor control and should be considered as control activities.

psychological research and gives a benchmark for a industrial work tasks classified humane.

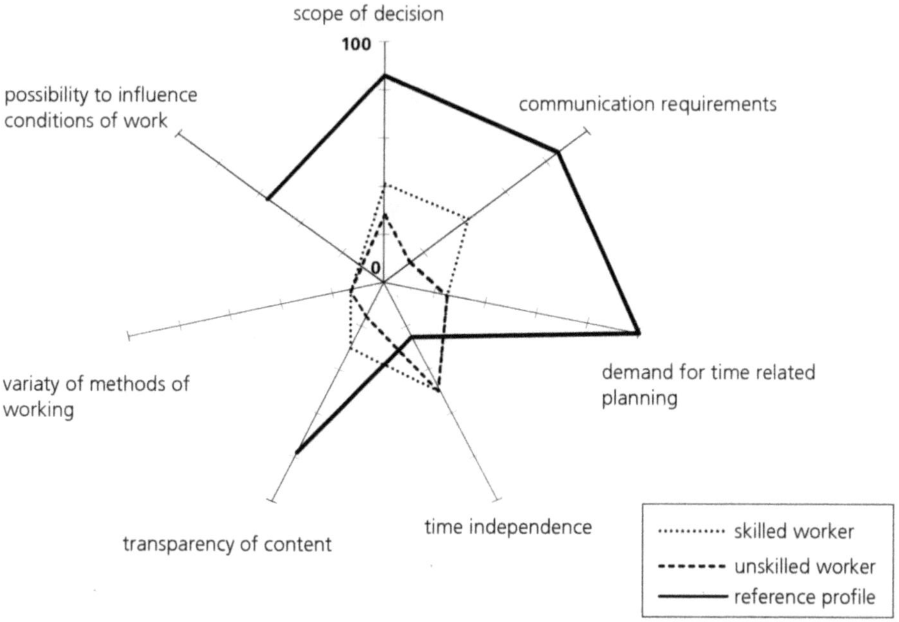

Fig. 7 Analysis results for shop floor control tasks of skilled and unskilled workers.

Close analysis of tasks yielded the finding that the floor supervisor in particular did not receive, or had only difficult access to, information necessary to task fulfillment (Figures 8 and 9). The handling of standard work orders, which as a rule may be planned in advance – that is, proactively – progressed smoothly.

Work tasks		Proportion of time
1	Management of shop floor order processing	~ 30 %
2	Coordination with other departments	~ 30 %
3	Personnel management	~ 30 %
Collateral tasks		
1	Quality assurance	< 5 %
2	Carrying out qualifying measures for employees	< 5 %
3	Participation in reorganization projects	< 5 %

Fig. 8 Work activities of the floor supervisor: tasks and proportion of time spent.

Task	Frequency	Information required	Information available	Information Means	Computer support
Control of orders on hand in the department	continuously	Type and number of orders	Type and number of order releases or orders being processed	Print-out: 'order overview'	none
Control of order progress	continuous	current processing status of all orders	orders in progress	Print-out: 'order overview'	none
Scheduling of rush orders	2 - 3 per week	Capacity load of shop floor, consequences of rescheduling	Type and number of released orders or orders being processed	Print-out 'order overview'	none
Intervention when internal problems arise: trouble-shooting	event-determined (as a rule, daily)	dependent upon problem	comprehensive knowledge of total order handling in the department	none	none
Coordination with other segments (logistics, commissioning, manufacturing)	event-determined (as a rule, daily)	dependent upon the issue: the situation in the particular segment	Knowledge on general operations relationships within order processing	none	none

Fig. 9 Planning and control tasks of the floor supervisor (excerpt).

Although the department has no information aids to support planning, the existing simple and inexpensive manual procedures showed themselves to be effective. In spite of numerous component groups and finished products, produced in small-size production, complexity of work order processing is rather low due to the small number of operations per order and the infrequent occurrence of changes. This results in general adherence to planned processing times and to due date fulfillment which only at certain times falls below tolerable values. The fact that planning is as a rule done by the foreperson given a positive evaluation, as this leaves the floor supervisor the freedom to concentrate upon directing shop floor operations.

In the planning area, that is in situations where it is necessary to deviate from standard procedures in order to react appropriately to events on the shop floor, the findings show a clear weakness. The shop floor supervisor does not have access to the information he requires for scheduling rush orders and KANBAN orders or to reschedule due to particular situations. The reason for this is mainly because these tasks do not have sufficient computer support. For the shop floor supervisor transparency on the shop floor is low, as control of characteristic values, such as order progress, is only possible with difficulty and in a limited way.

5.3. Measures

Analysis led to the following diagnosis: (1) much information is not easy to access or is missing, (2) the division between planning and execution of activities causes information loss and unnecessary interfaces, (3) a lack of work-related communication causes loss of knowledge, i.e., there is only communication on what to do and not how to do it. This diagnosis resulted in the redesign of order processing and related tasks in many respects. The prime measure was to reduce hierarchies to two levels. Supervisor and foreperson were combined into the role of 'floor controller', and all workers formed teams of 10 to 12 people. The key task within each group was the distribution and assignment of jobs. In the old procedure, each job was split up according to its routing and assigned the relevant machines. In contrast, with the new procedure jobs are assigned to the workers themselves. This was possible since there was no specific domination of machines as a restricting resource. Distribution of jobs through staff assignment is now conducted in daily meetings and encourages a more object-related perception by the workers. Figure 10 illustrates the minimization of hierarchies.

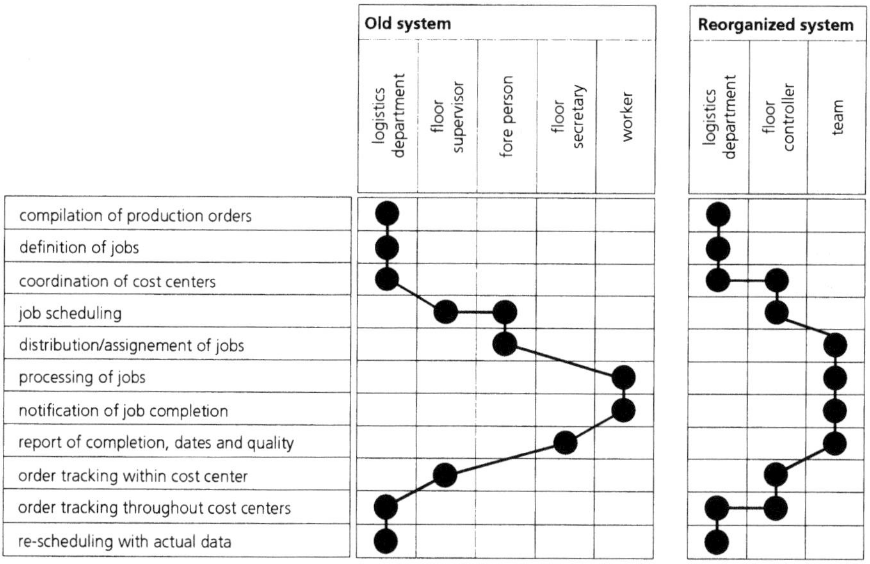

Fig. 10 Allocation of shop floor management-related tasks.

To support staff assignment as a key task, leitstand functionality was slightly adapted to treat human resources as key capacities in scheduling, and they were presented in the GANTT chart. Further contrastive analysis of human and information technology potentials led to the definition of the leitstand functions to be utilized, Figure 11. Through the leitstand and the allocation of job assignments within the teams, it is possible to reduce proactive planning and scheduling tasks of the shop floor controller. This allows more time for further management tasks and coordination with related segments. Reactive scheduling was improved through additional information provided by the system.

	leitstand functions	scheduling						report.	operating		information/ documentation			
	tasks	automatic scheduling	manual scheduling	GANTT chart (monitoring)	checking ressource availability	checking plausibility of schedule	simulation of schedules	reporting work progress	definition of scheduling priorities / objectives	relating dependent orders (order network)	information on schedules of other domains	automatic reporting on finish of jobs	printing out of bill-of-materials etc.	listing of orders on a terminal
	compilation of production orders									●				
standard process (make-to-order)	scheduling of standard production orders	●												
	re-scheduling / optimization of schedule		●	●			●							
	surveillance of due date compliance					●								
	release of order to supply raw material				●									
	supplying material												●	
	allocation of responsible team													●
	controlling work progress of specific jobs			●										
	recording of job completion, dates and quality							●						
	updating schedule represantation (GANTT)							●						
	reporting to MRP II system											●		
exceptions	scheduling of KANBAN orders		●	●			●							
	scheduling of urgent orders		●	●			●		●					
coordination	coordination with central business units													●
	communication with related production units										●			

Fig. 11 Leitstand functions and supported tasks.

6. Reflection

Several approaches to restructuring organization describe an enterprise as an organism and propose the formation of decentralized and partially autonomous segments, such as teams, as a major means to increase productivity (e.g., WARNECKE 1992). Thinking in processes and customer focus establish strategic guidelines for straightforward reengineering of organizational units at the operational level like the shop floor. Modern information technology thereby offers an extensive potential to support business processes. Information technology provides the information necessary to perform activities successfully and supports and controls the overall work flow. Distributed systems enable cooperation and communication among different business segments, teams and distributed production sites without the necessity of direct and personal contact. It is possible to obtain information previously difficult to access and to utilize it to make proactive decisions and decisions based on broader knowledge. Information technology seemingly offers a broad variety of design possibilities. Simultaneously it leads to the danger of unnecessary and inefficient automation. With regard to decision making and the execution of single process activities, it therefore is still necessary to consider the individual intelligence and communication skills only human resources provide within the overall system.

Strategic decisions to reengineer an enterprise's organization, e.g., to flatten hierarchies, establish customer-oriented business segments or introduce distributed production management software, always have to consider the potentials and limitations at the operational level. In the case of shop floor management, large scale restructurization has to be supported by adequate and realistic task design which takes individual qualifications into account. It also has to consider the basic conflict between physical and informational tasks as performed by most employees at an operational business level such as the shop floor. Therefore human resources and information technology equally have to be treated with expertise and conceptual methodology to form a holistic system. This does imply that a technologically feasible solution is not necessarily the solution required for positive system performance (KÖTTER & VOLPERT 1993, p. 131). Contrastive task analysis methods like KABA make it possible to design tasks and related activities within a distributed business process framework with an human resource perspective. To ensure the lasting efficiency of the newly realized organization, permanent organizational development is necessary. Employees have to have the opportunity to reflect upon their behavior and performance as teams and individuals and to adopt personal goals for their work. This includes the possibility to question the tasks and regulations set and to change them adequately (see MARKS 1991). The ability to reflect on processes and organization from the local perspective of the shop floor level ensures permanent improvement and flexibility within the process framework for production management.

7. References

BAINBRIDGE, L. (1983): Ironies of automation. Automatica, 19:6, p. 775-779.

BAITSCH, C. (1985): Kompetenzentwicklung und partizipative Arbeitsgestaltung. Frankfurt am Main.

BENOY, M., DEWILDE, P., VOET, M. (Eds.) (1994): Finite Scheduling: State-of-the-market. Ernst & Young Belgium, Brussel.

BULLINGER, H.-J., HIRSCH, B. E. (Eds.) (1994): Leitstände für die Werkstattsteuerung. Schriftenreihe der Bundesanstalt für Arbeitsschutz, Fb 705. Wirtschaftsverlag NW, Bremerhaven.

CURTIS, B., KELLNER, M. I., OVER, J. (1992): Process modelling. Communications of the ACM, 35:9, p. 75-90.

DUNCKEL, H. (1989): Contrastive task analysis. In: LANDAU, K., ROHMERT, W. (Eds.): Recent developments in job analysis. Taylor & Francis, London. p. 125-136

DUNCKEL, H., VOLPERT, W., ZÖLCH, M. ET AL. (1993): Kontrastive Aufgabenanalyse im Büro – Der KABA Leitfaden, Grundlagen, Manual, Arbeitsblätter. vdf/Teubner, Zürich/Stuttgart.

GROTE, G., WÄFLER, T., WEIK, S., ZÖLCH, M. (1994): KOMPASS – Komplementäre Analyse und Gestaltung von Produktionsaufgaben in soziotechnischen Systemen. Unpublished manuscript. Institut für Arbeitspsychologie, ETH Zürich.

HAMMER, M., CHAMPY, J. (1994): Reengineering the corporation. paperback edition, HarperBusiness, New York.

HUCKVALE, T., OULD, M. (1993): Process modelling: why, what and how. In: SPURR, K., LAYZELL, P., JENNISON, L., RICHARDS, N. (Eds.): Software assistance for business re-engineering. John Wiley, Chichester. p. 81-97.

KLEIN, M. M. (1994): Reengineering methodologies and tools. Information Systems Management, 11:1, p. 30-35.

KÖTTER, W., VOLPERT, W. (1993): Arbeitsgestaltung als Arbeitsaufgabe – ein arbeitspsychologischer Beitrag zu einer Theorie der Gestaltung von Arbeit und Technik. Zeitschrift für Arbeitswisenschaft, 47:3, p. 129 -140.

MANDL, C. (1993): Computerintegrationsprojekte – Die Logik des Misslinges. IO Management, 63:12, p. 54-58.

MARKS, S. (1991): Gemeinsame Gestaltung von Technik und Organisation in soziotechnischen kybernetischen Systemen. VDI Verlag, Düsseldorf.

MORRIS, D. C., BRANDON, J. (1994): Re-engineering your business. paperback edition, McGraw-Hill, New York.

OESTERREICH, R., VOLPERT, W. (1986): Task analysis for work design on the basis of action regulation theory. Economic and Industrial Democracy, 7:4, p. 503-527.

OESTERREICH, R., VOLPERT, W. (Eds.) (1991): VERA Version 2. Arbeitsanalyseverfahren zur Ermittlung von Planungs- und Denkanforderungen. Technische Universität Berlin

PICOT, A., MAIER, M. (1993): Interpendenzen zwischen betriebswirtschaftlichen Organisationsmodellen und Informationsmodellen. Information Management, 8:3, p. 6-15.

SCHEER, A.-W. (1994): Business process engineering: Reference models for industrial enterprises. Springer, Berlin.

ULICH, E. (1994): Arbeitspsychologie. 3rd Ed., vdf/Poeschel, Zürich/Stuttgart 1994.

VANSINA, L.S., TAILLIEU, T. (1995): Business process reengineering or socio-technical system design in new clothes? In: WOODMAN, R. W., PASMORE, W. A. (Eds): Research in organizational change and development. Volume 9. JAI Press, Greenwich. p. 81-100.

WARNECKE, H.-J. (1992): The Fractal Company – A revolution in corporate culture. Springer, Berlin.

WEIK, S., GROTE, G., ZÖLCH, M. (1994): KOMPASS – Complementary analysis and design of production tasks in sociotechnical systems. In: KIDD, P.T., KARWOWSKI, W. (Eds.): Advances in agile manufacturing. IOS Press, Amsterdam. p. 250-253.

ZÖLCH, M. (1992): Partizipation contra kriteriengeleitete Augabenbewertung – eine Scheinalternative. In: LANGENHEDER, W., MÜLLER, G., SCHINZEL, B. (Eds.): Informatik cui bono? Springer, Berlin.

ZÖLCH, M. (1997): Analyse und Bewertung von Bürotätigkeiten: Die Kontrastive Aufgabenanalyse (KABA). In: STROHM, O., ULICH, E. (Eds.): Unternehmen arbeitspsychologisch bewerten. vdf Verlag, Zürich. p. 223-244.

Guidelines for Systems Design for Operational Production Management

Franz Stuber

Competent production management on the shop floor requires efficient support by information systems. Along five guidelines, the paper works out the necessary requirements of such systems. The guidelines range from a simple basic model to visualization of competing planning goals to functions which allow work experience to be acquired and applied.

1. Introduction

In the context of new concepts of production organization, the deficiencies of production planning and control systems on today's market become apparent. Those responsible for carrying out reorganization which (once again) plans for relevant planning and control tasks at the shop floor level report that algorithms up to now calculated by the computer are being removed, as they are found to hinder new forms of order planning and work organization. The systems concepts gaining in importance are those which emphasize greater degrees of freedom for activity at the shop floor level as opposed to the detailed requirements of strategic production management. Production teams become subjects in operational production management, which, from the viewpoint of strategic production management are seen as a 'black box'. For this reason, research and development are faced with the issue of creating appropriate support for shop-integrated planning and control tasks through new Computer Aided Production Management[1] (CAPM) systems.

1 The term 'Computer Aided Production Management' here is used synonymously for Computer Aided Shop Floor Control.

But first a word on the use of the term 'guidelines'. Software development and application takes place within a field of tension between the postulation of *scientifically derived criteria* – as well as of 'standards' in the art of engineering – on the one hand and the demand for *participation*, user involvement in the development and improvement process on the other. The guidelines should bridge these two, at first glance contradictory, approaches. Thus in the following, requirements are formulated which scientists regard as significant for the design of future CAPM systems. The term guidelines, however, emphasizes at the same time that the results must be interpreted and transformed – by those affected and involved. Guidelines which are formed in this manner represent a first step in engaging users in a productive dialogue and in a cooperative realization of the system.

CAPM systems for shop floor integrated operational production management are one additional aid among other means and resources for task fulfillment. They are not oriented towards full-time planners and controllers, but towards skilled workers, who as part of a team plan, execute and control varying tasks with shared responsibility. In terms of temporal use, utilization of the systems is part-time. This means that guidelines which have been developed within research on software ergonomics, and which have in part become established as ergonomic norms, should be taken into particular account. With regard to part-time use, the following two requirements established in the field of software ergonomics are significant:

- System design should provide relief – if possible through automation – from activities which do not directly serve task completion, but which are caused by peculiarities of the tool computer, such as data security, reorganization measures or updating data bases between system units.
- Part-time utilization absolutely demands that systems be simple to use. Features of the user interface —e. g. editors, graphics, icons – should stem directly from the type of work and be highly recognizable. In short, the knowledge required to use the system must be reduced to a minimum to free workers minds for their real tasks.

The focus of the following guidelines is not, however, at the level of the ergonomics of the user interface. Research at that level, at least traditionally, presupposes a certain type of model building and modeling[2]. In this paper, a model is developed that is an *alternative* to established MRP (Material Requirements Planning) systems concepts.

> „The interface is important, but if we change only interfaces and not the systems behind them, we will only be able to scratch the surface.“
>
> (FISCHER 1995a, p. 23)

---

2 Although now boundaries are beginning to blur, as evident in the program title of a 1993 Conference on Software Ergonomics held by the German chapter of ACM: „From Interface Surface to Work Design“ (see RÖDIGER 1993).

2. First guideline: Simple work documents

The first guideline calls for work documents which are modeled in a simple manner and which can be used by production teams as an aid to production management. 'Work documents' is a metaphor which VOLPERT (1992) introduced into the debate over appropriate views of the computer. VOLPERT means to say that software units should function as memory supports which „ ... in an efficient way make the needed information, preliminary work, etc. available, without controlling or limiting the actions of the competent expert“ (VOLPERT 1992, p. 150).

In the simplest case, already existing work documents can function as such aids to memory, such as drawings and/or structural representations, e. g. bills of material, of the product to be manufactured. For many reasons, however, this can be only *one* option: first, these externally produced documents must be completely available – and of this there is no guarantee in practice. Secondly, the usefulness of such documents for production management depends to a great extent on the type and degree of repetition of the product. Thirdly, production teams should be allowed to develop their own evaluative criteria and forms of representation for the particulars of work tasks relevant to planning.

2.1. The structure of work documents

Going beyond the use of externally produced work documents, the CAPM system should make provision for the production of descriptions of planning particulars and connections of work tasks. Team members should be able to draw up and utilize noteworthy elements of work tasks as information units. These units should not be programmed into the system as fixed. The CAPM system should only provide a shared definition of such. General determinants of the production process serve as the basis for organizing these information units:

- In the place of setting an 'optimal' routing at the single operation level, the perspective of innovative production management finds a planning model appropriate which is oriented towards *work tasks.* The tasks for which production teams share responsibility are the link between work organization and technology (ULICH 1994).
- Every work process processes one or more *objects of work* which are processed and changed to reach the targeted result. This holds true independently of the degree of automation of the work process. Work objects may be, for example, raw materials or assembly components in the context of order processing, machines or equipment in maintenance work, or also non-material objects such as plans, lists or CNC programs.
- Successful processing requires further the implementation of one or more *means of work,* whether machines, workplaces, tools or factory supplies. Work objects and work means are not distinguished absolutely, but rather will

depend on the particular task at hand: a device or a CNC program may be a work object for one task in order to function as a means of work for another.

- In innovative production concepts, production teams organize their time programs autonomously as appropriate to tariff and agreements within the enterprise. To aid presence and absence planning, therefore, team members require the opportunity to manage their own time accounts. With these, each worker should plan his/her presence times to coincide with upcoming tasks. For planning task processing, these personnel time plans, at the same time, provide information with regard to who will perform the tasks.
- These considerations result in four general information units, which the CAPM system represents as elements of simple work documents: tasks, objects of work, means of work and personnel plans of the task holders. Figure 1 shows these in an overview[3].

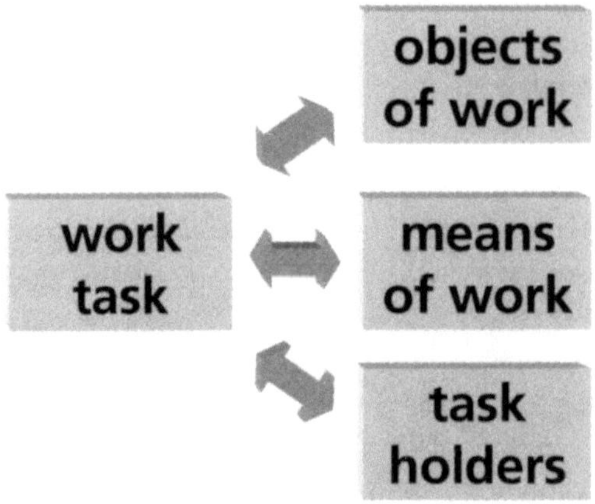

Fig. 1 A simple model for computer-based work documents.

3. Second guideline: Choice of amount of detail

The autonomy of production teams to make decisions must extend to include the choice of the degree to which acquired planning and control knowledge should, and should not, be aided by formal representation in the CAPM system. The design of the system should support this freedom to decide. It must be the users

[3] In the sense of object-oriented systems analysis, this corresponds to a representation of the four information units as *classes*, to which team members handle special *objects*. Even though the 'state of the art' as well as our considerations closely approach object-oriented specification and realization, we will maintain our use of social scientific terminology. This is not only so that non-engineers will follow the discussion, but rather also because the methods of 'Requirement Engineering' are not strictly *methods of analysis*, but rather *means to describe* (formally and graphically) *that which has undergone analysis.*

themselves who directly determine the degree of concretization of computer-based representation of work tasks.

With this principle, our model departs from the traditional concept in MRP II systems, whereby a plan must represent the complete description of a work task. For new organizational concepts, demands for detailed and exact —though as a rule not feasible – work plans for single parts and orders, which were made from the perspective of planning departments far from the shop floor, are now obsolete. The system concept must therefore take into account that when the planning subject changes, both the requirements and the type and breadth of relevant planning information must also change. For example, the orders received by production teams consist only of due date, technical specifications, quantities and perhaps external comments, such as quality requirements, but no instructions concerning scheduling or resource allocation of individual order components. A 'sparse' work document such as this may under certain conditions suffice to allow the order to be distributed and completed in terms of time, persons and resources. In such a case, the job of the CAPM system consists in creating, for all team members, a transparent level of information on the orders to be completed in mutual responsibility. A practiced team may, on this basis, perform internal task division and material provision simply – through votes of acclamation and short discussion – as well competently predict feasibility.

3.1. Details

With this guideline for freedom of choice, there is the option, if needed, to establish who will, when, and in what sequence work on processing, and where, or when, materials and resources are to be used. If task distribution and coordination is decided on and accepted in this way by team members, there is no need for a detailed computer based representation of work tasks. There may be various reasons, however, which make it desirable to have more detailed descriptions. Such reasons include the following:

- Information about special activities in setting up the machines, in the procurement of tools or about CNC programming,
- documentation requirements, for external or quality reasons, of operations and
- the need to have an overview of units where tasks are complex – involving, for example, many different resources, or complex components.

With these demands it becomes important to be able to produce detailed work documents. Freedom of choice, as demanded by this guideline, must therefore also mean that individual tasks are not necessarily assigned to be realized autonomously, but that assignments can contain indications of other, upcoming tasks. For this reason, the planning system must allow consistent detailing of *part tasks*. To avoid the restrictions of traditional MPR systems, the following should be taken into account: Detailing *one* work step in a task at one work station should not result in work station-specific planning of the *total* task. This would restrict the autonomy of the teams to make their own decisions in this regard. If work means and materials need to be assigned to perform the superordinate task, it must be possible to plan exact time schedules or other specifics with regard to one of the means —for example, a bottleneck machine – without affecting the general assignment of all work means. Expressed another way, it has to be possible that one planning element be detailed without other elements automatically being handled in the same way.

3.2. Grouping

Up to now the model assumed that required work objects and the utilization of work means can be established for tasks and part tasks. Assigning individual concrete resources, etc., can, however, result in undesirable restrictions to flexible use of a CAPM system. This occurs when there are several options with regard to the use of various materials or alternative machines, or when more than one worker may be assigned responsibility for task processing. For such demands, the system must allow degrees of freedom with regard to the assignment of resources and responsibility.

For this, the machine grouping approach, well-known in MRP development, will be examined and expanded. The purpose of grouping in the planning process is to assign resources not in terms of specific machines or materials, but rather in terms of a specific group (or 'class', see STUBER 1992) of many same or similar resources. To assure the necessary flexibility, it makes sense to use this method also with regard to the qualifications and competencies of the workers. Production teams should be free to determine that no, one or several workers will perform individual tasks, or to assign responsibility for a task to a certain set of workers. The most appropriate means to do this is to categorize team members in qualification groups, whereby each team member may be a part of several qualification groups.

4. Third guideline: Checking and informing

The first two guidelines are concerned with the structural requirements of simply and flexibly utilizable work documents as elements of shop floor-integrated CAPM systems. The third guideline now directs attention to demands for appropriate algorithms.

First of all, it is clear that workers, in meeting the disparate and contradictory challenges within dynamically changing situations, must not be hindered by the system's algorithms. For this reason, in contrast to many current approaches, optimization is understood not as the *function of computation methods*, but rather as the *task of the planning subjects*, who should be supported by calculation methods.

This orientation seems much more appropriate in view of the field of action which contains several, simultaneously targeted conflicting goals determined by technical, economic, and social variables (see SCHERER 1996 as well as the contribution of SCHERER in Part One of this book). For system design, this means that algorithms should be used where calculation tasks are routine or require excessive attention. Secondly, the main emphasis in their use should be the demonstration of the consequences of alternative decisions.

4.1. Task monitoring

The need is for task processing to be monitored, and the aim is to utilize supportive computer checks and calculations in so far as work documents contain quantitative information. Some examples of this are the following:

- If work documents contain merely a target date as well as planned length of processing time, the CAPM system might make continuos checks of the possible realization of the date and notify the users of possible delays.
- Where work documents contain the assignment of work objects the system can monitor demand coverage and suggest, or automatically initiate, any necessary logistical changes.
- If work means have been assigned planned machine times, this will allow capacity calculations to be made and warning information to be issued by the system with regard to bottlenecks.
- Where task descriptions include the assignments of particular workers, their work times can be monitored. This can be accomplished using the plan for personnel attendance times or also through comparison with other personnel planning.

This third guideline as developed so far stands in a certain relationship of tension with the first two guidelines, and certainly conflict can arise: the extent to which monitoring functions can be applied will depend completely upon the details of the work documents. If only drawings are used as work documents, with processing times and due dates, only a conflict between processing time and target date can be monitored. In addition, the degree of detail in work documents regarding resource utilization, for example, will determine the extent to which availability checks can be carried out or logistic actions initiated. The crucial factor is that the reason for modeling some information is not that they are prerequisites for the functioning of the system. Rather, the concrete working out of the details within this field of tension remains a matter which is in the hands of the production team, whose members may come to their own agreement as to the extent to which the progress in task fulfillment should be monitored by the computer system.

4.2. Demonstration of consequences

In addition to the functions of task monitoring sketched out above, computer checks can be used explicitly to demonstrate the consequences of planning decisions in terms of the conflicting goals inherent to production management. Where modifications in planning are to be made, whether experimentally, due to unforeseen events or due to changes in goals, it is useful to have the CAPM system demonstrate repercussions. Relevant information is, first of all, the degree to which sequence changes will become necessary for the planning object, and secondly, how other tasks and goals will be affected by the modifications. If dates are to be altered, for example, the system should calculate and demonstrate how this measure would affect the general situation with regard to due date delivery, utilization of work means, supply of work objects or the processing times of other tasks. In order to obtain a good basis for decision making and/or decision revision, it makes sense to execute the modifications as various simulations. Using the demonstrations of the effects of the modifications on the course of the task and on goals, the decision can then be made as to the most appropriate variant in full knowledge of its affects elsewhere.

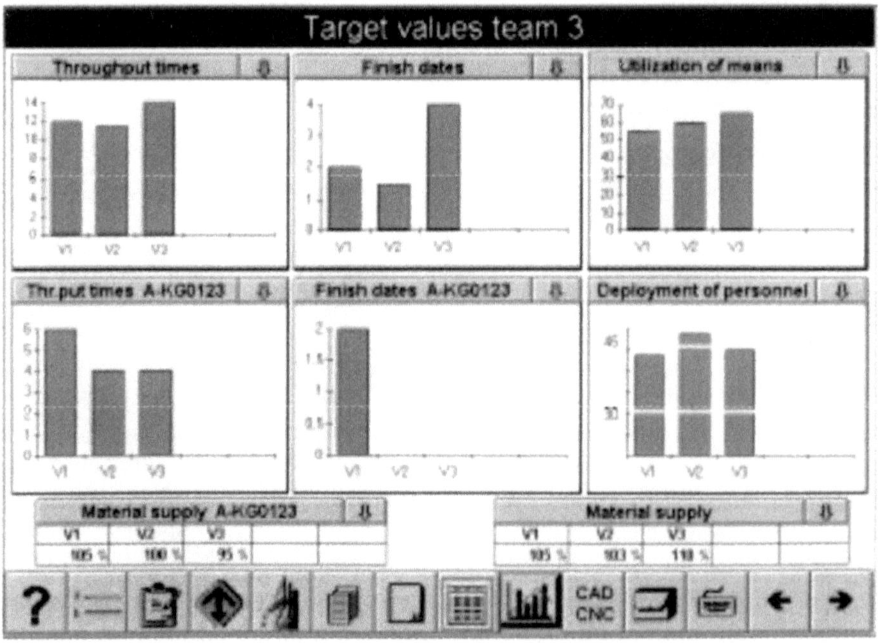

Fig. 2 Simulation of the effects on planning goals (from STUBER 1997).

Here it also holds that the possibilities and the meaningfulness of the calculations will result from the amount of detail in the work documents. It is crucial that there is no attempt, through computer processing, to cancel out the real contradictions inherent in planning by means of computer methods. Instead, the objective is to demonstrate changes of the quantitative relations among goals and, thus, to make evaluation and decision possible.

5. Fourth guideline: Tutorial character

All guidelines up to this point have made it clear that new system design opens windows of diverse decisions involving learning and problems, which are worked out in the process of communication among team members and can be stored in the CAPM system. A prerequisite for such flexibility of action is that new workers be introduced to the new system in an uncomplicated manner. Therefore, the requirement is for shop floor-integrated introduction and „learning on demand" (FISCHER 1995b). This pertains to:

- ease of learning the *use* of the CAPM system,
- learning the *task contents* of production management.

In order to support these learning processes, the CAPM system must possess tutorial characteristics. This is not a demand for independent instructional programs. The path needed is different from concepts which focus upon multimedia presentation of existing MRP system functionality in the sense of a 'computer-based training' instruction program which is *separate* from the work system. Here, rather, specific characteristics of the work system itself are addressed (compare KASVI ET AL. 1994, RAUNER & SPÖTTL 1995).

5.1. Elements of tutorial support

The first tutorial element is already contained in the concept of the simple system model. On this basis, there is a need for *context specific* help functions such as those already familiar in advanced systems in other application areas. For each screen window, icon to be selected or menu choice, there should be an explanatory text or graphics available whenever not self-explanatory to the novice. Such helps are self-understood today and require no further elaboration here.[4] Particularly helpful is the integration of hypertext elements and hypermedia structures with the helps: by cross-referencing further explanations, which, when needed, may be selected, the help system adapts to the varying degrees of knowledge and experience of each user.

A third element of tutorial system design are functions which support *exploration*. These are an aid to implicit learning in that they allow risk-free exploration and examining of the system functions: in first commercial standard systems, so-called 'UNDO' functions and freezing points ease the learning of system utilization where the user is not yet sufficiently familiar with the entire

[4] It is even in this that many CAPM technologies on the market today are gravely lacking.

system or the effects of individual operations (see CARROLL 1982). With this, however, the limits of explorative functions are revealed: explorative functions do support the learning of system functions, but from the start, it is either assumed that the user already has task-related competence or such competence is not promoted. Developers of the explorative approach readily admit to this state of affairs (see HAMBURG & PAUL 1993).

For the system to have true tutorial qualities, therefore, further elements must be integrated which are required by multimedia-aided, formalized learning processes. These are in part used today. For the tasks which are to be executed by the CAPM system, there should be built-in *learning sequences* which, when needed, may be activated during system use. The learning sequences may range from hypermedia explanations and animations to short video sequences on the task formulations that can be processed by the system. The design of such learning sequences, however, both in terms of content and teaching method, cannot be a complete given[5]. This kind of measured implementation makes use of the advantages of a combination of various presentation and simulation media for individual learning, according to need, of cost planning – without being taken in by the exaggerated claims and illusions raised by current-day multimedia euphoria.

A further benefit of tutorial system quality is that success in the learning process can be aided by the fact that:

- learners do not only adapt given knowledge, but actually *reformulate* this knowledge according to their own style of expression, and
- suitable help features stand ready to explicate what has been implicitly learned.

To support these processes, system design must build in a way for learners to make notes and explanations. Reformulations and explication of learned task elements should be manageable, both user-specific, individual notes as well as team-wide, collective notes on system functions of planning tasks. Figure 3 shows the individual elements of learning support in this type of tutorial system design.

It is important that the presentation of all elements of a tutorial system design may be selected according to choice, appropriate to various subjective and environmentally-determined learning conditions. For example, the degree of detail of information should be flexible, as selected by the user, and it should be possible for notes created during learning to be either 'privatized' or 'published' for all team members.

---

5 BINDER, FISCHER & NILSSON (1997) describes the results and experiences of participatory development projects within the manufacturing industry.

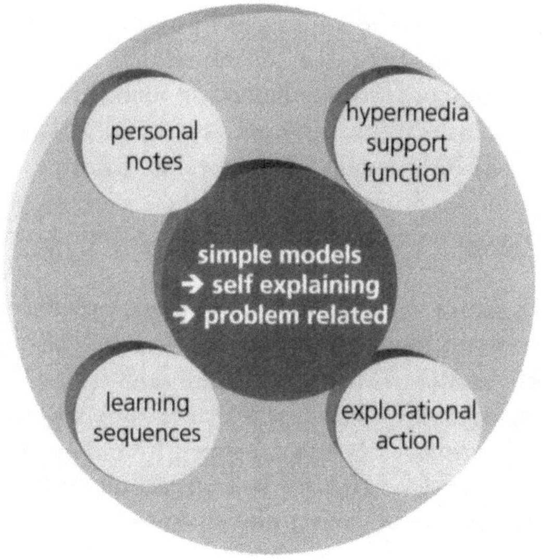

Fig. 3 Elements of tutorial system design.

6. Fifth guideline: Making previous experience accessible

In many ways, the work tasks of production teams consist of similar and partly repetitive instances. This means that there are good conditions for the development of competence in meeting the challenges of new planning tasks on the basis of previous experience. Computer-based work documents reveal their special usefulness in building a bank of work process knowledge when they are not specified anew for each new planning task, but rather make available experience gained from earlier tasks. In order to achieve this, the CAPM system allows for the documentation of experience in production management for later evaluation and future tasks. The system must therefore function as a medium for the gathering and utilization of work experience.

6.1. Planning patterns

To fulfill this guideline, previously realized work tasks must be documented and edited to serve as patterns and prototypes for current planning. In the simplest form, after task completion, the system must allow information to be entered with regard to deviations between targeted and actual time, between planned and actual material utilization or with regard to other difficulties or unexpected successes. At the same time, certain notes and instructions required only during actual processing and instructions related to particular moments in time should be deletable.

Continual documenting is known to be considered a great burden, in that it hinders workers in getting on with the next assigned tasks. It is here that a suitable field for automation of individual processes of documentation is found. On the basis of monitoring algorithms, various changes, such as in times or quantities, can be instituted automatically by the CAPM system. Other events worthy of documentation, however, such as unforeseen difficulties in the mastering of processing techniques or problems with material quality, would remain a matter for the workers.

A complementary form would consist in the extraction of relevant experience from several completed work documents to form a complied *task group*. Which of the two variants the team selects, and to what degree, should not be predetermined by the system. Certain information could be presented as decisional aids, such as the parts family to which a product belongs, its repetitive character, technical variances, etc. But the teams must be free to develop their own view, which will comprise many influencing factors that can only be evaluated within the particular shop floor situation.

Such patterns, based upon work experience, can be utilized in many ways. First, they may be used as the starting point for current planning as 'mother copies' which are then modified interactively to fit current requirements. Secondly, the use of task groups can allow more exact specification of items collected in the task group. Thirdly, any peculiarities of earlier task processing documented in the patterns can be used for purposes of statistical evaluation in order to search for the causes of any recurring problem situations. Figure 4 illustrates the development and use of patterns in the CAPM system.

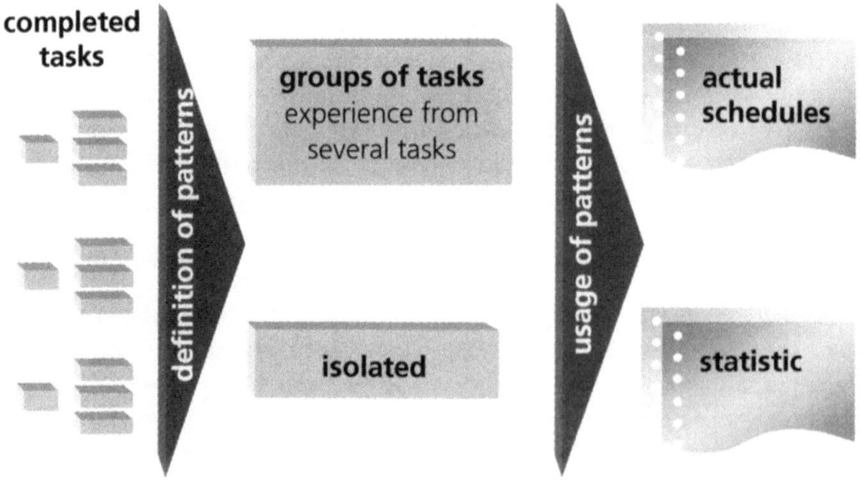

Fig. 4 Patterns as re-usable experience data bases.

6.2. Selection rules

The uses of patterns can be supported by complementing them with algorithmic elements that the teams produce themselves. Here production teams have the option of utilizing their knowledge of relations and interdependencies among individual planning instances from earlier experience by editing such knowledge in a form that can be processed by the computer.

Simple forms of the 'end-user programming' (see also EISENBERG & FISCHER 1994) can be utilized at many points within production management. For example, if a task group comprises several optional resources, selection rules establishing preferences for the one or other variant under varying conditions can be set up. Unambiguous selection criteria will be required from the start. Further, calculation formulas, such as for raw material dimensions in dependency on construction features, can be established. Quantitative relations among work objects, the work means to be utilized and qualification requirements may be formulated. Figure 5 shows an example of a raw material dimensioning derived from a simple rule.

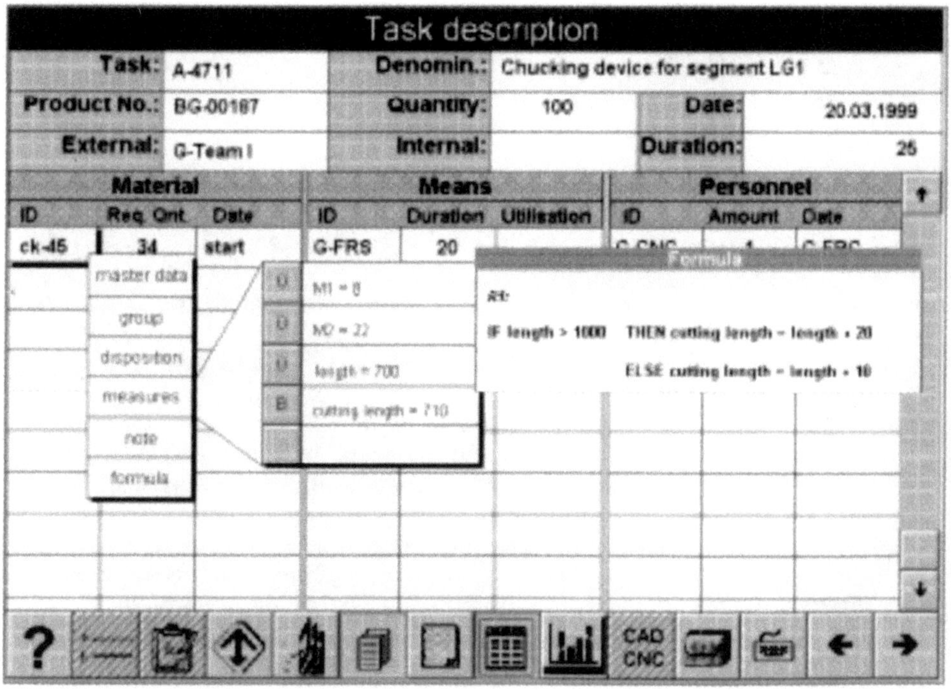

Fig. 5 Rules for dimensioning in work documents (from STUBER 1997).

The decisive point is that computer evaluation represents an *intermediate step* – for example, between the interactive selection of a certain task group and testing and complementation by the production team. The origins of computer-transmitted selections or calculations must be transparent at all times, and the option to accept or reject must always be given. Data in the task groups and formalized relations in selection rules need to be studied and evaluated in view of the current planning situation. The degree to which inference-based processing methods of knowledge-based technology can be implemented will depend on an evaluation of concrete planning problems. This is a field of conflict between declarative modeling on the one hand and the maintenance of consistency and transparency on the other (see BONSIEPEN & COY 1989).

7. Outlook

The guidelines sketched out suggest a design for computer-aided operational production management, formulated from a scientific perspective within new production concepts. In order to further implement the suggested design, a project consortium of users and software developers in dialogue will be formed. To succeed, it will be critical to embed the software development project within simultaneous redesign of planning organization. In addition, future users must remain permanent contributors to further developmental steps. A prototype described in detail by STUBER (1997) will be used at the initiation of the dialogue.

8. References

BINDER, TH., FISCHER, M., NILSSON, J. (Eds.) (1997): Learning with artifacts. Special issue of AI & Society. Springer, London.

BONSIEPEN, L., COY, W. (1989): Erfahrung und Berechnung – Kritik der Expertensystemtechnik. Heidelberg.

CARROLL, J. M. (1982): The adventure of getting to know a computer. IEEE Transactions on systems, man and cybernetics, 11.

EISENBERG, M., FISCHER, G. (1994): Programmable design environments: Integrating end-user programming with domain-oriented assistance. In: Human Factors in computing systems, CHI ´94 Conference Proceedings. Addison Wesley, Amsterdam.

FISCHER, G. (1995a): New perspectives on working, learning and collaborating and computational artifacts in their support. In: BÖCKER, H.-D. (Ed.): Software Ergonomie ´95. Mensch-Computer-Interaktion, Anwendungsbereiche lernen voneinander. Teubner, Stuttgart.

FISCHER, G. (1995b): Learning opportunities provided by domain-oriented design enviroments. In: DISESSA, A. A., HOYLES, C., NOSS, R. (Eds.): Computers and exploratory learning. Computer and Systems Sciences, Vol. 146. Springer, Berlin.

HAMBURG, I., PAUL, H. (1993): Beispiele für arbeitsorientierte Benutzungsoberflächen von DV-Systemen für Ingenieuraufgaben. In: KONRADT, U., DRISIS, L. (Eds.): Software-Ergonomie in der Gruppenarbeit. Leske + Budrich, Opladen.

KASVI, J., PULKKIS, A., VARTIAINEN, M., NIEMINEN, M. (1994): Building an information system to support shop-floor employees. In: LEVONEN, J. J., TUKIAINEN, M. T. (Eds.):

Proceedings of the interdisciplinary workshop on complex learning in computer environments (CLCS94). Helsinki University of Technology.

RAUNER, F., SPÖTTL, G. (1995): Berufliche Bildung und betriebliche Innovation als Moment des europäischen Strukturwandels, diskutiert am Beispiel der FORCE-Sektorstudie zum Kfz-Gewerbe. In: DYBOWSKI, G., PÜTZ, H., RAUNER, F. (Eds.): Berufsbildung und Organisationsentwicklung. Donat, Bremen.

RÖDIGER, K.-H. (Ed.) (1993): Software-Ergonomie ´93. Von der Benutzungsoberfläche zur Arbeitsgestaltung. German Chapter of the ACM, Berichte 39. Stuttgart.

SCHERER, E. (1996): Koordinierte Autonomie in verteilten, heterogenen Produktionssystemen. Ein Beitrag zur Flexibilisierung der Werkstattsteuerung. vdf, Zürich.

STUBER, F. (1992): Projektieren und Konfigurieren mit Standardsoftware für PPS. ZWF, 88:2, 80-83.

STUBER, F. (1997): Rechnerunterstützung für arbeitsprozeßnahes Planen. Software-Innovation im Kontext von Ökonomie, Organisation und beruflicher Bildung. Donat, Bremen.

ULICH, E. (1994): Arbeitspsychologie, 3rd Ed. Vdf, Zürich.

VOLPERT, W. (1992): Wie wir handeln ñ was wir können. Ein Disput als Einführung in die Handlungspsychologie. Asanger, Heidelberg.

On the Road to Agility: Changing Organizational Structures

Part Six

Making Teams Work – A Key for Effective Shop Floor Control

Albert Hurtz

Teamwork has become a major means to improve the flexibility of an organization. Accordingly the introduction of teamwork in manufacturing can increase the flexibility and quality of shop floor control. This paper aims to examine the factors which lead to successful teamwork. The experience which the Praxis für teamorientierte Arbeitsgestaltung (PTA) [Practice for Team-oriented Work Design] has gained in implementing teamwork in more than 20 industrial enterprises shows that certain prerequisites must be fulfilled if teamwork be implemented with successful results. In view of shop floor control, light will be thrown upon only a few significant factors. Many aspects, such as the significance of the behavior of management or issues of qualifications, must necessarily be neglected, as well as consideration of teamwork in non-productive areas.

1. Teamwork – A contradictory concept?

The issue of teamwork has been the subject of intensive discussion since the late 1980s (see WOMACK, JONES & ROOS 1991). Teamwork was seen as the measure which would rescue industry and lead it back to the path towards success. Teamwork, it was believed, would considerably increase production and cut costs. The idea became widespread that workers would simply need to be grouped in teams and given the opportunity to talk among themselves, and success would be automatic.

Teamwork went on to become an indispensable instrument in the modern manager's toolbox. Production managers were instructed short-term to institute teamwork as well as follow the goal of reducing manufacturing costs by 20%. The consequences of proceeding in this way are undeniable: teamwork is now commonly held to be an instrument of rationalization and job downsizing. Numerous enterprises have announced the failure of their experiments in teamwork and have returned to the old system of divided-labor manufacturing. In other enterprises, teamwork has brought a measurable degree of success, and workers are more satisfied and active at work than before. Many a company would not exist today had it not restructured and introduced teamwork.

2. Ways and instruments of effective teamwork in manufacturing

The term teamwork is very variously defined today. Here teamwork is understood as semi-autonomous team work. In Figure 1, the main features of semi-autonomous teamwork are presented schematically. The team is responsible for their own area of work, which is unambiguously defined and determined. The team organizes itself, plans and controls the processes and tasks within the team. Quality control is also conducted by the team. The greatest possible number of team members must be capable of performing several tasks within the team; in the ideal case, each member can do all things. This allows the greatest possible flexibility (also see HURTZ 1994).

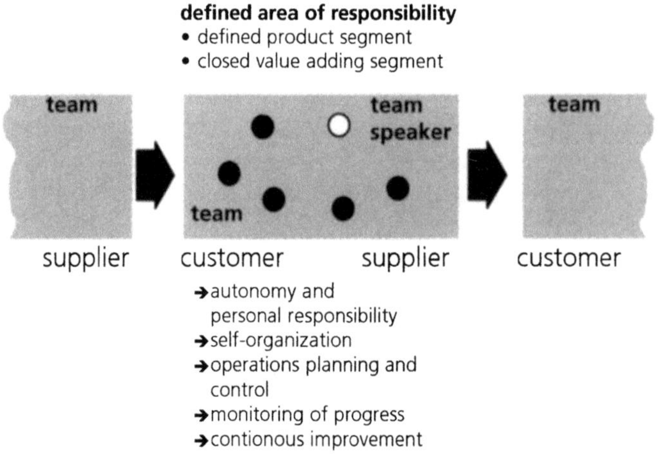

Fig. 1 Basic structure of teamwork in manufacturing.

The relationship among the teams is that of customer-supplier. Each team views the team which follows in the manufacturing process as its customer, who must be supplied to fullest satisfaction. The teams preceding are seen as suppliers, who are expected to supply the team on schedule. The teams communicate among themselves directly, with no detours via managers or other outsiders. In interaction with others, each team is represented by a chosen team speaker, who is equal in status to the others and a productive worker. In no case should the team speaker carry out all non-manufacturing tasks alone; rather, team tasks should be distributed among the greatest possible number of team members.

As a rule, it takes a long time for a team to become able to work independently and responsibly. If rapid success in increasing productivity and cutting costs is desired, teamwork is the wrong means. Teamwork leads to success only in the long term. In Figure 2, the most important steps towards successful teamwork are outlined.

Fig. 2 Steps toward successful teamwork.

„The fundamental prerequisite for successful teamwork is the building of trust."

From the start, the team needs trust from the organization. If managers and bosses have no confidence in the team, it will achieve little or nothing. A team granted no freedom to organize itself or make plans will not develop the ability to organize itself independently. A team which learns to organize work within the team and to divide tasks in a meaningful way will also become capable of improving the manufacturing process. It will begin to discover wastes of time, materials, transportation routes and so on, and find ways to eliminate them. With time, the team will learn to build well-ordered relations with customers and suppliers. It will base its actions on the criteria of fulfilling the needs of its internal customers optimally, and through the course of time, the external customer, who buys the products, will become the focus of interest.

Team activity will be regularly examined, weak points discovered and improved approaches developed and tested. Although tasks and areas of responsibility will of course vary considerably from team to team within manufacturing, there are the following central activities, which are usually carried out by manufacturing teams (Figure 3).

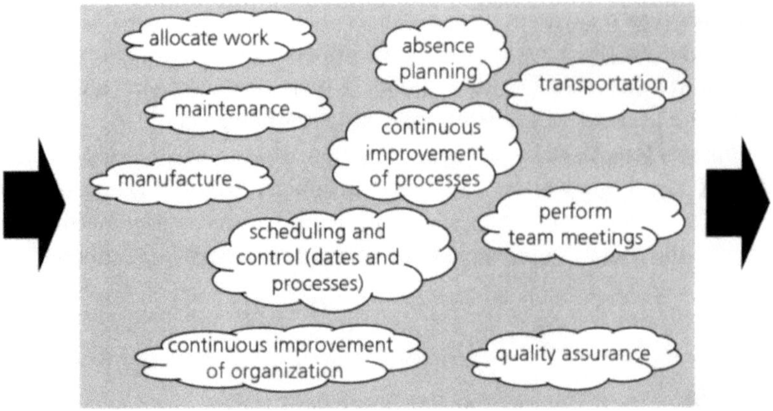

Fig. 3 Manufacturing team tasks and areas of responsibility (example).

The main task of every manufacturing team is to produce in due time parts, components or machines of optimal quality. To do this involves quality control as well as the optimal planning of processing and work division within the team. The crucial factor in implementing teamwork successfully is involvement of the team in design and decision processes. For this reason, mutual exchange of information on the situation, planning and new developments is extremely important; only these allow true involvement of the coworkers. Team communication is the most important instrument of teamwork. At team meetings information is exchanged within the team, and problems and possible solutions are discussed. Superiors or representatives from other teams are included in the talks to discuss and handle any problems transcending the individual team, to present new proposals and determine proceedings. Once a team is built, it goes through a process of development which, in its ideal type, consists of three phases (Figure 4).

In a first phase, team members seek their place and role within the team. Normally, superiors, production management, have developed a concept of the team's task area and have proposed the make-up of the team. Future team members must first deal with this proposal from the outside. The team will develop well only if its members accept the team's composition and its tasks. If the team as planned is not comfortable with the proposal, a better alternative should be sought by management and the prospective team together. Once workers and executives have reached a consensus, the team will be able to develop itself.

Consensus within and among teams is one of the most important determinants of the success of teamwork, from formation to 'mature' team. When team members have been appointed, they frequently 'wall themselves off' from the outside. As relations among the team are still unclear and there is uncertainty as to how internal cooperation will be achieved, the team will fend off any intrusions or examinations from the outside. Only once relationships within the team have been set.

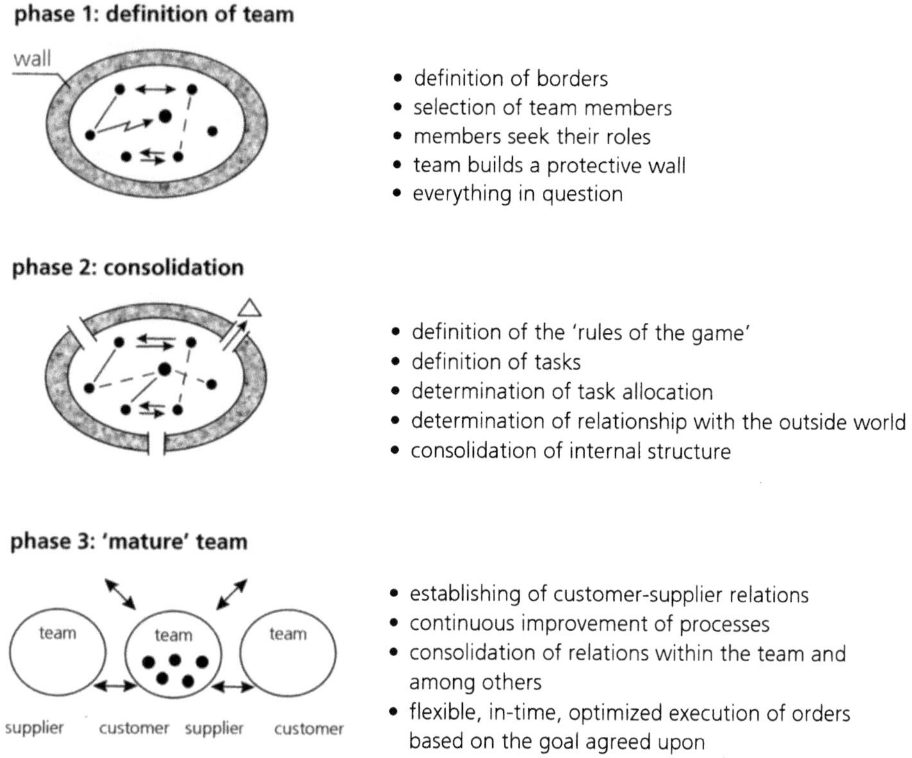

Fig. 4 The process of team development.

The 'mature' team is characterized by exchange with other teams, with which they stand in a customer-supplier relationship. The team works on the basis of their set goals to continually improve processing. Problems and conflicts within the team and among teams are settled in an independent manner. The team will be included in any decision-making which affect the team. Figure 5 presents the characteristics of and steps in development towards the 'mature' team.

The experience of many enterprises shows that teams can be supported in their development by a number of instruments: through goal setting, plans of activity, visualizations and qualifications planning. Goal setting brings together the objectives of team and management level. Goals are reasonable only to the extent that they are realizable and accepted by all. Team goals should be set for a comprehensible time period of, for example, three months. Teams and management meet regularly to determine the status of goal achievement, institute additional measures or, if need be, adjust goals.

Characteristic of a 'mature' team	Developmental steps
1. Team's internal rules of the game are brought to life	▪ rules of the game developed, ▪ internal rules adhered to, ▪ internal rules changed or extended.
2. Activities and responsibility are divided among team members.	▪ team representative functions mainly as mouthpiece, ▪ each member has developed a sense of responsibility.
3. Team members execute all activities within the team.	▪ each activity can be performed by any of several members, ▪ work is conducted according to systematic rotation.
4. Regular, independent team meetings are held.	▪ team meetings are held at least every two weeks, ▪ a team member moderates the meeting, ▪ teams regularly review what was well done and not well done and how work may be improved.
5. Each team member is accepted and integrated into the team.	▪ the opinions of all are respected, ▪ no member is excluded, ▪ the team shows tolerance towards its members.
6. Team members identify themselves with their team.	▪ team decisions are supported and carried out by all.
7. Team members are open and honest with each other.	▪ each member expresses opinions openly, ▪ no one speaks behind another's back, but rather addresses the other openly, ▪ disturbances are reported immediately.
8. Team conflicts are addressed, worked through and resolved.	▪ conflicts are brought into the open, ▪ conflicts are discussed rationally, ▪ conflicts are resolved in a lasting way.
9. The team can visualize the aspects most important to their work.	▪ team maintains information boards, ▪ important variables are given concrete illustration, ▪ team results are presented in graphs (in continuous fashion).
10. Team exchanges information and measures with other teams.	▪ team sets up rules of dealing with clients and suppliers (conditions of transfer, for example), ▪ problems with other teams and divisions are directly addressed, ▪ team finds lasting solutions to problems with other divisions.
11. The team works to continually improve work procedures and to avoid waste.	▪ waste is eliminated.
12. The team makes and follows plans of activities and seeks to implement measures quickly.	▪ the team institutes measures which lead to improvements, ▪ the team sets due dates and appoints those responsible for implementation, ▪ the team controls adherence to due dates.

Fig. 5 Characteristics of a 'mature' team.

A plan of activities is a representation of the issues and measures with which the team must deal. This plan is posted visibly (in large format) and kept up-to-date. Figure 6 is an example plan of activities of a work team. In addition to plans of activities, a team requires a number of further means of visualization in order to be permanently informed of their achievement status, to have access to information from management or to make internal agreements visible. Figure 7 presents an example of an information display. Each team work out a qualifications plan and follows it closely. The qualifications plan (Figure 8) comprises the tasks that each team member can execute or would like to learn.

plan of activities					
sheet no.:					week:
no.	issue	necessary measure	responsible	agreed completion date	degree of completion
1	operators have to bend down too much to get parts	lift conveyor belt by 12 inches	Mark	9/29	
2	tools are in no order and difficult to find	install board for tool storage	Peter	10/18	
3	team room is to noisy	install door	Peter	9/20	
4	air conditioning out of order	contact maintenance	Steve	11/19	
5	to view members can drive the forklift	get license for all team members	Mr Burnes/ driver's school	to be determined	

Fig. 6 Example plan of activities.

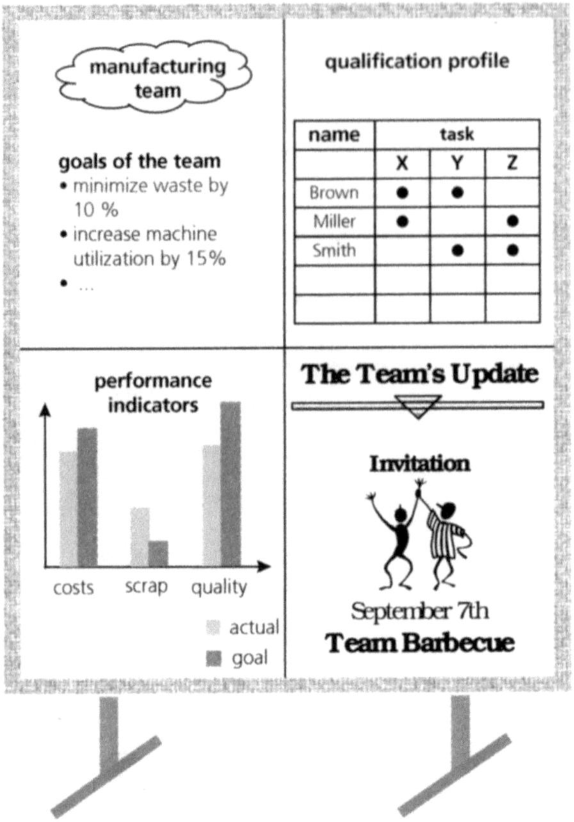

Fig. 7 Example design of a display board.

	name									
team: / **task:**	Miller		Smith		Brown					
store material	●		●		●					
lathe set-up			●							
fixture adjustment	◔		●		◓					
optical inspection	○		●		◕					
...										

○ member to learn the task
◔ member has been present during processing
◓ member has performed the task at least once
◕ member can perform the task under supervision
● member can perform the task independently

Fig. 8 Example qualifications plan.

3. Determinants of successful teamwork in manufacturing

The introduction of teamwork does not automatically guarantee its successful implementation. Rather, there are requirements which must be met for teams to develop well.

„Teamwork is implemented successfully only if the entire enterprise is involved in the restructuring."

If teamwork is introduced into manufacturing alone, while other divisions continue to work as before, the desired effects will not – or to a limited extent only – be achieved. Inevitably, teams will run into stonewalls if there is no behavioral change in their internal partners. It is not necessary that the entire enterprise work in teams, but within the enterprise new behaviors must develop which support and not hinder the teams' engagement and growing independence.

„Without active support from management, teamwork will remain half-hearted."

It is not enough for management to introduce teamwork and to make the necessary funds for its institution available. It will also not suffice for management to give verbal support to teamwork. What is required is that management embody the 'spirit' that marks teamwork: openness, willingness to change and rapid implementation of improvement approaches. Only when management recognizably demonstrates that which the team must achieve in its area will team members show lasting motivation to further develop their team.

„Responsible action of team members will develop only where there are degrees of freedom for action and decision-making; without these, the team will not develop."

Without opportunities to act with responsibility, a team cannot learn to take on responsibility for its work. If superiors control the team, members remain dependent and do not develop the motivation to be actively involved.

„Teamwork is successful only where management is appropriate to teams."

Leadership of teams differs considerably from managing individual workers. Management has been accustomed to dealing with workers directly, without having to consult with anyone else, as, for example, when a worker is moved from one workplace to another. With teamwork, however, the situation is much more difficult, because the team must be informed of the changed situation, and the decision may only be made with the team's agreement. Their involvement in decision-making is very difficult for some managers.

The form of leadership appropriate to the teamwork system is leadership through goals, that is, goals must be set with the team and goal attainment controlled. Experience has shown that many managers need to be specifically trained and accompanied in order for them to fulfill their new task appropriately.

„Successful teamwork requires motivated workers, who work on the team in a responsible and engaged manner.“

A team cannot be forced to take on responsibility. It will, however, show effort and work on improvement if active involvement in teamwork would bring positive advantages.

„Lasting motivation on the part of workers requires a good work environment as well as the institution of effective incentive systems.“

If a team does not identify itself with the enterprise and its products, its motivation to make extraordinary achievements with not be lasting. Achievement must be rewarded, and good achievement will last only if there are incentives. In addition to money, there are further incentives for teams such as praise and recognition, the experience of success and grants of extended authority.

„Work teams in manufacturing can develop all the better the closer they come to the ideal of the 'closed segment of the value-adding chain' and are responsible for a product or a parts family.“

For example, teamwork does not make much sense in a conventional lathe shop, because the degrees of freedom and possibilities are too few. The more the value-adding steps that are executed within the team, the greater the degrees of freedom for good teamwork. The optimal situation for teamwork is one whereby the team manufactures completely one product, one sub-assembly or one part.

4. Planning and control in manufacturing teams

This aspect will be examined particularly, due to the special importance of planning and control in efficient teamwork. Manufacturing teams which work 'to the beat', have no freedom to plan or schedule, and produce according to an externally ordered program have little opportunity to develop or improve. Such teams experience helplessness and constraint, and motivation for teamwork quickly dissipates. The probability that this team will advance on its own initiative is low. At some point team consciousness will fade and each member will once again be a lone struggler. On the other hand, the existence of freedom to plan and schedule by no means ensures that a team develop; the degrees of freedom must first be utilized in a meaningful way. There are many roadblocks to effective utilization of degrees of freedom to plan and schedule:

„The team is unsure of what it may and may not do.“

Right at the outset, irritations often result from uncertainty about what is expected of the team. The team is not sure that what is does is correct and prefers to ask the foreman. The foreman is all too willing to help, agrees that is hard to get the big picture, and tells the team to leave the planning to him. It makes much more sense to prepare the team carefully for planning tasks. The team requires training in the use of planning and control instruments and should have an advisor at its disposal at the beginning.

„Planning and control systems are often too complex, too unwieldy, and too rigid for adequate finite scheduling and control in manufacturing teams."

Most planning and control instruments used today are not suitable for implementation in teams. The biggest problem is their production of untranslatable data, which makes rapid and error-free decision-making very difficult to realize.

Teams tend to circumvent the shortcomings of control systems by means of their own solutions. This may help the team, but often this leads to breakdown in the total system. These problems will be resolved in the long-term only through the implementation of simple planning and control instruments designed to meet the needs of manufacturing teams. Teams in manufacturing as a rule require only certain bits of information in order to control processing and to plan work schedules of the workers, such as the following:

- quantities,
- completion dates,
- planned manufacturing lead time,
- delivery reliability of internal and external suppliers.

„Due dates are constantly rescheduled, making reliable due date scheduling in the team impossible."

There is nothing more frustrating to manufacturing teams than constant interruptions by rash express orders, which throw over the day's or week's plan. Team efforts to work effectively and to reduce set-up times by skilled planning are destroyed by these interruptions. A certain amount of 'interference' caused by the rescheduling of due dates can be managed by the team without difficulty, and in fact, it is here that the advantages of the team system come to the fore. But when interference becomes the rule, nothing works.

„The material needed for further processing is not available."

But it is perhaps missing parts which roadblock the team even more than constant, short-term due date rescheduling. Missing parts interfere with work flow, the team gets out of rhythm and production is disrupted. Teams feel helpless when they can no longer move forward and can do nothing to influence the situation.

„Teams often have no understanding of how planning data is generated, so that the available data is not correctly interpreted. As a result, planning and control within the team are beset with wrong conclusions and poor decisions."

Teams must be given intensive training in the use of planning systems. It is not enough to qualify one team member, but rather, everyone on the team should know how planning is done in the enterprise. Particularly at the beginning, teams require regular feedback on their planning activities. Successful planning attempts should be acknowledged, and errors and breakdowns should be discussed and analyzed in detail, so that the teams gradually learn to plan realistically and well and to control processing.

5. An outlook: The future of teamwork in manufacturing

The initial euphoria regarding teamwork has dissipated; hardly anyone believes that teamwork alone can solve the problems facing manufacturers. Exaggerated expectations of productivity increases and cost-cutting have given way to a much more realistic view of teamwork as a form of organization. In the last five years, we have gained a more differentiated understanding of the areas and divisions in which teamwork can be implemented successfully and of the conditions which must be met.

In all cases where segmentation of production has led to the formation of closed segments of the value-adding chain, and where relatively complete processing of parts or assembly of components, machines or tools takes place, teamwork – in the form of semi-autonomous team work as described – is possible. Although it was at first assumed that teamwork lent itself to best realization in serial production, it has become evident that teamwork works just as well, or even better, in single-item or small-sized production.

For successful introduction of teamwork, what is far more important than the number of parts manufactured by batch is (1) a surrounding with a structure that is adapted to the requirements of the team and the workplaces and (2) the creation of sufficient degrees of freedom for the team. And not least, whether or not a team fulfills expectations will depend upon the behavior of management.

If the tendencies recognizable today towards the creation of smaller and more flexible organizational units gain general acceptance, teamwork will become an indispensable element of the organization of work – the germ of self-organization and self-responsibility. In such powerful organizations oriented towards the rapid implementation of ideas, work in teams will be self-understood. The dividing line between mental and physical labor will be erased to a great extent. No engineer or business manager will find it belittling to work in cooperation with the 'worker' on solving problems or on the development of new products.

In enterprises which maintain the previous hierarchical structures and behaviors based in subordination and superordination, however, teamwork will have a difficult time. It is in any case questionable whether such enterprises will remain competitive on today's market.

It is between the two extremes that teamwork will develop in future. In those enterprises which create a firm foundation for teamwork, teams will gradually grow towards self-regulation. Teams will be provided with planning and control instruments which are designed for their needs. Many firms have already begun to design programs, using standard software, which produce the data which the team actually requires. The advantage of such programs is that they may be adapted quickly and easily as requirements change. And they are inexpensive to boot. There are teams which have developed their own solutions to planning and control within the team and with which they work successfully. Teamwork becomes more viable with such solutions, teams become more flexible and can better contribute their share towards economic and client-oriented production.

Teamwork will prove itself in those cases where team members find that it is to their own personal advantage to work in a team and where managers succeed in setting and achieving goals in cooperation with the team. In future, there will be few industrial organizations in which individual competitors are needed, because work will proceed in teams not only in manufacturing, but on all levels and in all divisions within the total enterprise. Teamwork will fail where a team orientation is not consistently applied. Work must become teamwork, for the challenges of the future can only be met together!

6. Literature

WOMACK, J.P., JONES, D.T., ROOS, D. (1990): The machine that changed the world. Rawson Associates/Collier MacMillan, New York/Toronto.

HURTZ, A. (1994): Qualifizierung für Gruppenarbeit. Theoretische Konzepte und ihre Umsetzung in der betrieblichen Praxis. in: ANTONI, C. (Ed.): Gruppenarbeit in Unternehmen. Psychologie Verlags Union, Weinheim. p. 81-99.

Shop Floor Control as a Task of Organizational Development

Steffen Weik and Eric Scherer

Flexible, self-regulating, (semi-) autonomous, decentralized cells, fractals or teams, in which independent, intrinsically motivated individuals enterprising in thought and in action are current concepts intended to insure that manufacturers will be able to react adequately to the increasing dynamic nature and complexity of production conditions. Management theories based upon the principles of „Cybernetics I" (see STAEHLE 1994, p. 41), which aim to eliminate variations and disturbances by means of preventive systems oriented towards stability and balance, have increasingly given way to theories built on „Cybernetics II". The premise of these is that only by increasing the system's tolerance for failures and instabilities can an adequate strategy for mastering the problems facing management be found (STAEHLE 1994, p. 41, see also SCHERER 1996). The goal of modern management cannot be the reduction, or avoidance, of complexity, but rather complexity must be mastered by means of appropriate system design. Stated simply, for shop floor control this means that planning and control tasks are no longer carried out in central, higher-level departments uncoupled from manufacturing operations, but rather, that these tasks must be executed whenever possible by the manufacturing islands themselves. Only this can guarantee that there are sufficient degrees of freedom in manufacturing for complex events to be mastered. Where there is no such freedom, manufacturing has to depend upon complexity-reduction internal to the system in the form of, for example, deterministic systems production planning and control systems (PPC).

Upon the background of this currently popular position, those working within practical in-plant reality, as well as external consultants, are continually faced with the issue of how such logistics concepts can be implemented in work systems which up to now have lived and worked according to contrary concepts. A case study of a Swiss vehicle manufacturer will illustrate a possible path leading to pragmatic, practical implementation of visionary, theoretically-derived, well-founded logistics concepts – a path which both operations and management can develop and uphold together.

1. Changes in sociotechnical systems

The following considerations start out from the sociotechnical systems approach which is going back to studies by TRIST & BAMFORTH (1951) – established that: Organizations are made up of various work systems which all have a social and technical component system. The social system is composed of „members of the organization, with their individual physical and psychological needs, in particular their demands on their work and their knowledge and abilities.“ (ALIOTH 1980, p. 26). The technical system constitutes „the assets, equipment and their layout; in general the technological and spatial work determinants which form the demands on the social system“ (ALIOTH 1980, p. 26). The social and technical subsystems are subject to different laws, which must be taken into account in designing or redesigning a work system. While in the „technical world“ (see SUSMAN 1976, p. 146) the rational principles of efficiency rule, in the social, or „phenomenological world“ (see SUSMAN 1976, p. 146), in which humans experience their daily lives, psychological principles of cognition and emotion hold. Particularly the latter has received ever less attention within modern reorganization concepts. WOMACK (1995, p. 100), for example, accuses the 'reengineering prophets' of an unwillingness acknowledge „human and organizational realities“.

> „Nowhere in the now abundant literature on reengineering is there an indication that attention is given to the legitimate needs of the people employed. People appear in various categories: leaders, redesigners, people who ought to be flexible, committed to continuous learning or as subject to be committed, to develop shared values and goals; and as an average of 20 percent who need to be moved out.“ VANSINA & TAILLIEU (1996, p. 94).

These painful experiences were felt by the company – or the employees – which is described in the following case study. All at once, in a move called by the workers the 'Kristallnacht' (in reference to Crystal Night of the 1938 Nazi pogrom), 20% of the employees were let go practically overnight. This experience has had lasting effects on the workers and continues to influence the company's attempts at reorganization. The assumption that organizations – or the processes constituting the organization – can be separated from the people who execute them, viewed and redesigned separately, is falls emphatically short. „Thereby they neglect the very basis of any organization: the pattern of relations between people who make the system work.“ (VANSINA & TAILLIEU 1996, p. 97).

1.1. Expert versus process consultation

When we take on a sociotechnical perspective and explicitly incorporate the social system into our considerations on reorganizing work systems. The question is raised as to the most suitable concept regarding consultation. SCHEIN (1968) differentiates between two concepts of organizational consulting: expert and process consultation. In the first, a client system (organization/corporation) typically delegates a problem – reorganization of shop floor control, for example – to a consultant or team of consultants and contracts them to suggest problem solutions. The second assumes that a client system, through the aid of the consultant, must become capable of solving its own problems. Here the consultant does not propose solutions, but rather has as a goal that solutions develop within the client system itself. While the role of the consultant in expert consultation is that of the doctor or fireman, who knows exactly what to do when and how, the process consultant is more of a moderator, or therapist, who seeks to broaden the client's basis for decision–making by means of self-searching questions. The process consultant helps the client towards self-help, without aiming to directly influence problem solutions. (see, for example, KÖNIG & VOLMER 1993, WOHLGEMUTH 1991).

It thus becomes clear that justice can only be done to the social system through way of process consultation. A crucial part of process consultation is involvement of those affected in the process of change (participation). Only through participation is there a guarantee that those involved are in a position to register their legitimate interests and needs with regard to the redesign of their working conditions. Participation's significance with respect to the individual and the individual's development of competence, to the utilization of existing knowledge and experience and to handling resistance – which is always part of the process of change – will be discussed later in this paper.

With this as a background, let us look at the issue of how expert knowledge which is generated outside the individual enterprise – in universities for example – must be handled. Even dedicated process consultants have no interest in keeping this knowledge from their clients. KÖNIG & VOLMER (1993, p. 47ff) hold to the view that process consultation may certainly include phases of expert consultation – such as informing about possible measures – but not, however, with the goal of supplying packaged problem solutions, but rather to give the client system supplementary information to aid the decision-making process. Within sociotechnical systems theory, there is agreement that strategic goals, such as changing from a central to decentralized PPC system, can certainly be triggered 'top-down', or by an external expert. But the realization of structural changes must in any case be achieved 'bottom-up', with consistent involvement of those affected (see VANSINA & TAILLIEU 1996, 95). Expert or process consultation – the question is not one of dogma, but one of dosage and emphasis!

1.2. Changing social systems and individual development processes

Processes of change in social systems – just as in individual learning and development – follow certain, psychologically-founded laws, which reorganization planning in social systems must take into account. However, in the context of purely economically-motivated change, social scientific knowledge is seldom given any regard. An apt metaphor characterizing a method frequently implemented is the 'bomb throw' (see WOHLGEMUTH 1990). Changes – in most cases downsizing or other rationalization measures – are thrown out into the company like a bomb, that is, an irreversible blow without warning. Management and expert consultants rely on members of the organization to improvise to fill in the resulting organizational gaps. This assumption was also made in the manufacturing enterprise we will study below – under significant sacrifice, however. Reactions from the employees to new changes are understandably strong, even if the present form of organization has had evident failings. Employees continue to be caught up with digesting the crucial event and the loss of control that accompanied it. They are not willing to enter into new adventures. Loss of control means that a person is confronted with events and conditions which are neither transparent nor predictable, never mind controllable. Loss of control is accompanied by feelings of being threatened, vulnerable, helpless and dependent (see ULICH 1994, p. 400f).

A strategy of change oriented to social scientific knowledge and attempting to prevent the above is 'organization development'. ROSENSTIEL (1992, p. 418) describes it as a procedure which „must be understood in the light of the entire organization, must take place with the active participation of those involved, whereby planning has as its goal to serve both the productivity of the organization and the development of its individual members". Organization development assumes as a principle that employees are themselves the actual experts with regard to redesign of their work conditions and that they are able and willing to do their utmost – in so far as they are permitted to. Participation does not mean, however, that decision trees in the enterprise are rejected, but rather that there is change in the preparation of decision-making. Different, or better, decisions become possible, their practical implementation is easier, and they are more readily accepted (see Frei et al. 1993, p. 175ff).

Along with participation, time plays a crucial role in the organization development process. Organization development is a long-term process, more evolutionary than revolutionary in nature (WOHLGEMUTH 1990, p. 41). Social systems, like individuals, progress through stages of development, which must progress in a meaningful relation to each other (see also FREI ET AL. 1993).

Following LEWIN (1952, p. 228ff), three phases of change for social systems of all kinds are characterized: unfreeze – move – freeze. Each phase must be completed for lasting and workable change to be ensured (Box 1).

Box 1: The three steps of change of social systems (see FREI ET AL. 1993, p. 30ff).

Unfreeze

Social systems exist – when not currently undergoing change – in a quasi-stationary state of balance. This means that system-maintaining and system-changing forces check each other in the balance; they neutralize each other in such a way that they seem relatively stable. Even if this state is not 100% satisfactory to all participants, it does have a certain advantage: it assures security and makes experiencing control possible. Disturbance of the balance, in any case, causes anxiety and insecurity, which inevitably lead to resistance. Resistance is thus not a personality problem but a normal reaction of those involved and indicates that their needs are not being sufficiently attended to. 'Unfreeze' therefore means that fears and biases must be reduced, old habits overcome and trust in the new formed. Careful attention must be devoted to participants' expectations. Expectations not fulfilled in the later course of change trigger frustration, resignation and renewed resistance. For this resistance to be overcome, it is important that there be true participation and the creation of an atmosphere of communication among all involved. The standpoints of each side must be laid out openly – contradictions in the system have to be addressed and discussed, because the more those involved experience contradictions in the present situation, the greater their readiness to change.

Move

The openness to change developed in the unfreeze phase must now be transformed into a need to change. Timing is very important – precipitate, headlong action can be just as counter-productive as inappropriate hesitation. Over-extending people with too much of the new must be avoided as well as resignation due to too much passivity. This can happen only when employees and their individual development are taken seriously and when project directors orient their planning in accordance with these (both individual learning processes and the development of individual willingness to participate will be discussed below). It is important that procedures be implemented in meaningful stages. Having goals for each stage – and these may be worked out in a participative manner – makes evaluation of the steps to change possible. Undesirable developments may be checked, or feedback on their progress may be given to those involved.

Freeze

The goal of this phase is to achieve a new, quasi-stationary balance which allows participants to integrate the changes into their daily work. The new must become the norm, the temporary loss of control must yield to regained security. Should this fail to occur, everyone is in danger of falling back into old patterns and habits. This phase of consolidation must be recognized as such by all in a conscious manner. Some kind of symbolic finale to the 'move' phase (official announcement, celebration) may ease this process and signal that one now return to daily tasks. A social system needs ever recurring phases of quasi-stationary balance in order to come to rest and once again be productive – a phenomenon which is often difficult for some change-happy managers to accept. When changes follow one another in too rapid succession, there is the danger that workers become so insecure that they begin to show resistance to any, however slight, restructuring.

The 'three steps' make clear that social systems must be carefully prepared for change. Of course, the steps can not always be clearly separated from each other or planned independently within the framework of a concrete project of change; they will overlap. This is the case not least because different social systems – or their members – are at different phases in the development.

Just as for many social systems, FREI ET AL. (1993, p. 28ff) also differentiate among three phases with regard to individual learning processes: growth, differentiation and integration. In the growth phase, the individual is confronted in a previously unfamiliar situation with new experiences, must take on new tasks and accept changed conditions. In the differentiating phase, the individual begins to – triggered by the confrontation with the new – see the old, familiar things and procedures in a new light. He or she now perceives subtle differences and begins to make new judgements of the familiar. In the integration phase, finally, the individual is in a position to view various aspects of a problem area in a holistic manner. Connections, which were previously not perceived, are recognized, and they form the basis of new suggestions for problem solution. The central factor, however, is that such individual developments never occur in isolation, but always in dealing with the environment. This may be the working conditions an employee faces, which – by means of appropriate task design, for example (see ULICH 1994, p. 141ff) – make it at all possible for the individual to come into contact with the new. In addition, communication and cooperation with other members of the social system have great significance. All exist with their own, individual 'learning curves' (growth – differentiation – integration) and influence each other mutually, in that they exchange feedback on their own attitudes and behaviors, which – all together – make up the very dynamics of the social system. Within the framework of a innovation project, we must therefore consider the interaction between individual learning curves and the development stages of the social system. FREI ET AL. (1993, p. 134) speak in this connection of a „double helix of competence development" in an attempt to convey a picture of the two as mutually determined and entwined. It is characteristic of the unfreeze phase, for example, that individual learning curves begin more and more to point in the same directions.

> „[The individual members of the system] may begin to look at things differently; they many start to question why things 'have always been done this way,' and imagine something different than what is 'normal' in their social system. They may begin to recognize in themselves an in others some initial motivation to change. However, at this point people only begin to *think* differently, while they still behave as usual". FREI ET AL. (1993, p. 31).

The same holds for the phases of change and consolidation; here again, individual learning and developmental processes must be coordinated with the changes that the social system as a whole is undergoing. Prerequisite, however, is that – as mentioned variously above – all employees participate in the change, not just during implementation of the solutions, but in the preceding phases as well.

VANSINA & TAILLIEU (1996, 83) caution that not all employees have achieved a stage where they explicitly demand to participate in company matters. They refer to PASMORE & FAGANS (1992), who speak of a continuum of participation readiness of the ego. This continuum is composed of five developmental stages, to which the employees of a social system can be assigned. It is important to note that the position within these stages has nothing to do with personality structure of individual employees, but rather has a great deal to do with – is decisively influenced by – the organizational form in which they live and work (Table 1).

Individual's readiness to participate	Organizational form
Ego regressed	*Control*
Readiness towards active involvement in decisions which influence one's own future is not (any longer) present	Lower hierarchical levels are excluded from participation
Ego potential	*Sympathy*
Readiness to participate in decisions which conform to present system and involve little risk	Openness to ideas which do not question present distribution of power or goal orientation of the system
Ego prepared	*Alliance*
Active involvement in controversial decisions,, which requires a certain degree of self-definition	Recognition of the interests and values of all hierarchical levels
Ego involved	*Mutual Development*
Willingness to contribute to changes in the system. Desire to understand better the needs of self and also others	Invitation to all hierarchical levels to contribute to change of the system
Ego committed	*Transcendence*
High degree of commitment to change important systems in a longer-term and lasting way	Willingness to examine the system as a whole; even, for example ,to question its right to exist

Tab. 1 Readiness to participate and form of organization (based on PASMORE & FAGANS 1992 and VANSINA & TAILLIEU 1996).

A common objection to change in organization towards 'mutual development' contends, namely, that the coworkers themselves are not willing to bring forward active involvement, and this serves as reason enough not to attempt change from the start. But here we must note that coworkers' readiness to participate and the form of organization are variables within a typical cause-and-effect loop:

„No variable in a loop controls other variables without itself being controlled by them. You can start the sequence anywhere you want by changing any variable. That change initially looks like a cause that triggers subsequent events in the loop. But the changes within the loop keep happening and eventually come back and modify the initial change which you intentionally made. And any cycling back means that what was originally the cause is now suddenly an effect.“ WEICK (1979, p. 77).

Informally put, this means that each organization can count on as much active involvement from its coworkers as it itself extends, or more pithily 'every organization has the workers it deserves'.

How and in what manner all these considerations can be included within the framework of a concrete project for change will be illustrated by means of the following case study. The project was initiated in April 1995 and is at present in full development – in other words, rather than summing up a completed project, this is a report on the steps that have been taken up to now and their consequences, as well as on the experiences which have been made along the way.

2. Case study

The enterprise described in this study is a vehicle manufacturer with approximately 900 workers, 450 being in production. Of these 220 are employed in parts manufacturing. Production is divided into parts manufacturing, pre-assembly and final assembly, and covers the greater part of the value-adding chain. The focus of parts manufacturing lies in sheet metal parts and prismatic as well as rotation-symmetric parts. Electrical systems, hydraulics motors and similar parts are purchased as finished product modules. Due to increasing dynamics of the market with an accompanying product diversification, the demands on the enterprise, both in terms of logistics and production technology, have changed drastically. While for decades series of 5 to 40 pieces were produced almost exclusively, the new product is produced in a form modified by client specifications and, as a rule, in the quantity of one piece. Demands have increased greatly, mainly in the areas of delivery times and flexibility.

In order to cut costs, the enterprise eliminated 20% of its workforce all at once. This resulted in a weakening of departments, which now only through all kinds of temporary measures were able to perform their work. The reduction of jobs led to improvement in terms of the cost situation, but at the same time, worsened work in terms of quality and longer-term efficiency. For these reasons, it became necessary to reexamine the entire order processing system from the ground up and to restructure in view of the new situation. Within the framework of a comprehensive project, the firm wishes to completely reorganize its logistics, computer-aided, by developing a new PPC system.

Initial analysis of the situation by the authors (called external coaching in the following) revealed that, among other things, the planning departments could not, despite numerous attempts and considerable expenditures, put together a clear and relatively realistic production schedule for one week. The main problem was the large number of released orders – up to 700% above the theoretical limits of capacity – and the simultaneous poor quality of calculated manufacturing due date. For example, some manufacturing orders were put through as rush orders even though assembly of the orders had been scheduled for much later dates. These were some of the reasons why the project 'independent parts manufacturing' – which will be described here – was put into effect. The project aimed to show how, in the future, manufacturing orders should be executed by parts manufacturing, how pre- and post-manufacturing areas must proceed, and how all these activities can become computer-aided. In addition, procedures and their sequence within individual areas of parts manufacturing should be examined in detail and re-conceived in view of the comprehensive whole. Table 2 shows an overview of the course of the project as well as the happenings leading to its initiation.

Date	Event	Actors
Beginning 1992	In order to rescue the economic viability of the firm, 20% of the **workers are laid off** with short notice ('bomb throw'). This triggers a plan to develop a concept for reorganization. ➔ The operations level is forced to organize itself, and many provisional solutions arise.	External economic rehabilitator, directors of the firm, management consultants
Spring 1994	**Start of the comprehensive reengineering project 'PPC 2000'** aiming to improve order processing and to support this with the aid of a modern, integrated PPC system. The project is managed by the logistics department. The project team is made up of high and mid-level cadre.	Project team 'PPC 2000'
Spring 1994 - Spring 1995	Numerous **detailed analyses and concept meetings,** training measures, consultation with system salesmen, system prototyping, pilot projects in many areas. ➔ Employees bear increasing burden caused by the project. Numerous problems in data processing area. Economic and organizational problems in the day-to-day.	Project team 'PPC 2000'
April 1995	**Initiation of the project 'Independent Parts Manufacturing'** in the form of a two-day start-up workshop. Founding of a project team composed of directors of production, shop supervisors, representatives from logistics department and external coaches. ➔ Project participants on the side of the enterprise rather passive. Activity stems mainly from the external coaches. Firm expects to be given a concept which solves all problems - in the sense of expert consultation.	Project team 'Parts Manufacturing'

Tab. 2 Course of the project.

Date	Event	Actors
May 1995	**All reorganization projects except 'Independent Parts Manufacturing' are terminated.** ➔ Management shows extensive lack of clarity with regard to future proceeding. The project still stands only because the participants are interested in it in principle and believe that a solution can be gained from the outside.	Directors of the firm
May/ June 1995	**Gross analysis of the situation (strengths/weaknesses):** Informal conversations with shop supervisors, foremen and department heads on tasks, processing, informational aids and problems in parts manufacturing. ➔ Problems are articulated and hope expressed. However, confidence in success is still low.	External coaches
Sep-tem-ber 1995	Three-part **workshop to asses and rate current situation** with members of the project team. Extended discussion on strength/weakness analysis. Summary of results in the form of ISHIKAWA diagrams. ➔ Mutual consciousness of the problems arises. There is now agreement that changes must be made. A first improvement of planning process is recognized as an urgent goal, in order to improve the totally overloaded, unrealistic weekly production schedule. No clarity as yet as to further proceedings or concept.	Project team, external coaches
11/ 1995 - 1/ 1996	**Sociotechnical Systems Analysis performed according to MTO method** (Interviews, see Box 2), Interviews with 18 employees in the planning department ➔ The interviews result in a rise in problem consciousness in those involved, in informal conversations in the framework of the interviews some initial principles of 'holistic work and task design' are transmitted.	External coaches
Feb-ruary 1996	**Development of a basic concept of 'total logistics'**. The goal is to create the organizational conditions necessary for extensive self-control by operations workers in parts manufacturing.	External coaches
March 1996	**Extensive discussions on the basic concept** and suggested strategic approach, approval of basic concept, discussion and resolution on immediate measures and further proceeding. ➔ Extensive discussion strengthens mutual problem consciousness further. Contradictions recognized and named explicitly. The project team generates numerous ideas and suggestions for implementation.	Production director, project team, external coaches

Tab. 2 Course of the project (continued).

Date	Event	Actors
April 1996	**Establishment of a decentralized control center** in parts manufacturing by employees in the planning department and parts manufacturing. Weekly production schedule now made on site. ➔ The project shows the first concrete results and thus becomes tangible to operations workers in parts manufacturing as well. Reconsideration of previous tasks and procedure leads to rapid improvement in terms of work expenditure and quality of the weekly schedule.	Project team, workers in planning department and in parts manufacturing
April/ May 1996	**MTO analyses in parts manufacturing** (interviews and workplace observations). Inventory of the spectrum of parts and the machines to prepare for segmentation. Presentation of the analyses to production management, shop supervisors and foremen. First group discussions regarding future design of work islands. ➔ Temporal lags between analysis, conception phase and implementation become shorter, more and more spontaneous ideas are generated, and the will to implement grows, particularly in the production manager and some shop supervisors.	External coaches
June 1996	**Resolution made to implement segmentation** and the formation of semi-autonomous manufacturing islands in the milling and turning section (as a pilot).	Production management
July 1996	**Kick-off meeting for forming of manufacturing islands**, general information meetings for all involved. ➔ Workers initiated into topic of islands in group discussions. Stimulated discussions ensue. Initial ideas about later island design are formulated.	Production management, shop supervisors, foremen, workers and external coaches
July 1996	External coach assigned to be in charge of group formation processes in parts manufacturing. ➔ The central role of operations workers is finally acknowledged. Initial measures for qualifications training in social competence for future group members planned.	Production management, shop supervisors, foremen, workers and external coaches

Tab. 2 Course of the project (continued).

For an initial overview, informal interviews and conversations were carried out at all levels of the enterprise at the outset of the project – in the context of a first, broad analysis of the situation – which were subsequently reviewed and systematized in a workshop (September 1995) held with those responsible and those involved from parts manufacturing and representatives from the specialist departments of logistics and computer systems. In the workshop, which lasted a total of nine hours distributed over a period of three days, 6 persons and two moderators took part. Figure 1 gives the results of the cause-and-effect analysis in the form of ISHIKAWA, or fishbone, diagrams (see ISHIKAWA 1985). In the figure, the strength of the arrows indicates the subjective weighting of the individual factors. The figure is the product of the workshop. Logical disagreement resulted from the fact that the entries were not all examined in detail.

Figure 1 reveals that the reasons given for the failure of the control system examined were very diverse in nature and in some cases contradictory. The dense linking of the reasons for failure mean problems for manufacturing reality. Workers having planning tasks within the planning department have, as a rule, only a limited perspective of the total problem. They attempt to base their decisions on their knowledge of the problems without any proximity to happenings at the level of operations. This leads at best to localized optimization of the system which may very well be based upon incorrect assumptions. An example of this typical failing is the much described vicious circle of due date and capacity planning (IBM 1975, p. 73). Due to low delivery reliability, expediters release orders even earlier in order to – in the mind of the expediter – guarantee delivery. But the consequence is over-extension of the workplaces, and thus an increase waiting times and overall cycle times. Another example is the issue of the degree of detail in planning, which is also revealed in Figure 1 (gray background): in the view of the planning department, too little exactness and detail in planning leads to an overloaded weekly production schedule, while the shop floor sees no degrees of freedom in the already too detailed plan. Here again, no simple and clear solution is found.

Goal of the situation analysis and the workshop was to achieve an initial 'unfreeze' of the predominant quasi-stationary balance. All were dissatisfied. with this state, but due to numerous determinants internal to the system, suggestions for change had always 'died in the bud'. The 'bomb throw', which had occurred four years previously, was still being felt and was often brought up in the interviews. There was no will to risk altering the controls which had subsequently been worked out under great difficulty. It was the ISHIKAWA diagrams which in the main stimulated broader consciousness of the problems and confronted workshop participants in the context of extended discussion with the new idea of independent parts manufacturing. Independent parts manufacturing were described as capable of - independently of an implemented PPC system – self-organization within a defined temporal frame and local handling of variations and disturbances.

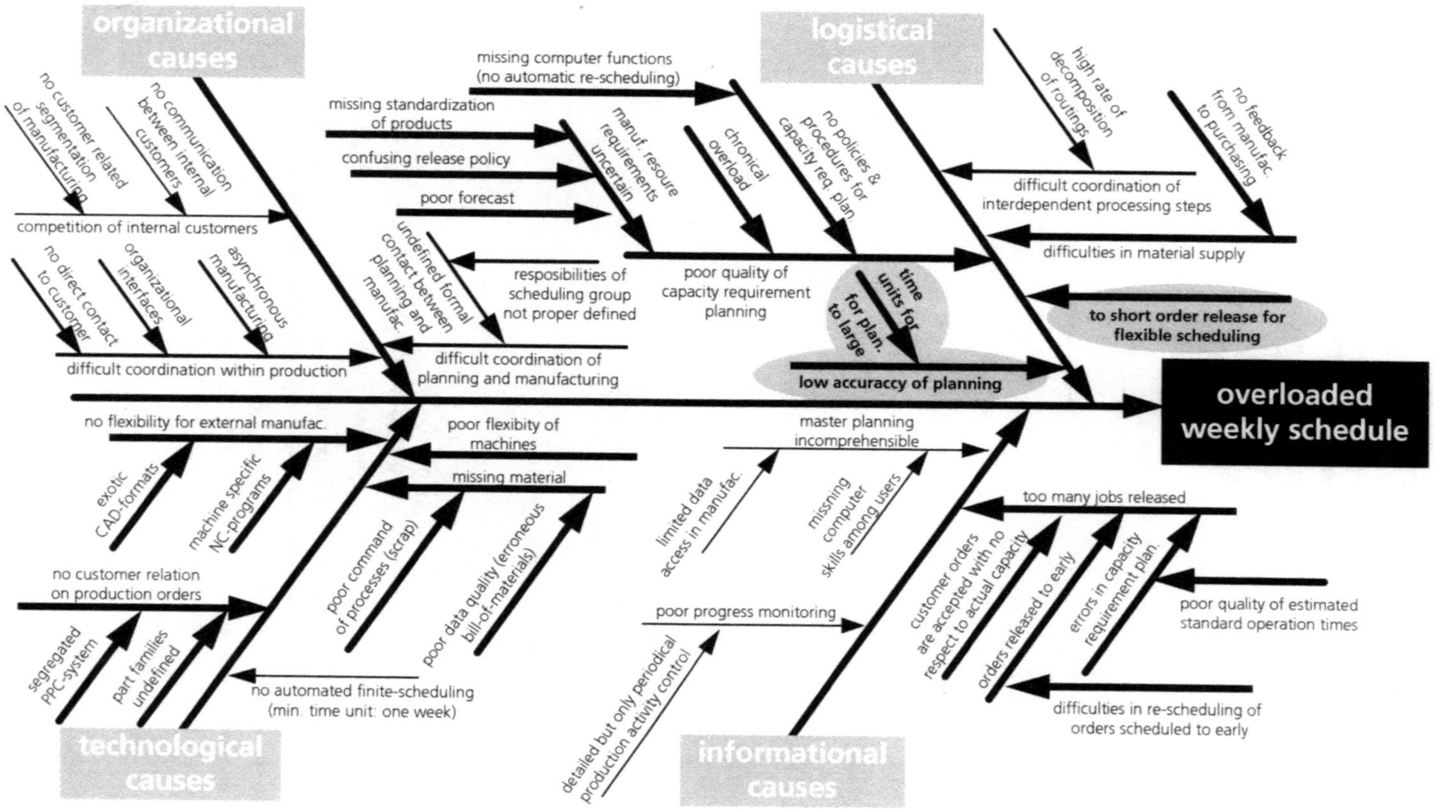

Fig. 1 Causes of the logistics problem 'overloaded weekly schedule' in the form of an ISHIKAWA diagram.

2.1. Developing a concept for independent parts manufacturing

On the basis of the initial experiences, we resolved to proceed with a three- phase plan (Table 3). The aim was to prepare the organization, to allow it to gain necessary readiness and competence for change. In particular, the plan aimed to boost the participation readiness of many coworkers, which could be assigned to the levels 'regressive' and 'latent' (see Table 1), successively towards 'active' participation. This always means that all steps in Table 3 had to be executed with consistent involvement of those affected.

Phase 1 ('unfreeze'): Clearing up operational organization along the boundaries of the parts manufacturing system. (improved interface design between parts manufacturing – to become independent – and material suppliers, planning information and assembly as internal customer). **Goal:** Assuring availability of all necessary resources (planning information, orders, engineering documentation, raw materials, for example), creation of organizational flexibility to control parts manufacturing.
Phase 2 (further 'unfreeze' and initial changes, or 'move'): Designing internal procedures and work organization in parts manufacturing (division of labor, orders responsibility, formation of islands, complete processing, etc.). **Goal:** Optimal processing of the work plan, increase in flexibility, and increase in total quality of work.
Phase 3 (further change and consolidation): Supporting processing by means of computer-based shop floor information system and final clearing of the system boundaries (that is, full assumption of any responsibilities not yet turned over). **Goal:** Optimization and stabilization of execution.

Tab. 3 Phase plan of the project.

Phase 1 required that a total logistics concept be developed with shop floor control as an integral part of the concept. For this design options for independent parts manufacturing were to be suggested and the necessary steps to realization defined. To develop the concept for Phase 1, standardized interviews were conducted in all areas connected with parts manufacturing during a period of three months. An interview method for sociotechnical systems analysis developed by the Work and Organizational Psychology Unit (IfAP) at the Swiss Federal Institute of Technology (ETH) in Zürich was used (Box 2).

Box 2: MTO Analysis

The reorganization of shop floor control must fit into a holistic and comprehensive design approach. ULICH (1994, 290) formulated work-oriented structural principles for the various levels of organization of enterprises.

Work-oriented structural principles for levels of the organization (ULICH 1994, p. 290).

Organizational level	Structural principle
Enterprise	Decentralization
Organizational unit	Functional integration
Group	Self-regulation
Individual	Qualified production work

The structural principles suggested form the basis of a comprehensive organizational analysis taking human resources, technology and the organization into account ('MTO' analysis in the original German[1]), a form of analysis developed in recent years at the Work and Organizational Psychology Unit (IfAP) at the Swiss Federal Institute of Technology (ETH) in Zürich and presented by STROHM & ULICH (1996). „MTO" analysis is a battery of methods; sociotechnical systems analysis serves to analyze work systems. On the basis of the guidelines for interviews with workers, aspects of the work system such as primary tasks, processing and structural organization, jobs and work stations, work methods, information networks, meetings, boundary regulations, variations and disturbances and general problems are assessed. The interview guidelines are very comprehensive and may be modified in accordance with the particular problem case to be studied. The length of the interview ranges from 2.5 to 4 hours.

On the basis of knowledge gained from the interviews, a relative evaluation is carried out according to the following criteria (see ULICH 1994, p. 156):

Independence of the organizational unit
Is the unit of organization - due to the completeness and independence of the task assigned to it – in a position to deal with variations and disturbances right then and there where they arise, without these disturbances spreading uncontrolled to other organizational units?

1 MTO refers to the German 'Mensch - Technik - Organisation'.

Task integration within the organizational unit

Are the part tasks in a work system related in content? Are task-related communication and cooperation necessary to fulfillment of the task? Is there a common and shared understanding of the task?

Congruity of product and organization

To what degree can work products be attributed to the organizational unit – both quantitatively and qualitatively? Can, for example, mistakes or quality deficits be attributed directly to the work system? The congruity of product and organization allows identification with one's own product.

Polyvalence of workers

To what degree are the workers within a system qualified to fulfill or execute the various part tasks of the work system, so that they may support each other mutually and represent each other? The degree of polyvalence is a significant indicator of the flexibility of a work system.

Technological-organizational convergence

Are work organizational and technological conditions finely tuned to each other? To what degree are the demands resulting from work organization met by technology, and to what degree is available technology actually used?

Evaluation of the individual criteria is relative and not absolute, that is, grounds for positive and negative evaluation are gathered, and on the basis of discussion (consensus validation), degree of fulfillment is rated on a 5-point scale (degree of fulfillment 1= low, 5= high). The method thus differs from most other work psychological analytical methods, which commonly work with normed scales. As a consequence, the results in different enterprises are hard to compare to each other. But the advantage is that the method may be adapted flexibly to the enterprise studied. While negative evaluations using normed scales quickly trigger expressions of displeasure by those workers involved and those responsible, the relative evaluation approach is more dynamic and, in the experience of the authors, much more apt to led to discussion and further communication.

The focus was placed on analysis of process organization and work organization in the logistics department, as it was de facto responsible for shop floor control. Here we must note that parts manufacturing itself had no formal delivery responsibility at all, and thus was excluded from shop floor control de jure. This also means that between shop floor control and its client – assembly – there was no regular contact, as this was taken over by the logistics department.

The logistics department itself was subdivided into three planning groups (each of 4 to 5 members) for production planning, who were responsible for the two main product groups and one for all parts supplied to external customers (LP1, LP2 and LP3). Logistics planning groups were responsible for materials planning, assembly planning and total supervision of due dates for parts manufacturing and subsequent assembly. They were assigned a separate manufacturing scheduling group (LPM), which like an service center derived the finite weekly schedule for parts manufacturing on the basis of plans set for assembly. (Figure 2).

One fact proved to be particularly difficult: the actually relevant finite scheduling for shop floor control (machine planning, exact daily plans) took place months ahead in an remote office. The planning procedure was, due to the previously used computer system, very demanding and tedious. Once planned, any changes in an order were possible only under very difficult conditions, as the system was not coupled to material planning. Manufacturing orders were based on customer orders (lot size from 5 to 20), but once issued, the manufacturing orders could not be clearly distinguished as the client orders. In consequence, due dates were not met as a rule, and for shop floor workers, they were far from realistic. Actual control was carried out by the 'delayed job hunters', who run after hundreds of orders and 'pushed' them through production. These 'hunters' were assigned directly to the logistics department and relatively unpopular in the production areas. It could happen parts missing in assembly could lead to a completely new manufacturing order of a lot size of 1 or 2, which would be rushed through manufacturing. This was due to the non-identifiable nature of the manufacturing orders. Again and again, even very expensive A-parts would be manufactured in excess of orders, at the same time jamming capacities. While the store prepared the raw materials for the whole weekly production schedule set out by the information system, working through of the orders occurred only when the orders were marked 'very urgent' on the priority list or when 'delayed job hunters' pushed an order through. As the entire responsibility for delivery reliability lay with the planning groups, production was not particularly motivated to attempt to work through the orders in a logistically optimal way.

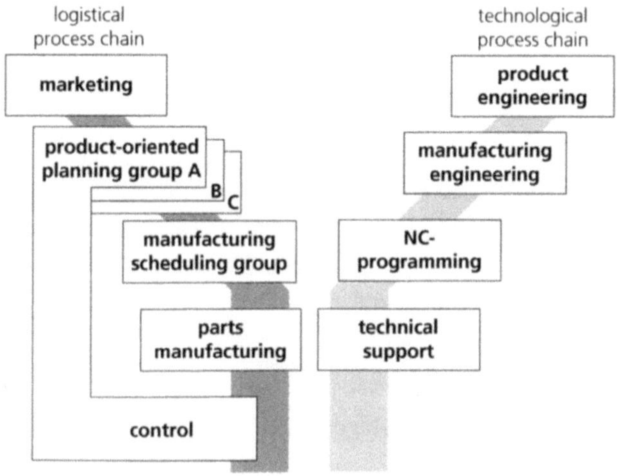

Fig. 2 Existing organization of PPC and shop floor control.

The results of the sociotechnical systems analysis are presented in Figure 3 for one planning group, the manufacturing scheduling group and one manufacturing group. What showed clearly, was that according to sociotechnical criteria, the planning groups formed an reasonable organizational unit, while the manufacturing scheduling group functioned as a mere dependent service to the planning groups.

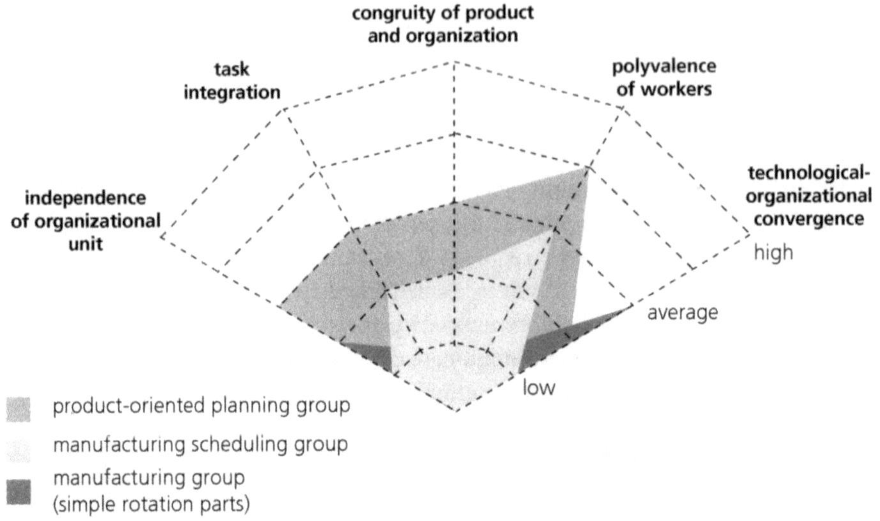

Fig. 3 Results of sociotechnical system in the planning groups.

The manufacturing department was mainly structured in job shops The different job shops (e.g. turning, milling, sheet metal processing, welding) in parts manufacturing struggled with work conditions which, from a sociotechnical systems analysis perspective, were totally inadequate. In addition to strong dependency upon the planning groups, but also upon the NC programming, no inner task interrelation was recognizable. Along with this, workers were also highly specialized in terms of a particular machine or control type, so that polyvalence was low. In spite of their rather high qualifications (all workers had a vocational training) and the fact that there were no problems of language communication within parts manufacturing (all spoke German), workers were not included in any planning or preparatory activities. Workplace analyses yielded the finding that the demands of the individual jobs in terms of thinking and planning lay far below the level called for skilled workers.

Sociotechnical analysis revealed clearly that the logistics department, despite its formal responsibility for shop floor control, cannot fulfill this task at all. The reasons lie both in the time perspective of its other tasks – the planning of raw materials having delivery times of nine months, for example – and in the lack of organizational and spatial proximity to production. On the other hand stood parts manufacturing, which with a formal overload up to 700% and order instructions marked 'rush order' of up to 50%, had lost all belief in the validity of the due dates from the logistics department. Consequently, an informal shop floor control developed with a certain 'cheekiness' – sequences are set up at the discretion of the shop floor and only then, when a 'delayed job hunter' appears on the scene, is there an attempt to act as a total system. Accordingly, manufacturing is full of half-completed manufacturing tasks, which are left in a depot to await further processing until they show up on a due list as 'very urgent'.

Already at the start of the analyses, it was evident that some tasks in the planning groups and the manufacturing scheduling group highly related to shop floor control had to be shifted into parts manufacturing. This meant that the previous organizational units of the logistics department should be maintained in a meaningful form, but that at the same time, parts manufacturing should have their own task area of shop floor control. Here the main goal was to transfer related tasks to an organizational unit so that the criteria of 'independence', 'integrated tasks' and 'congruity of product and organization' might be better met.

With the results of the analyses as a basis and with these goals in mind, strategies (Table 4) were derived to serve as guidelines for the fundamental arrangement of new organization and as a basis for project management. The strategies were conceived and discussed extensively by interest representatives from logistics, production, engineering and information systems together with the external coaches. This process proved to be enormously time-consuming and at times very trying, but it resulted in the valuable creation of a durable basis for the further proceedings of the project as well as for management behavior in the context of day-to-day work.

The sociotechnical analyses, as well as the mutual creation of strategies and the organizational concept – which will be described below – have proved to be extremely fruitful with regard to the entire process of change. In particular, the sociotechnical analyses yielded results which allowed all those involved to truly grasp where the deficits in their organization lay. Presentation in a marked form of the contradictions within the system served to transmit the latent desire for change some members already had to other members, and a commonly shared need to change developed. After all, the more that those involved experience the contradictions inherent in their situation, the greater the readiness to make changes. If many experience the contradictions together, it is more likely that they will begin to face new ideas and enter into a phase of growth.

The bases of the concept	Repercussions for project management
■ Within the enterprise, production is the area which is value-adding. All operational planning activities must be aligned to production's possibilities and capacities.	■ In every reorganization project, those affected must become involved participants.
■ Control should proceed in a proactive rather than reactive manner; for this, as much information as possible must become readily available to all.	■ Personnel and organizational development are permanent tasks of enterprise management. This ensures lasting flexibility and the ability to learn of both the organization and its workers.
■ Planning tasks must be understood as decision aids serving to guarantee effective operations.	■ Existing know-how must not be lost through reorganization, but rather should be utilized more effectively.
■ Responsibility for delivery reliability must be distributed along the temporal axis. Production is responsible for meeting deadlines in the context of their own commitments.	■ The project must promote independent and economic thinking in all workers.
■ Priority evaluation of an order should be made within the situation and directly prior to its possible realization. Priority can thus be determined only short-term and not 'planned' far in advance.	■ Agreement on basic matters is necessary and must be supported by management.
■ Cooperation among the various organizational units should be a major and meaningful consideration starting already at the level of task definition. Task definition should foster cooperation.	

Tab. 4 Strategies of reorganization.

The new strategies were discussed in the context of a basic concept for reorganizing the shop floor supplied by the external coaches. This basic concept was oriented along the lines of the considerations sketched out above regarding the local mastery of complexity through self-regulating organizational units. The goal was to transform parts manufacturing from its previous hierarchically-delegated and strongly labor-divided organization to self-controlled organization based upon production islands. The considerations, and thus the basic concept as well, were granted fundamental acceptance by the project team. The major consideration here was that any area, such as parts manufacturing in this case, can function effectively only if surrounding areas cooperate in a meaningful way and are more attuned to the needs and possibilities of their particular internal customer.

In the new organization proposed (Figure 4), the previous planning groups are clearly strengthened in their role as universal 'information turn-tables' for all logistics activities. They gather, compile and make available all planning information. Their focus is the fulfilling of external client orders under simultaneous optimal utilization of the company's internal resources.

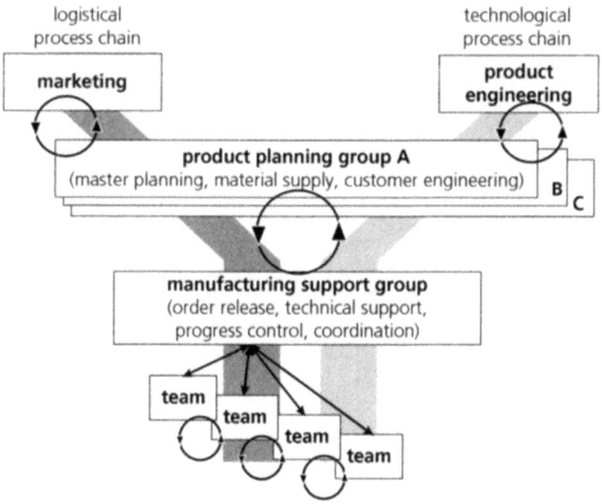

Fig. 4: New organizational concept of PPC and shop floor control.

Additionally, planning groups take on the task of materials procurement; this is necessary due to some suppliers' long lead times. Standing facing the planning groups is a stronger shop floor control in the production area, which flexibly and efficiently controls the processing of orders at the operations level in parts manufacturing. It procures the required raw materials from the supplier or from inventory. Through the creation of production islands, or groups with a large degree of self-regulation, it becomes possible to reduce present long wait times and to utilize more continuously currently unused potential. Production islands are assisted by a manufacturing support group (or 'support island'). This support island is assigned to production and will be staffed by two coworkers formerly with logistics (former LPM) and a (formerly) job shop supervisor. Also joining its staff will be the previous job shop secretary and a representative from manufacturing engineering.

The planning island is, in a first step, responsible for shop floor control. Approximately four weeks prior to the manufacturing due date for parts – that is, the starting date of assembly – the planning island takes over from the logistics planning groups and is now fully responsible for the order. Within this period, the planning island thus has responsibility for meeting deadlines.

The longer-term goal is to transfer this responsibility to the production islands themselves. This transference must proceed step-by-step and only with successful qualification of the production island workers. Common meetings held between logistics planning groups and the planning island as well as meetings between the planning island and the production island should serve to provide the best possible information flow and, thus, guarantee proactive control.

Following a one-year 'unfreeze' phase, which provided particularly the members of the project team with an opportunity to attune their learning curves (growth – differentiation – integration) to each other, the ground had been worked and prepared for the first real change: the installing of a decentralized scheduling and control group in parts manufacturing. This relatively small step – physically moving the former manufacturing scheduling group from its remote office into parts manufacturing and upgrading it to a support island – seemed at first glance to take a very long time indeed: an entire year! Here it is valuable to note that had this change been dictated and forced through right from the start, great resistance would have been encountered. Certainly innumerable rational reasons would have been given to show how impossible such an undertaking would be, and disproving these objections would have been difficult if not futile. But, to a degree astonishing to the authors, the groundwork of examining the old system together and the shared efforts at developing a vision of more efficient shop floor control, with small doses of external expert input, had led ultimately to very positive dynamics. Even before workplace analyses in parts manufacturing, being performed in order to prepare for future task design of the production islands, had been completed, the production manager surprised us with the request that islands be set up immediately. It is very important that in the change phase, the process consultant be capable of hooking up with the enthusiasm that has developed in the 'unfreeze' period and helping it towards implementation and realization of the vision. If this significant moment is not grasped and is lost, the new dynamics – which involved overcoming resistance and hesitation – are in danger of dissipating and sinking into deep resignation ('once again, we were promised something that did not come through').

A subsequent informative conference was held to call upon all coworkers to participate actively in the formation of the islands. There will naturally be some resistance here, which needs careful attention and handling. Just as given to the members of the project team, workers in parts manufacturing must be granted the opportunity to understand the nature of the coming changes and to influence them by helping with their design. For the processes of group formation within the islands planned, a coaching concept is worked out (training in social competence), which is carried out with the aid of external specialists. While coworkers in parts manufacturing are still in the 'unfreeze' phase, project team members and, in particular, the manager of the planning island are in the midst of the phase of change. This manager is now finding all kinds of new possibilities to make weekly scheduling more realistic. For example, at his own initiative he has begun to work together with a representative from the computer department to modify data processing lists, and the power of his new technical authority has led to some tangible improvements. Theoretically, he might have proceeded in this way a year earlier, but at that time he would neither have had the confidence to do so nor any support from others. Today, these efforts and measured are welcomed.

With regard to the third phase – consolidation of the changes – we can at present say only this: no decisions on systems for computer-aided control will be made until the new organizational measures have set in to a degree that a new quasi-stationary balance is achieved, and all coworkers involved have had a chance to become accustomed to and accepting of the new situation.

Organizations must have reached a state of rest before new technologies can be implemented successfully, because modern information technologies are particularly unsuited to serve as strategic instruments for organizational innovations (see BRINCKMANN 1991). This is important to note in the face of certain expert consultant efforts which hold the contrary and throw integrated PPC systems like bombs into organizations which, in obvious consequence, become thoroughly over-taxed.

3. Enterprises on the road to change

Enterprises are again and again faced with the task of reorganization. With the ever narrowing number of personnel and increasing burdens of daily business, reorganization projects themselves become a burden and are increasingly difficult in execution. In past years, support has been offered mainly by consulting firms which sell their expert knowledge. For cost reasons, enterprises have as a rule sought quick solutions which are easy to implement. Almost always, the search was for the 'ideal concept' which would lead directly to economic success. The consultant as expert was supposed to give support - simply and as quickly as possible. Implementation itself, and the expenditure which it incurs, have generally been evaluated only from the negative perspective of considerations of cost.

Such 'ideal concepts' frequently do not solve the real underlying problem, which - as evidenced by many reorganization projects - proves to have significantly more facets and to be much more dynamic in nature than could be handled with one simple measure. This has made consultation mandates which work out a theoretical concept 'externally', but leave implementation solely to the enterprise itself, increasingly problematic. The very process of change itself must stand at the center of all efforts of any organization and at all levels. Messages such as 'the way is the goal' are gaining particular significance within corporate reorganization projects. In order to achieve success, reorganization efforts must view the *path* towards the implementation of a concept as *a process of change* which cannot be commanded nor ruled, but rather must be accompanied.

4. References

ALIOTH, A. (1980): Entwicklung und Einführung alternativer Arbeitsformen. Verlag Hans Huber, Bern.

BRINCKMANN, H. (1991): Technological and organizational innovation. in: VAN DEN BESSELAAR, P., CLEMENT, A., JÄRVINEN, (Eds.): Information system, work and organization design. Elsevier, Amsterdam. p. 121-130.

FREI, F., HUGENTOBLER, M., SCHURMAN, S., DUELL, W., ALIOTH, A. (1993): Work design for the Competent Organization. Quorum Books, Westport.

IBM (1975): Executive perspective of manufacturing control systems. Technical publication. IBM Corporation, Atlanta.

ISHIKAWA, K. (1985): What is Total Quality Control? Prentice Hall, Englewood Cliff.

KÖNIG, E., VOLMER, G. (1993): Systemische Organisationsberatung – Grundlagen und Methoden. Deutscher Studien Verlag, Weinheim.

LEWIN, K. (1952): Field theory in social science. Tavistock Publications, London.

PASMORE, W. A., FAGANS, M. R. (1992): Participation, individual development and organization change: a review and synthesis. Journal of Management, 18:2, p. 375-397.

ROSENSTIEL, L. VON (1992). Organisationspsychologie. 3rd Ed. Schäffer-Poeschel, Stuttgart.

SCHEIN, E. H. (1968): Process consultation: Its role in organization development. Addison-Wesley, Reading.

SCHERER, E. (1996): Koordinierte Autonomie und flexible Werkstattsteuerung. vdf, Zürich.

STAEHLE, W. H. (1994): Management – Eine verhaltenswissenschaftliche Perspektive. 7th Ed. Verlag Franz Vahlen, München.

STROHM, O., ULICH, E. (Eds.) (1997): Unternehmen arbeitspsychologisch bewerten. Ein Mehr-Ebenen-Ansatz unter besonderer Berücksichtigung von Mensch, Technik und Organisation. vdf Verlag, Zürich.

SUSMAN, G. I. (1976): Autonomy at work. A sociotechnical analysis of participative management. Praeger, New York.

TRIST, E. L., BAMFORTH, K. W. (1951): Some social and psychological consequences of the longwall method of coal getting. Human Relations, 4, p. 3-38.

ULICH, E. (1994): Arbeitspsychologie. 3rd Ed. vdf/Poeschel, Zürich/Stuttgart.

VANSINA, L.S., TAILLIEU, T. (1995): Business process reengineering or socio-technical system design in new clothes? In: WOODMAN, R. W., PASMORE, W. A. (Eds): Research in organizational change and development. Volume 9. JAI Press, Greenwich. p. 81-100.

WEICK, K. E. (1979): The social psychology of organizing. 2nd Ed. Addison-Wesley, Reading.

WOHLGEMUTH, A.C. (1990): Erfolgreich eine neue Struktur einführen. IO Management Zeitschrift, 7, p. 39-44.

WOHLGEMUTH, A.C. (1991): Das Beratungskonzept der Organisationsentwicklung. 3rd Ed. Haupt, Bern.

WOMACK, J.(1995): Book review on Hammer & Stanton and Champy. Sloan Management Review, 36:4, p. 99-100.

Work-Process Knowledge and the School-to-Work Transition

Martin Fischer and Franz Stuber

To promote the development of modern organizational forms in production, vocational and technical training must first of all introduce new concepts of transmitting knowledge of organizations and planning. Based upon experience gained in a pilot project in schools for vocational training in trade and technology in Germany, it will be shown how the transition from school to the world of work can be designed by means of teaching by examples which allows the student to acquire vocational work-process knowledge.

1. The Dynamics of Work Organization

New concepts of organization ...

Current debate on economic and social modernization confronts us with many proposals and concepts which present new organizational models in enterprises and contain a new understanding of the role of qualified production work. 'Cooperation between different professions', 'product- and process-oriented reorganization', 'intra- and intercorporate networks' and 'the learning organization' are some key words in the debate. With regard to the development of human resource qualifications, two types of efforts are particularly relevant:

- The traditional separation of planning and organizational processes as practiced in production concepts based on Taylor and Ford are being subjected to fundamental criticism. Company-wide and functionally organized planning departments are to be disbanded in favor of 'product areas', 'cost and profit centers' or 'production units'. The units execute self-responsible planning and production and cooperate in flexible networks with other 'units'. With this, work area- and work area-transcending cooperation gains increased significance.

- The strict division between planning and controlling on the one hand and production on the other, which was also typical of Taylorism and Fordism, is being revised. New team and team work concepts are designed to give planning and organizational tasks back to direct production. For example, production teams are to be given 'controlled autonomy' within a manageable area of responsibility.

Figure 1 illustrates the developmental trend of new organizational approaches.

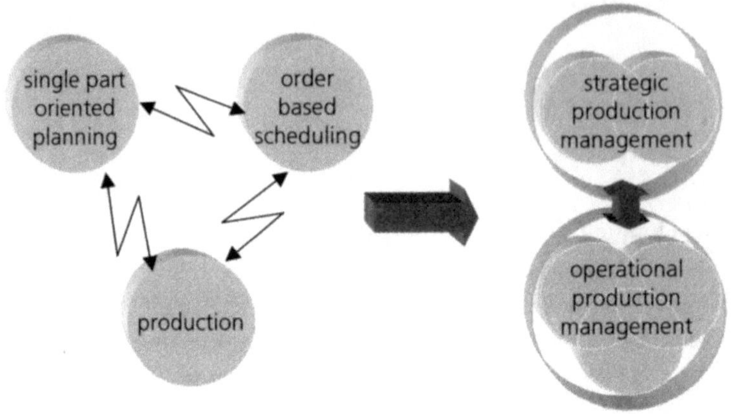

Fig. 1 From departmental organizational to increasing operative production management.

Shop floor workers, traditionally degraded to performing simple motions and work according to precise instructions, are gaining a new role in production. According to the European Centre for the Development of Vocational Training (CEDEFOP 1993), the new status of skilled work may be characterized as follows:

- Recognition is now being given to the types of knowledge and experience which were not utilized in traditional forms of organization. The fact that directives from planning must frequently be modified, as well as the fact that the objectives followed are contradictory, has always called for experience and improvisational talents on the part of shop floor workers (see FISCHER 1995).
- Through the inclusion of tasks from previously upstream and downstream areas, workers' areas of competency are being expanded. The integration of planning, control, maintenance and quality assurance tasks is not simply the addition of isolated skills. Instead, this integration represents a qualitative change in individual and collective task profiles.
- This means that novel skills and competency must be acquired. For example, the demand to optimize an order requires not only knowledge of the logistics

chain leading to complete order processing, but the development of new competency within a field of activity beset with contradictions. The demand is thus for planning and organizational knowledge which will aid the worker in evaluating and weighting the various objectives in relation to the conditions and constraints of their realization. With regard to the division of labor, for example, having the most experienced workers always perform a certain demanding task ensures that the task will be done quickly and reliably. However, this deprives other workers of an opportunity to further qualify themselves in the work process – which runs counter to the increasingly important goal of flexibility among personnel. Going beyond this, new competency is also required if social cooperation and communication is to take place within team-oriented work contexts or in the context of new supplier-manufacturer-customer relations.

... and the reality of the shop floor

Organizational concepts which systematically promote and utilize the competencies outlined above remain the exception rather than the rule in the industrial world of work – despite proclamations of the 'learning organization'. In particular, the vertical division of labor in planning, control and execution has remained virtually unrevised. A recent study of a number of industrial branches in Germany concludes:

> „The sum total of the creativity of companies with regard to organization – with a few exceptions – is mere modification of Taylorist principles by adding reintegration of control activities, ergonomic improvements in small details and work teams with internal rotation of works which is still short-cycled." (SCHUMANN ET AL. 1994, p. 645).

In spite of the fact that companies are currently turning to team work, many go only half-way when it comes to delegating decision-making authority and responsibility in decentrally operating units. Team work concepts follow, as a matter of priority, the pattern of adding various executing activities and of integrating, for example, quality assurance control tasks into manufacturing teams (see MOLDASCHL 1996). However, if the advantages of increasing operative production management are to come into full play, Taylorist principles of vertical labor division will have to be revised in a consistent manner. This is clear if one looks at an example such as set-up time minimization. Giving workers the opportunity to minimize set-up time will not suffice if there is, at the same time, no decentralized responsibility to participate in decisions within materials logistics and further loading plans. Namely, whether or not an order sequence may be changed will depend upon a sufficient supply of materials, among other things, and consideration of set-up aspects will affect subsequent planned processing dates (see STUBER 1997).

2. Work-process knowledge as a mediator between new concepts of organization and production experience

This raises the question of how skilled workers have dealt with the problems in the past. If we analyze the type of knowledge that characterizes shop floor workers in trade- and technical occupations, we discover the following phenomenon: this knowledge is work experience – experience with concrete machines, work orders, situations and people – which enables the worker to master problem situations at work. The importance of experience is demonstrated even within computer-aided manufacturing (see BÖHLE & MILKAU 1988; BÖHLE & ROSE 1992). This fact has been shown and analyzed in Germany within research efforts in the field of „computer-aided, experience-guided work" (called CEA) (compare MARTIN 1995; FLEIG & SCHNEIDER 1995, also see the contribution by FLEIG & SCHNEIDER in this volume); it has also been confirmed by our own work (FISCHER 1995, for example).

The relevance of work experience has meanwhile been increasingly emphasized in the research literature, and discussion centers around terms like skilled workers' experience knowledge, experience-guided work activities, and so on. Work experience is held to be the determining element in the acquisition of competency and mastery in industrial work. There is now even talk of redesigning work organization in order to more strongly promote training on the job – which would possibly make costly and extensive occupational training systems, like Germany's, superfluous.

But there is something overlooked in the discussion: all investigations on the subject of the significance of work experience over recent years have studied skilled workers – persons who had had an average of three years of technical training and thus had acquired specialized knowledge. Work experience does not lead to competent work activity automatically. Rather, in our investigations it was found that woven into the experience-guided work activity of skilled workers, there is also always – rudimentary or otherwise – specialized theoretical knowledge. Our studies within the areas of maintenance and computer-aided production planning and control also pointed up the significance of work experience for occupational activities in the trades and technical fields (see FISCHER, JUNGEBLUT & RÖMMERMANN 1995; FISCHER 1995). But, in order to prevent the misunderstanding that this might be completely unschooled experience, we use the term *work-process knowledge* to characterize this kind of competence.

WILFRIED KRUSE suggested the use of this term many years ago. With it, he referred to the knowledge of the context of the production process which is characteristic of shop floor workers. However, it is becoming more and more difficult for such knowledge to be acquired, due to complexity and due to the networking of computer-aided machines and facilities (see KRUSE 1985, 1986). According to our understanding, work-process knowledge can be described as follows:

- It is that knowledge which is a direct necessity within the work process (in contrast to knowledge which is structured systematically according to corresponding scientific fields).
- It is usually acquired through the course of the work process itself – for example, by learning through experience – which does not, however, exclude the possibility that theoretical knowledge also be applied.
- It comprises an entire process, in the sense of planning, execution and evaluation of one's own work in the context of internal company processes.

Even though work-process knowledge in production will differ from the construction and planning knowledge of engineers, it does transcend the semi-skilled worker's simple knowledge of how to run a machine. It is not just knowing single motions or repetitive part-tasks, but rather also knowledge of how the various part-tasks are bound together in the factory context. Figure 2 illustrates the importance of technical work-process knowledge in production practice.

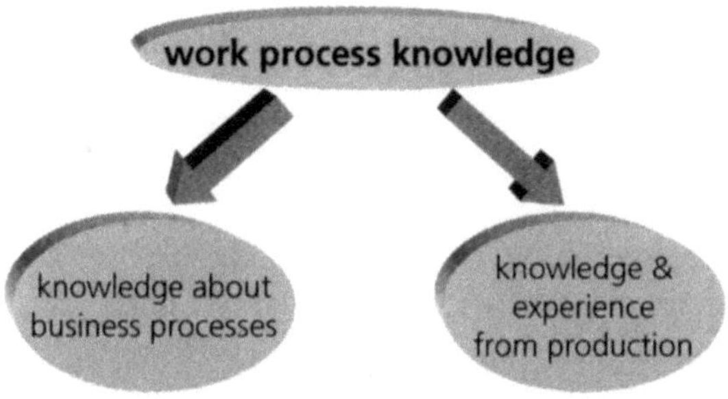

Fig. 2 Work-process knowledge in industrial production.

Occupational work-process knowledge is, of course, not secondary knowledge – derived from (scientific) academic knowledge by means of didactic reduction – but rather has its own independent quality. It imparts the relation between conceptual models of work organization and planning and the real world of work, between the artifacts constructed by engineering and the real situations in which the artifacts are utilized. With this, all the discrepancies and gaps become visible which are the order of the day in modern industrial work:

- between technology's intended functioning and its quirks,
- between industrial-economically optimized planning and its processing in production,
- between the official organization of the enterprise and informal cooperation during so-called 'unexpected' events.

Work-process knowledge is important to innovative company concepts, because responsibility is not limited to the individual locality of the workplace, but extended to the processes to be mastered by the work team. The term 'work-process knowledge' expresses the fact that industrial-economical business process knowledge is here complemented by knowledge of actually existing (or desired) technological and social degrees of freedom, constraints and developmental perspectives. The demands of cost-profit calculations are placed into relation to factual prerequisites and the means of their realization.

The true starting point for the activation and acquisition of work-process knowledge is the 'problem situation'. With this is meant not only interferences in planned work processes and in technical facilities, but also novel tasks for which the worker has no action plan. Qualified work activity gains in significance under the condition of computer-aided production: on the one hand, a large part of the job is machine work, automated and planned in advance by information technology. The option of allowing semi-skilled and unskilled workers execute such planned work activities, on the other hand, has not been put into practice – at least not in Germany – to the extent that had been predicted by industrial sociologicists (compare, among others, KERN & SCHUMANN 1984). Instead, even those jobs highly defined and determined by technology are given to skilled workers. From these workers it is expected, as self-understood, that they will master any problem situation within production.

ULICH & BAITSCH (1987) demonstrate in a longitudinal study that whether or not problem situations will result in the accumulation of work-process knowledge will depend upon the interaction of three critical factors:

1. It is necessary that the worker become conscious of the contradictions within a given work situation and that new objectives for this situation, based on a newly developed or previously latent motif in the work activity, be formulated.

2. So that new objectives may be developed, information must be accessible which will aid the drafting of alternatives.

3. Practical execution of the new activity which resolves the contradictions or serves the new goals is that which ultimately leads to the development of competencies. (ULICH & BAITSCH 1987, p. 516).

On the basis of our own findings, we can add the following: The above points should not be understood in a deterministic-logical sense, along the lines of 'if one is aware of the contradictions and if information is accessible, then in every case problem-resolving activity will result, which will automatically result in new competency'. This is certainly not the case. Problem situations in work are experienced by human beings in very various ways, and for some, they represent a threat. As a prerequisite to the acquisition of work-process knowledge in the context of problem situations, therefore, workers must have a sense of occupational identity which enables them to act independently. Without occupational training, this identity would hardly be achieved (see GEORG 1996).

3. Problems in the school-to-work transition and a new approach

Vocational and technical training is organized in Germany in a multitude of ways – in vocational schools within companies, unattached to specific companies and in public educational institutions. Traditionally, the core consists of dual elements: private in-company training combined with training in a public professional school (on one to two days per week). The problem is how in-house training in a company can best be combined with technical school training (compare LIPSMEIER 1987, p. 56 ff.; DEDERING & FLEIG 1996, p. 392 f.). The traditional assumption has been that first, the technical school will provide the learner with theoretical occupational knowledge, while second, in-house training will allow practical skills to be learned, and third, the learner himself will somehow be able to link theoretical knowledge with work experience and the demands of work in practice.

This traditional perspective of two-pronged technical training is, however, no longer viewed as unconditionally valid. On the basis of recent investigations, which confirm our own hypotheses, the belief that knowledge acquired in the training school is and can be utilized automatically on the job has been shaken. HEINZ MANDL and coworkers pointed out that in the training institutions of vocational school or the technical college, what is taught is often so-called „inert" knowledge, which is not utilizable in other contexts (MANDL, GRUBER & RENKL 1993). The authors demonstrated that advanced students of industrial engineering – in spite of their great knowledge of the field – did no better than a control group of students (studying education) in controlling a computer-simulated company (see RENKL ET AL. 1994). The authors concluded, with reference to commercial training, that:

> „A main reason for the development of 'inert' knowledge is the transmittal of knowledge in educational courses which is non-applicable, abstract, and artificially systematized and which does not do justice to the complexity of the real commercial world." (STARK ET AL. 1996, p. 23).

This problematic aspect is not restricted to the dual system of training in Germany. Certainly, in the Anglo-Saxon speaking world, skilled workers are trained and qualified in systems very different from that in Germany. But there also exists a dualism of education (which takes place in schools) and training (which is the task of companies). This dualism actually hinders the process which would combine theoretical knowledge with work experience, a process necessary if work-process knowledge is to be acquired (see RAUNER 1996). There is no grounds for the assumption that in that system as well students will transfer schooled knowledge effortlessly to the work context. In fact, the opposite is more the case: current findings of research on learning (see LAVE & WENGER 1991; HOLZKAMP 1993) have shown that students learn that which teachers intend to transmit as *content* to the least extent. What students learn best is to interpret the learning *situation*. JEAN LAVE's Californian research group has demonstrated repeatedly that all learning is bound to situation and context[1].

For this reason, consideration must be given to contexts which would promote the acquisition of work-process knowledge. Shop floor workers require a field of experience in which they may subjectively work through the results of their training and further education – their knowledge of the field. This field of experience was traditionally the job itself: possibilities to learn within the work process began initially with one's first training, while later, problem situations on the job were the main factor allowing, and requiring, learning at work.

Although this field of experience has not disappeared completely in today' world, it has changed in many respects and has become limited in computer-aided production. The situation is characterized by such key words as the non-transparency of technical phenomena and processes, the chaining and networking of facilities, the amalgamation of metal-, electro- and computer issues. Moreover, learning during the work process is not only made more difficult for those who have already completed training. Even during training, the possibilities to learn within the work process have become restricted: in particular, the domination of book knowledge within training and further education in large enterprises has resulted in a situation where trainees are not allowed to gain any experience using the costly computer-aided systems.

This raises the issue of just where the fields of experience can be located that learners require in order to apply their theoretical knowledge of their areas. This represents a challenge not only to industry, where learning- and experience-promoting reorganization is needed, but also to vocational schools and technical colleges.

Upon this background, curriculum research and development must define as its task the development and testing of new learning concepts for technical training. These must be learning concepts which support the skilled worker in moving towards competent execution of operative production management and with this, towards participation in processes of change in operations. It is obvious that the concepts should be oriented to work-process knowledge within production.

The development of work-process knowledge in training and further education in schools is thus confronted right away with particular difficulties. Traditional school learning processes lack that significant factor which is the pivot and hinge of

---

1 The following findings of diverse studies and empirical investigations illuminate this assertion: For example, in studying how the learning and application of (elementary) mathematics is applied in school and outside of school, e.g., in the supermarket, in the kitchen, etc. LAVE (1988) found that good math achievement at school does not lead automatically to good calculating abilities at the supermarket, while „poor" math students can very well achieve flawless results at the supermarket. Conversely, it was found that students who achieve correct answers in math lessons at school often do not utilize the calculating techniques taught by the teacher at all. Instead, they continue to use – unknown to the teacher – their own techniques, which they have mastered in order to avoid shaming themselves in math class (for example, regrouping, addition and subtraction instead of multiplying and dividing).

the activation and acquisition of work-process knowledge, because traditional, didactically reductionist technical schooling is at an abstract level removed from application contexts. Thus, it does not contain the typical problem situations which constitute job-relevant learning: ranging from interferences a planned work process to novel tasks which must be fulfilled in the context of real conditions and constraints.

The pilot project 'Work Organization in Technical Training' was designed to meet the challenge of discovering the prerequisites of the acquisition of work-process knowledge within technical training and education. A conception was needed which would make the fields of production management activities transparent and which could be implemented within training in technical schools and colleges. Therefore, the concepts of the project focused upon the *development of fields of experiences in schools* which would systematically link new learning content with practical activity.

3.1. A teaching company for the training of skilled workers

The concept of a 'teaching company' was tested for basic vocational training and education. With the teaching company, relevant prerequisites of learning through example to acquire work-process knowledge – through experimental contact with various production management tasks – were created. The concept is based upon the following considerations:

- First, we wanted to counteract the difficulties which arise when students in traditionally organized training programs are supposed to participate in planning department and planning station tasks. Such difficulties are reinforced due to the apprentice workshops found in large firms, which are completely separate from productive work processes. Moreover, traditionally, apprentices in small firms as well are given little opportunity to familiarize themselves with planning.
- Knowledge of the inherent conflicts among the competing economic, technological and social planning objectives with all of their repercussions is normally not an open topic within firms, let alone openly discussed with those in training. Through the teaching company, 'background knowledge' on company planning goals and methods becomes part of the curriculum. Further, the educational context allows for appropriate thematization of issues, such as the consequences of improving delivery reliability with respect to economic objectives, or the effects of postponing due dates on work hours and qualifying of workers, etc.
- The educational field of experience of the teaching company also, and not least, allows for greater degrees of freedom in dealing with computer-aided planning and organizational technologies. On the one hand, experience with these planning aids may be gained, and on the other hand, existing technologies may be critically examined and issues raised, such as: To what

extent are the technologies suited to supporting planning and organizational tasks at the level of skilled work?

The teaching company in the pilot project is organized in two, overlapping areas for study: the planning office and the production island. The two areas are themselves divided into various task areas, which consist of one or more learning workplaces. In the teaching company, alternative organizational scenarios can be played out. For example, tasks in material planning for production orders, work sequence planning and CNC programming, as well as procurement and materials logistics, may be dealt with and evaluated from the perspectives of both the planning office and the production island. Figure 3 presents an overview of the structure of the two areas of learning.

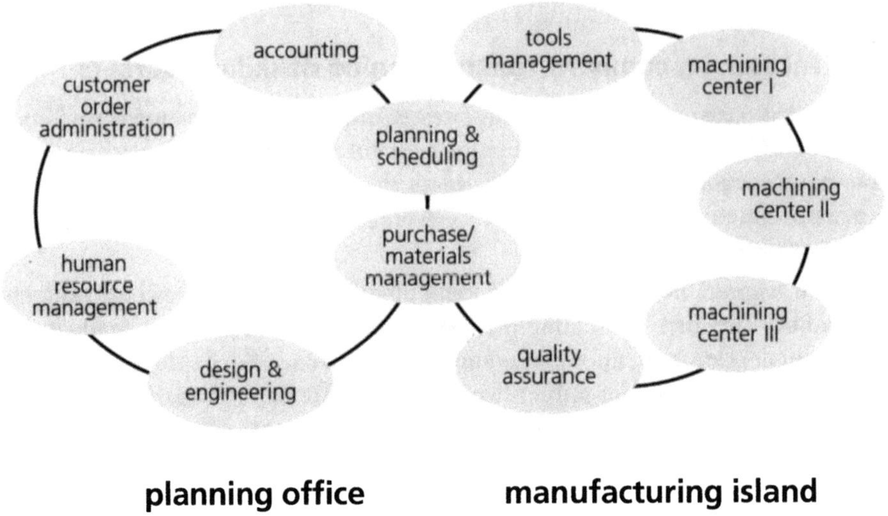

Fig. 3 Learning and work areas in the teaching company.

Each task area is equipped with documents as well as specially designed method manuals. Individual workplaces are linked via an internal network. As computer-based tools, data base applications developed within the framework of the project are utilized, and, in addition, a PC-based production management system is tested. Teaching is not organized according to the traditional class group. Instead, several learning and task groups are formed. Learners in the introductory and learning phase initially work on the same tasks. Subsequent to this phase, there is cooperation in a 'production phase' among cooperative groups divided along task lines. Teachers follow the objective of gradual and increasing withdrawal from the traditional teaching role, and they move on to moderating the groups or working within them themselves.

3.2. Interdisciplinary 'linked tasks' and internships for technical training

For vocational training of technical trades, various teaching plans were developed within the pilot project which based upon the concept of so-called 'linked tasks'. Through the aid of this concept, for the first time in the technical school, various aspects of production management in different fields were 'linked' via the tasks they share in common. Didactic concepts such as 'parts production on customer order', 'project organization in systems construction 'organization of the CNC environment' are treated by specialists in different fields, working in cooperation. In presenting the common task, students are transmitted knowledge of technical, economic and organizational contexts; in addition, students are meant to gain a certain practice-relevant understanding which will help them to evaluate alternative forms of organization from multiple perspectives. The project group found this to be a good point at which to break through the strong individual scientific disciplines within technical schooling and to increasingly orient teaching towards business and work processes. Students work within groups and organize their cooperation among themselves in a largely independent manner.

This interdisciplinary approach to learning based upon complex, job-relevant tasks finds continuation in the framework of specially-guided cooperative efforts between the technical school and regional businesses. Groups of 3 to 5 students are assigned tasks dealing with business and work organization as a two-week internship at a local factory or business. The students work closely with managers and coworkers within the companies. They are also put under the care of a teacher at the technical school, with whom they meet as their schedules allow. The tasks set are formulated by the companies themselves, who also provide the necessary documents and means of work. Internship results – suggestions for improvements and solutions – are presented by the students both in the companies and at school, where they are evaluated. These solutions may have to do with internal materials logistics and shop floor layouts, assembly methods, or the ergonomic redesign of workplaces. Companies which participate in the program welcome in particular the unconventional, non-run of the mill ideas presented by this external source. The aspiring technicians themselves are stimulated and motivated by the serious and real-world character of their tasks, because it gives them the opportunity to plumb the depths of their strengths and weaknesses with respect to future occupational roles.

4. Results and evaluation of the pilot project

The concepts guiding the pilot project are intended to make a contribution which would allow trainees and technicians to grasp existing company structures and planning tasks and master the challenge of innovative production management.

Their abilities to analyze a system, to develop their own suggestions for design and to evaluate alternatives upon the basis of acquired work-process knowledge, however, are impeded by some continuing obstacles. The obstacles cannot be overcome through content and methodological-didactic conceptions alone:

- One crucial difficulty lies in the fact that the learning processes initiated by the project run *crossways* to established disciplines. The concepts seen by the project team as meaningful balk at learning within purely defined and separate disciplinary areas. This incurs not only additional cost and effort – within the means of a government-supported project of course – in curriculum planning, but also provokes considerable conflict with the traditional learning organization. Under the given circumstances, the most a dedicated teacher can do is to practice the art of inserting the content shared from area to area, whereas what is really called for is the breaking down of boundaries and barriers (for a detailed evaluation, see FISCHER & STUBER 1997).
- Our experience with planning and organization technologies currently used indicate, moreover, that there are deficits with regard to support of the planning tasks of operative production management. There continues to be a lack of simple and learning-promoting systems, and the current orientation of market-leading systems towards traditional forms of work organization is lamentable. It was found during the course of the project that the production management system in use was actually utilized less than we had originally assumed (see discussion in STUBER 1997 and contribution by STUBER Part Five in this volume).

All in all, however, the approach of using example learning tasks to foster the acquisition of work-process knowledge proved to be successful. The concepts tested

- contribute to resolution of the problems posed by the school-to-work transition,
- expand the design competence – including technical, organizational, and social competencies – of future skilled workers and technicians,
- meet the growing need for production-economics qualifications of vocational-technical workers,
- increase the competence of trade- and technical schools in their role as contributors to regional innovation.

With this, the concepts developed could serve as an orientation aid to future curriculum development in Anglo-Saxon technical (vocational) high schools and community colleges. Putting the results into practice, as well as meeting the problems revealed through the project, will require the cooperative efforts of scientists, politicians, training and further education institutions and innovation-oriented businesses.

5. References

BÖHLE, F., MILKAU, B. (1988): Vom Handrad zum Bildschirm. Eine Untersuchung zur sinnlichen Erfahrung im Arbeitsprozeß. Campus, Frankfurt a. M.

BÖHLE, F., ROSE, H. (1992): Technik und Erfahrung. Arbeit in hochautomatisierten Systemen. Campus, Frankfurt a. M.

CEDEFOP (1993): Europäisches Zentrum für die Förderung der Berufsbildung (Ed.): Die Stellung des Unternehmens im Prozeß der Qualifikationsvermittlung: Ausbildungseffekte der Arbeitsorganisation. Bericht der Konferenz vom 31.3. und 1.4.1993.

DEDERING, H., FLEIG. G. (1996): Berufliche Grundbildung. In: H. DEDERING (Ed.): Handbuch zur arbeitsorientierten Bildung. Oldenbourg, München. p. 377-396.

FISCHER, M. (1995): Technikverständnis von Facharbeitern im Spannungsfeld von beruflicher Bildung und Arbeitserfahrung. Donat, Bremen.

FISCHER, M., JUNGEBLUT, R., RÖMMERMANN, E. (1995): Jede Maschine hat ihre eigenen Marotten! Instandhaltungsarbeit in der rechnergestützten Produktion und Möglichkeiten technischer Unterstützung. Donat, Bremen.

FISCHER, M., STUBER, F. (1997): Arbeitsorganisation als Gegenstand beruflicher Bildung. Abschlußbericht. Universität Bremen, Institut Technik & Bildung, Bremen.

FLEIG, J., SCHNEIDER, R. (1995): Erfahrung und Technik in der Produktion. Springer, Berlin.

GEORG, W. (1996): Arbeitsorientierung in der Ausbildung von Berufsschullehrern und betrieblichen Ausbildern. In: DEDERING, H. (Ed.): Handbuch zur arbeitsorientierten Bildung. Oldenbourg, München. p. 817-839.

HOLZKAMP, K. (1993): Lernen. Subjektwissenschaftliche Grundlegung. Campus, Frankfurt a. M.

KERN, H., SCHUMANN, M. (1984): Das Ende der Arbeitsteilung? Rationalisierung in der industriellen Produktion. München.

KRUSE, W. (1985): Ausbildungsqualität, Arbeitsprozeß-Wissen und soziotechnische Grundbildung. In: Gewerkschaftliche Bildungspolitik 5/1985.

KRUSE, W. (1986): Bemerkungen zur Rolle von Forschung bei der Entwicklung und Technikgestaltung. In: Sachverständigenkommission Arbeit und Technik, Universität Bremen (Ed.), Perspektiven technischer Bildung. Bremen.

LAVE, J. (1988): Cognition in practice. Mind, mathematics and culture in everyday life. Cambridge University Press.

LAVE, J., WENGER, E. (1991): Situated learning. Legitimate peripheral participation. Cambridge University Press.

LIPSMEIER, A. (1987): Neue Technologien, Technikwissenschaft und Technikdidaktik in ihrer Relevanz für eine Didaktik beruflichen Lernens. In: Verband der Lehrer an berufsbildenden Schulen und Kollegschulen in Nordrhein-Westfalen (Ed.), Berufliche Schulen – Multiplikatoren technologischer Innovation. Frankfurt a. M., p. 42 ff.

MANDL, H., GRUBER, H., RENKL, A. (1993): Neue Lernkonzepte für die Hochschule. Das Hochschulwesen, 41/1993. p. 126-130.

MARTIN, H. (Ed.) (1995): CEA – Computergestützte erfahrungsgeleitete Arbeit. Springer. Berlin.

MOLDASCHL, M. (1996): Kooperative Netzwerke – Komplement und Alternative zur Gruppenarbeit. In: SCHERER, E., SCHÖNSLEBEN, P., ULICH, E. (Eds.), Werkstattmanagement – Organisation und Informatik. vdf, Zürich.

RAUNER, F. (1996): Lernen für die Arbeitswelt in den USA. In: H. DEDERING (Ed.): Handbuch zur arbeitsorientierten Bildung. Oldenbourg, München.

RENKL, A., GRUBER, H., MANDL, H., HINKOFER, L. (1994): Hilft Wissen bei der Identifikation und Steuerung eines komplexen ökonomischen Systems? Unterrichtswissenschaft, Nr. 22, p. 195-202.

SCHUMANN, M., BAETHGE-KINSKY, V., KUHLMANN, M., KURZ, C., NEUMANN, U. (1994): Trendreport Rationalisierung. Edition Sigma, Berlin.

STARK, R., GRUBER, H., GRAF, M., RENKL, A., MANDL, H. (1996): Komplexes Lernen in der kaufmännischen Erstausbildung: Kognitive und motivationale Aspekte. In: BECK, K., HEID, H. (Eds.): Lehr-Lern-Prozesse in der kaufmännischen Erstausbildung. Wissenserwerb, Motivierungsgeschehen und Handlungskompetenzen. Zeitschrift für Berufs- und Wirtschaftspädagogik. Beiheft 13. Franz Steiner, Stuttgart. p. 23-36.

STUBER, F. (1997): Rechnerunterstützung für arbeitsprozeßnahes Planen. Software-Innovation im Kontext von Ökonomie, Organisation und beruflicher Bildung. Donat, Bremen.

ULICH, E., BAITSCH, C. (1987): Arbeitsstrukturierung. In: KLEINBECK, U., RUTENFRANZ, J. (Eds.): Arbeitspsychologie. Hogrefe, Göttingen.

Index

The Authors

Volker Ahrens, Dr.-Ing.
KÖNIG METALL GmbH
Königstr. 1
D 76571 Gaggenau
Germany

Volker Ahrens is assistant of the general manager of König Metall GmbH. He studied mechanical engineering at the University of Hannover, Germany. After working several years as a research assistant, he recently finished his doctoral dissertation with Prof. H.-P. Wiendahl at the Institute for Production Systems (IFA) on 'Decentralized Production Planning and Control – Systems Foundations and potential Applications'.

Mirko Dobberstein, Dr.-Ing.
Hella KG Hueck & Co.
Rixbergstr. 75
D 59552 Lippstadt
Germany

Mirko Dobberstein is assistant of the general manager of Hella KG Hueck & Co. He studied electrical engineering at the Aachen University of Technology (RWTH), Germany, and at the Norwegian Technical Institute (NTH) in Trondheim. He holds a post graduate degree in business administration and received his Ph.D. from RWTH in the field of industrial management. Before joining Hella he has been a researcher at the Laboratory for Machine Tools and Production Management (WZL) with Prof. W. Eversheim.

Martin Fischer, Dr. phil.
Institute for Technology and Education (ITB)
Universität Bremen
Grazer Str. 2
D 28359 Bremen
Germany

Martin Fischer is research fellow at the Institute of Technology and Education (ITB) of the University of Bremen. He holds a degree in psychology and focuses his research on the question of work process knowledge in manufacturing and aspects of vocational training.

Jürgen Fleig, Dipl.-Wirt.-Ing.

Fraunhofer Institute Systems Technology and Innovations Research (FhG-ISI)
Breslauer Str. 48
D 76139 Karlsruhe
Germany

Jürgen Fleig holds a joint degree in engineering and management from the Technical University of Karlsruhe. After working for a software provider as project manager and trainer, he joined the Fraunhofer Institute for Systems Technology and Innovation Research (FhG-ISI) where he is member in the task force on 'production and innovation'. He recently finished his Ph.D. thesis on environmental protection in the enterprise.

Albert Hurtz, Dr.-Ing.

Practice for Team-Oriented Work-Design (PTA)
Händelstr. 25-29
D 50674 Koln
Germany

Albert Hurtz is managing director of the Practice for Team-Oriented Work-Design (PTA). After receiving vocational training as a mechanic, he studied mechanical engineering and pedagogy at the Aachen University of Technology (RWTH) and the University of Cologne. After several years as researcher in the field of work organization, he co-founded PTA. PTA is a major service provider and consultant for team-based work structures and has a vast experience in the field of change management and supervision.

Julia K. Kuark, Dr. sc. techn.

Burghaldenstrasse 25
CH 5600 Lenzburg
Switzerland

Julia Kuark studied mechanical engineering at Stanford University. She then worked in the field of robotics and cybernetics at the Norwegian Institute of Technology (NTH) at Trondheim. She holds a postgraduate degree in mechatronics and a Ph.D. in industrial management both from the Swiss Federal Institute of Technology (ETH), Zürich. Until 1997 she worked as a researcher at the Work and Organizational Psychology Unit (IfAP) of the Swiss Federal Institute of Technology (ETH) in Zürich and recently became an independent researcher and consultant. Her research interests include CIM strategies, their impact on work organization, and cybernetical and sociotechnical systems design in production management. Besides she specializes on gender aspects of industrial working life. Since 1994 she is president of the Swiss Union of Women Engineers.

Manfred Moldaschl, Dr. phil.

Institute for Social Research (ISF)
Jakob-Klar-Str. 9
D 80769 München
Germany

Manfred Moldaschl is a research fellow with the Institute for Social Research (ISF). He studied psychology, sociology and history at the Universities in Tübingen and Berlin. He has been working with a number of research institutions in the fields of work psychology and industrial sociology and has a long time experience in analysis and supervision of organizational change in manufacturing. Recently he started INNOVATOP, an independent consulting and transfer firm supporting technological innovation and organizational change. His interests include innovation strategies in industry, teamwork and cooperation, aspects of stress and quality of working life, artificial intelligence and knowledge based systems.

Kenneth Preiss, Prof. Ph.D.

Department of Mechanical Engineering
Ben Gurion University of the Negev
Beer Sheba
Israel

Agility Forum
Lehigh University
Bethlehem, PA 18015
USA

Kenneth Preiss holds a chair in Computer-Aided Design at Ben Gurion University of the Negev and is senior fellow of the Agility Forum at Lehigh University. Over his career, he has worked in enterprise engineering, computer-aided design and manufacturing, artificial intelligence and many more. He has held leadership roles in industrial projects in Israel and the U.S. Together with Steven Goldman and Roger Nagel he is co-author of "Agile Competitors and Virtual Organizations" (Van Nostrand Reinhold, New York 1995).

Eric Scherer, Dr. sc. techn.

Institute for Industrial Engineering and Management (BWI)
Swiss Federal Institute of Technology (ETH) Zürich
Zürichbergstr. 18
CH 8028 Zürich
Switzerland

Eric Scherer currently works in the field of business information systems and logistics management. He was senior research assistant and head of the research group on 'Shop Floor Management and Information Systems' at the Institute for Industrial Engineering and Management (BWI) of the Swiss Federal Institute of Technology (ETH) in Zürich. He studied mechanical and industrial engineering at the Universities of Stuttgart and Aachen, Germany, as well as at Rutgers University, New Brunswick NJ, USA. In 1992 he joined ETH where he received his Ph.D. in Industrial Management. His research interests focus on shop floor control and operations management with respect of holistic design principles, application of IT systems and methods to facilitate sustainable change.

Robert Schneider, Dipl.-Psych., Ing. grad.

Fraunhofer Institute Systems Technology and Innovations Research (FhG-ISI)
Breslauer Str. 48
D 76139 Karlsruhe
Germany

Robert Schneider holds a degree in mechanical engineering as well as in organizational and work psychology. After several years in various industries and research he currently is a researcher at the task force for 'production and innovation' of the Fraunhofer Institute for Systems Technology and Innovation Research (FhG-ISI). His research interests include team-based work organization, applicability of new technologies, e.g., CIM, concepts of experience and knowledge in manufacturing.

Heinz Schüpbach, Prof. Dr. phil.

Institute of Psychology – Work and Organizational Psychology Unit
Albert-Ludwigs Universität Freiburg
Niemensstr. 10, R 116
D 79085 Freiburg im Breisgau
Germany

Heinz Schüpbach holds the Chair for Work and Organizational Psychology at the University of Freiburg. He studied work and organizational psychology at the University of Berne, Switzerland, where he also received his Ph.D. Before joining the University of Freiburg he was researcher at the Work and Organizational Psychology Unit (IfAP) of the Swiss Federal Insitute of Technology (ETH), Zürich. In 1993 he received his Venia Legendi from ETH. His major research interests include the unification of cybernetical and work-psychological principles towards a holistic systems design.

Franz Stuber, Dr. phil.

Institute for Technology and Education (ITB)
Universität Bremen
Grazer Str. 2
D 28359 Bremen
Germany

Franz Stuber is research fellow at the Institute of Technology and Education (ITB) of the University of Bremen. He holds a degree in economy and has major experience from a long time involvement in commercial IT systems design for production management. His research interests include user oriented software design in production management and application of the concept of work process knowledge.

Elizabeth Szelke, Dr.

Computer and Automation Research Institute
Hungarian Academy of Sciences
Kende u. 13-17
1111 Budapest
Hungary

Elizabeth Szelke is a senior research associate at the Computer and Automation Research Institute of the Hungarian Academy of Sciences, within the 'Intelligent Manufacturing Research Group'. She holds a M.Sc. degree from Eotvos Lorand University, Budapest, in mathematics and physics. Her industrial experience include the development of shop floor scheduling and control systems, especially in the machine tool industry, and various pilot CIM projects in Hungary. Her current research interests lie in the application of AI technologies to shop floor scheduling and control, supervisory control and adaptive systems behavior in a dynamic environment.

Steffen Weik, lic. phil.

Swiss Bank Corporation
P.O. Box
CH 8010 Zürich
Switzerland

Steffen Weik is team leader in the human resource development group of the Swiss Bank Corporation. He studied psychology at the University of Zürich and for several years worked with the Work and Organizational Psychology Unit (IfAP) of the Swiss Federal Insitute of Technology (ETH), Zürich. He has been involved in several projects concerning organizational development, innovation and change in manufacturing and financial industry

Martina Zölch, Dr. phil.

Institute of Psychology – Work and Organizational Psychology Unit
Albert-Ludwigs Universität Freiburg
Niemensstr. 10, R 116
D 79085 Freiburg im Breisgau
Germany

Martina Zölch is research fellow at the Chair for Work and Organizational Psychology at the University of Freiburg. She studied psychology at Landau and Berlin and has been working with research institutions in Berlin and Zürich. She holds a Ph.D. in work psychology from the University of Potsdam, Germany. and has long time experience from various projects in industry and administration. Her research focuses on the design of man-machine systems and aspects of cooperation at operational business levels.

MIX
Papier aus verantwortungsvollen Quellen
Paper from responsible sources
FSC® C105338

If you have any concerns about our products,
you can contact us on
ProductSafety@springernature.com

In case Publisher is established outside the EU,
the EU authorized representative is:
Springer Nature Customer Service Center GmbH
Europaplatz 3, 69115 Heidelberg, Germany

Printed by Libri Plureos GmbH
in Hamburg, Germany